THE ENDLESS SPACE FRONTIER

*A History of the House Committee
on Science and Astronautics
1959-1978*

The Endless Space Frontier:

A History of the House Committee on Science and Astronautics, 1959-1978

By Ken Hechler
(An abridgement by Albert E. Eastman)

AAS History Series, Volume 4

A Supplement to Advances in the Astronautical Sciences

An American Astronautical Society Publication

AAS Publications Office
P.O. Box 28130
San Diego, California 92128

1982

Affiliated with the American Association for the Advancement of Science
Member of the International Astronautical Federation

ISBN 0-87703-157-6 (Hard Cover)
ISBN 0-87703-158-4 (Soft Cover)
ISSN: 0730-3564

Published for the American Astronautical Society
by Univelt, Inc., P. O. Box 28130, San Diego, California 92128

Printed and Bound in the U.S.A.

IV

Foreword

The History Series of the American Astronautical Society attempts to make available sound and important works neglected by the academic and commercial publishers. This volume is a distinguished addition to our growing list for it genuinely helps explain more fully the genesis and evolution of the space program of the United States. It does this by a rare insight into the workings of the responsible Congressional committee in unvarnished detail not found in the prints of hearings, reports, and other documents. Moreover, it was written by a professional historian by education and experience before he became a member of the Congress. He tackled his recent scholarship with both detachment as well as some first-hand knowledge and biases. His treatment seems strengthened by his personal involvement on matters not otherwise documentable.

As a history this volume has a breadth on urgent concerns, issues, and contradictory factors as well as strong personalities interacting over a period of time. These tend to be fractionalized rather than integrated by political scientists or skewed by politicians in their hindsight recreations. Dr. Ken Hechler, its author, earlier wrote a history of the German historical profession in Nazi Germany during world War II, lending appreciation for his careful awareness of the interplay of competing ideals and actions via political institutions. And indeed, the Science and Astronautics Committee of the U.S. House of Representatives, about which he writes here, was itself transformed into the Science and Technology Committee with space as only one of its concerns.

Importantly, the source of all funding authorization for the U.S. space program resided in the House Committee on Science and Astronautics from the shock of the sputniks into the post-Apollo let-down. Members were the overseers for the Americans who elected them. NASA and the Pentagon and other agencies had to explain what they were up to in the years ahead. The so-called "space race" with the Soviet Union, the practical benefits of space activities, the needs of science in space, and how they might best be achieved had to be equated with the budgetary askings of the White House from Eisenhower to Nixon. These are some of the prevailing themes wondrously assayed in this history. Space exploration and exploitation because of their costs and broad benefits require that the collective interests of the American people be served by their representatives on Capitol Hill. This volume deserves a wider audience.

We are appreciative for the author's contribution as well as the able historical service of Dr. Albert Eastman in ordering this volume in its present form.

Eugene M. Emme
Editor, AAS History Series

Preface

This volume is an abridgement calculated to focus on the fate of the American space program on Capitol Hill for two decades. Historians rarely fractionate major phenomena to focus on a single policy issue such as space science and applications. So novel, dynamic, and successful were U.S. space happenings in a brief span of time, readers and sponsors may recognize the usefulness of this abridgement of the massive opus of Ken Hechler's, *Toward the Endless Frontier: History of the House Committee on Science and Technology, 1959-79* (35-1200 GPO, 1980). The House Committee on Science and Technology is to be highly congratulated for placing this basic and massive work in the public domain.

Fidelity to the original has been maintained in the abridgement. Nearly all material concerning the space program and related matters has been retained. Chapters and sections concerning areas not closely related to space have been eliminated.

Critical histories of Congressional affairs are rare indeed, and we are in debt to the dedicated efforts of Historian Ken Hechler for the existence of the original work.

Thanks are due Robert H. Jacobs for coordinating the editorial work on this abridgement, including the painstaking task of changing pagination and adapting the Source Notes and the Index to reflect only material which appears in the current volume.

Albert E. Eastman
Editor

TABLE OF CONTENTS

Introduction

By Charles A. Mosher [1]

It is an easy error to assume that all congressional committees are alike. They differ significantly. Each takes insistent pride in its own uniquity. And in several ways the Committee on Science and Technology is the most interestingly different of all.

At age 20, this is very junior among the standing committees of the U.S. House of Representatives. It was born of an extraordinary House-Senate joint leadership initiative, a determination to maintain American preeminence in science and technology, reacting to the U.S.S.R.'s Sputnik. And throughout its difficult, formative years to its now increasingly complex, still evolving adulthood, this committee has exerted a forceful policy influence and oversight responsibility, a galvanizing role in an unprecedented era of scientific and engineering accomplishment.

Our modern age of tremendously increased Government support for research originated in World War II, rapidly accelerated and expanded in the early years of this committee's influence, the 1960's, and now is in a transitional stage to new directions and dimensions as yet not clearly identified.

Recognizing the historic significance of this great burst of creativity, and also recognizing the Science and Technology Committee's central innovative part in helping guide and shape much of that successful effort, it is appropriate and important that the committee's first 20 years now be recounted fully and accurately. Four distinctly different chairmen, each very creative in his own way, have directed the committee's activities. And it is particularly appropriate that this history was conceived by Chairman Olin E. Teague, whose decisive leadership in a changing period of increasingly larger, varied, more complex committee jurisdiction, included recently added responsibilities for research, development, and demonstration (R.D. & D.) in all the crucially important aspects of energy resources.

Early in 1977, anticipating the decision to publish a history of these 20 years and after conferring informally with other members and staff, Chairman Teague appointed Mrs. Bonnie Seefeldt to begin the necessary research of documents, to conduct a series of interviews and

---

[1] Ranking Republican member, House Committee on Science and Technology, 1971–77 and executive director of the committee, 1977–79; former Member, U.S. House of Representatives from Ohio, 1961–77.

to develop a basic chronological report of the committee's activities, preparatory to the final manuscript. He also instructed other members of the staff to cooperate in every possible way with Mrs. Seefeldt's efforts. At a meeting of the whole committee on September 20, 1977, Chairman Teague informed its members—most of whom already were aware and cooperating—that he had taken that responsibility, and he requested the advice and assistance of the members. Chairman Fuqua has likewise advocated publication of the history, and the committee members' consensus support for that goal has been evident throughout its preparation.

Who should the author be? One of the chairman's first assignments to me—when I returned to this committee temporarily as staff director in September 1977—was that of proposing the general character, shape, and thrust our history should take, and to identify the best available talent to write it. The latter task proved most difficult, taking more time than expected. While Bonnie Seefeldt continued her background research, for over three months we explored further our options for producing the final product. After a tentative start or two that proved inadequate, we became convinced the final manuscript must be written by one author rather than by several of us collaborating, and it was imperative that the author have a firsthand knowledge of the Congress and its committee process.

I confess to being somewhat hesitant when I first suggested that we try to persuade Ken Hechler to write our history. But in his typically decisive way, Chairman Teague immediately reached for his phone to learn whether our former colleague would consider the task. As we feared, though interested, he was not immediately available. We waited, exploring other options, then returned to Hechler and it finally was our great good fortune to obtain his consent. He began this work on June 30, 1978.

All of us recognize Ken Hechler's superb qualifications. He was by profession a scholar, a product of Swarthmore and Columbia University (A.B., 1935, Ph. D., 1940), long before he became a politician. He had taught successfully political science and history at Columbia, Barnard College, Princeton, and Marshall Universities. He had assisted in the preparation of Franklin Delano Roosevelt's public papers, had held important posts in the Truman administration, including speechwriting for the President and later for Adlai Stevenson, he was experienced as a radio-TV commentator, he had authored successful histories: *The Bridge at Remagen; Insurgency: Personalities and Politics of the Taft Era;* and *West Virginia Memories of President Kennedy*. He was first elected as Representative to Congress from West Virginia's Fourth District in 1958, and then in each succeeding election for nine consecutive terms.

Most importantly, in January 1959 as a freshman Congressman, he became one of the original members of this committee, participating very actively throughout all of its first 18 years, as chairman of important subcommittees for 15 of those years.

So why did I first hesitate to suggest Ken Hechler to be our author? First, there was the obvious question, whether one so intimately involved in the committee's history could recount it with sufficient objectivity. Second, there was the fact that Hechler, though always one of the most productive and stimulating of our committee members, exerting genuine leadership qualities, also was at times a stubbornly independent member on occasion provocatively at odds with the committee's decisions. (Example, in the 94th Congress, his ultimately successful opposition to the committee's bill which would have authorized funding by Government-guaranteed loans for private industry to construct large facilities, to demonstrate the feasibility of producing synthetic fuels on a commercial scale.) We enjoyed and respected him even when we disagreed with him; but would the chairman or other members think me facetious in suggesting that he was ideally the one to write our history?

Those doubts were quickly resolved. It indicates the wisdom and vital spirit of Teague's chairmanship that he and other committee members recognized Hechler's qualifications, trusted his judgment and integrity, ignored any old disagreements, and agreed he should write our history. It emphasized the objectivity of this work, our requirement that it not be a superficial puffery job, that the author has solid credentials as an independent critic. It also emphasized the nonpartisanship typical of this committee, that I—a long time Republican member—and a Democrat, Hechler, of independent spirit, were assigned prime responsibilities for the history.

As author, he was promised a free hand in recounting these 20 years. He was instructed only to be as accurate as possible, but to produce far more than a routine chronological report, to deal realistically with substantive issues, personalities, and interesting anecdotes, to tell this committee's story "warts'n'all," to try to identify the actual significance, strengths, and weaknesses of its role and impact during a most vital, changing period of American history. The facts, the emphases, the adjectives, and somewhat colloquial style, all these are Ken Hechler's own responsibility.

He has fulfilled that charge admirably. This is far different and better reading than the ordinary congressional report. Personally I find it more fascinating than I could have hoped, it abounds in meaningful incidents and details of which I was not aware. I have learned much, I have a very valued, better understanding of the inner dynamics and broader influence of our committee's efforts.

I am confident that knowledgeable critics will find this a remarkably honest, absorbing, and significant work. I am confident that I speak for other members, past and present, in expressing enthusiastic appreciation to our longtime colleague for reporting the actualities of our committee's role with such integrity and verve.

Acknowledgments

On my office door is a cartoon depicting a writer being told by a supervisor: "On first reading, Ken, I'd say cut back on the prickling insights and beef up the anecdotes."

In compiling this 20-year account of the origins and activities of the committee, we have been very fortunate to have available the personalized comments and recollections of many of the participants in these dramatic events. This has made it possible to bring the history alive with anecdotal material.

Chairmen Olin E. Teague and Don Fuqua, as well as executive directors Charles A. Mosher and Col. Harold A. Gould have insisted that this history be objective and nonpartisan, maintain professional standards, and pull no punches. Mr. Mosher has been my severest critic, while at the same time supplying numerous useful editorial suggestions and support. At no time has he attempted to change or soften any of my critical judgments on events involving the committee.

Present and former members of the committee and staff, as well as agency officials and observers, have enriched the history through letters and interviews.

The interviews were tape recorded by Bonnie Seefeldt, who also contributed invaluable assistance in research, editing, and helping to organize the final product. In fact, it would have been impossible to complete this work without the careful groundwork and consistently high-caliber input which she provided.

During the summer of 1978, a number of interns helped with the typing and documentary research, including David Orleans, Kathleen Miller, Michael Gentry, Laura Bartlett and F. Marion Harrelson. In the latter stages, valuable typing assistance was received from Kimberley Woodruff. We were also fortunate to draw on staff assistance from Melinda Bradley, Gail Mathias, Peggy Witzel, Julie Fisher, Lillian Trippett, Suzanne Gibson, Carol Rodgers, April Applegate, Mary Bly, Tim Lockett, Mary C. Jatkowski, and Michael C. Helmantoler, as well as advice from Herbert Wadsworth, Jr.

In the early stages of this effort, John L. Swigert, Jr., executive director of the committee, 1973–77, and Gordon L. Harris, former public affairs director of the Kennedy Space Center, were of great help. In supplying needed documentary materials, we are grateful for the cheerful cooperation of Dr. Monte D. Wright, Carrie E. Karegeannes and especially Lee D. Saegesser of the NASA History Office. At the

Library of Congress, Robert T. Ennis, Pat Ayers and their staffs at the Congressional Research Service were unfailingly helpful. Dr. Charles S. Sheldon II was a constant inspiration, and his staff at the Science Policy Research Division made many suggestions.

Others now or formerly with the committee staff furnished useful advice, including Philip B. Yeager, general counsel; Regina A. Davis, chief clerk; Helen Lee Fletcher, former executive secretary. Among those giving us good editorial criticism were Daniel G. Buckley, Dr. John P. Andelin, Jr., Robert C. Ketcham, Dr. J. Thomas Ratchford, Dr. John V. Dugan, Jr., Dr. William G. Wells, Jr., Frank R. Hammill, Jr., James W. "Skip" Spensley, Dr. Thomas R. Kramer, Nancy Mathews, Dr. John D. Holmfeld, Darrell R. Branscome, Dr. Radford Byerly, Jr., Paul Vander Myde, Daniel E. Cassidy, Thomas N. Tate, Gerald E. Jenks, James E. Wilson, Jr., Paul C. Maxwell, Alexis J. Hoskins, Ralph N. Read, James Jensen, David D. Clement, Sherman E. Roodzant, John G. Clements, James H. Turner, Patricia G. Garfinkel, and Anthony Scoville.

I want to express my admiration and appreciation to the dedicated crew at the Government Printing Office, whose professional handiwork is evident in the production of this volume.

We acknowledge with thanks the following individuals whose interviews and letters to Chairman Teague helped immeasurably:

Hon. Carl Albert (Democrat of Oklahoma), former Speaker, U.S. House of Representatives.

Dr. John P. Andelin, Jr., former staff director, Subcommittee on Advanced Energy Technologies and Energy Conservation, Research, Development and Demonstration, Science and Technology Committee.

William A. Anders, astronaut, later Ambassador to Norway.

Former Representative Leslie C. Arends (Republican of Illinois), former member of Select Committee on Astronautics and Space Exploration.

Neil A. Armstrong, astronaut.

Former Representative Wayne N. Aspinall (Democrat of Colorado).

Dr. Allen V. Astin, Director Emeritus, National Bureau of Standards.

Former Representative Perkins Bass (Republican of New Hampshire).

Alan L. Bean, astronaut.

Former Representative Alphonzo Bell (Republican of California).

Former Representative Bob Bergland (Democrat of Minnesota), presently Secretary of Agriculture.

Jules Bergman, ABC News.

Dr. Charles A. Berry, former Director of Medical Research and Operations, Manned Spacecraft Center, NASA, and personal physician for the astronauts.

Dr. Eugene W. Bierly, Head, Division of Atmospheric Sciences, Climate Dynamics Research Section, National Science Foundation.

Karol J. Bobko, astronaut.

Lt. Gen. Frank A. Bogart, retired, former Associate Director, Manned Spacecraft Center, NASA.

Frank Borman, astronaut.

Dr. Edward L. Brady, Associate Director for International Affairs, National Bureau of Standards.

Vance D. Brand, astronaut.

Dr. Lewis M. Branscomb, former Director, National Bureau of Standards.

Dr. Harvey Brooks, former member, National Science Board and former member, President's Science Advisory Committee.

Representative George E. Brown, Jr. (Democrat of California), chairman, Subcommittee on Science, Research and Technology, Science and Technology Committee.

Former Representative J. Edgar Chenoweth (Republican of Colorado).

Dr. Robert S. Cooper, Director, Goddard Space Flight Center, NASA.

Dr. Edgar M. Cortright, former Director, Langley Research Center, NASA.

Former Representative Paul W. Cronin (Republican of Massachusetts).

Dr. Edward C. Creutz, former Assistant Director, National Science Foundation.

Walter Cunningham, astronaut.

Former Representative Emilio Q. Daddario (Democrat of Connecticut).

Former Representative John W. Davis (Democrat of Georgia).

Dr. Bowen Dees, former Associate Director, National Science Foundation.

Joseph Del Riego, former minority staff member, Science and Astronautics Committee.

Former Representative Thomas N. Downing (Democrat of Virginia).

Dr. Lee A. DuBridge, former president, California Institute of Technology.

Charles F. Ducander, former executive director and chief counsel, Science and Astronautics Committee.

Former Representative Marvin L. Esch (Republican of Michigan).

Hon. George J. Feldman, former chief counsel and director, Select Committee on Astronautics and Space Exploration.

George W. Fisher, former administrative assistant to former Representative Olin E. Teague (Democrat of Texas).

Dr. James C. Fletcher, former Administrator, NASA.

Hon. Gerald R. Ford, former President of the United States, and former member of Select Committee on Astronautics and Space Exploration.

Representative Edwin B. Forsythe (Republican of New Jersey).

A. N. Fowler, Acting Division Director, National Science Foundation.

Former Representative Louis Frey, Jr. (Republican of Florida).

Dr. Robert A. Frosch, Administrator, NASA.

Representative Don Fuqua (Democrat of Florida), chairman, Science and Technology Committee.

Arnold W. Frutkin, Associate Administrator for External Relations, NASA.

Charles G. Fullerton, astronaut.

Dr. Robert R. Gilruth, former Director, Johnson Space Center, NASA.

Edward G. Gibson, astronaut.

Dr. T. Keith Glennan, former Administrator, NASA.

Richard F. Gordon, Jr., astronaut.

Col. Harold A. Gould, executive director, Science and Technology Committee.

Dr. Norman Hackerman, Chairman, National Science Board.

Frank R. Hammill, Jr., former counsel, Science and Technology Committee.

Dr. Philip Handler, President, National Academy of Sciences.

Representative Tom Harkin (Democrat of Iowa).

Henry W. Hartsfield, Jr., astronaut.

Dr. Leland J. Haworth, former Director, National Science Foundation.

Former Representative Brooks Hays (Democrat of Arkansas), former member of Select Committee on Astronautics and Space Exploration.

Rev. Theodore M. Hesburgh, president, University of Notre Dame.

Roger W. Heyns, former chancellor, University of California, Berkeley, Calif.

Richard P. Hines, former staff consultant, Select Committee on Astronautics and Space Exploration and Science and Astronautics Committee.

Dr. John D. Holmfeld, science consultant, Science and Technology Committee.

Richard E. Horner, former Associate Administrator, NASA.

Dr. Lloyd G. Humphreys, former Assistant Director for Education, National Science Foundation.

James B. Irwin, astronaut.

Dr. Leonard Jaffe, former Deputy Associate Administrator, Applications, NASA.

Former Representative Joseph E. Karth (Democrat of Minnesota).

Joseph P. Kerwin, astronaut.

Robert C. Ketcham, counsel, Science and Technology Committee.

Former Representative David S. King (Democrat of Utah).

Dr. Christopher C. Kraft, Jr., Director, Johnson Space Center, NASA.

William B. Lenoir, astronaut.

Representative Jim Lloyd (Democrat of California).

Dr. Alan M. Lovelace, Deputy Administrator, NASA.

James A. Lovell, Jr., astronaut.

Dr. George M. Low, former Acting Administrator, NASA.

Hon. John W. McCormack (Democrat of Massachusetts), former Speaker, U.S. House of Representatives and chairman, Select Committee on Astronautics and Space Exploration.

Dr. William D. McElroy, former Director, National Science Foundation.

Dr. Archibald T. McPherson, former Associate Director, National Bureau of Standards.

Former Representative Dale Milford (Democrat of Texas).

Former Representative George P. Miller (Democrat of California), chairman of the Science and Astronautics Committee, 1961–73.

Edgar D. Mitchell, astronaut.

Former Representative Charles A. Mosher (Republican of Ohio), former executive director, Science and Technology Committee, and former ranking Republican, Science and Technology Committee.

Dr. George E. Mueller, former Associate Administrator for Manned Space Flight, NASA.

Dale D. Myers, former Associate Administrator for Manned Space Flight, NASA.

Representative William H. Natcher (Democrat of Kentucky), former member of Select Committee on Astronautics and Space Exploration.

Dr. John E. Naugle, chief scientist, NASA.

Dr. William A. Nierenberg, director, Scripps Institution of Oceanography, University of California, San Diego.

Former Representative Leo W. O'Brien (Democrat of New York), former member of Select Committee on Astronautics and Space Exploration.

Brian O'Leary, astronaut.

Dr. T. B. Owen, former Assistant Director, National Science Foundation.

Dr. Lowell J. Paige, former Acting Deputy Director, National Science Foundation.

Donald H. Peterson, astronaut.

Dr. Rocco A. Petrone, former Associate Administrator, NASA.

Dr. E. R. Piore, former member, President's Science Advisory Committee and National Science Board.

Dr. Frank Press, Director, President's Office of Science and Technology Policy.

Former Representative Bob Price (Republican of Texas).

Dr. J. Thomas Ratchford, former staff director, Subcommittee on Energy Research, Development and Demonstration, Science and Technology Committee.

Former Representative Richard L. Roudebush (Republican of Indiana).

Former Representative J. Edward Roush (Democrat of Indiana).

Representative James H. Scheuer (Democrat of New York), former chairman, Subcommittee on Domestic and International Scientific Planning, Analysis and Cooperation, Science and Technology Committee.

Russell L. Schweickart, astronaut.

Dr. Robert C. Seamans, Jr., former Deputy Administrator, NASA.

Dr. Charles S. Sheldon II, former assistant director, Select Committee on Astronautics and Space Exploration and technical director, Science and Astronautics Committee.

Donald K. "Deke" Slayton, astronaut.

Philip M. Smith, Assistant Director, President's Office of Science and Technology Policy.

Former Representative Neil Staebler (Democrat of Michigan).

Lt. Gen. Thomas P. Stafford, astronaut.

Dr. H. Guyford Stever, former Director, National Science Foundation.

Richard H. Sullivan, former member, National Science Board.

Former Representative James W. Symington (Democrat of Missouri).

Thomas N. Tate, counsel, Subcommittee on Space Science and Applications, Science and Technology Committee.

Former Representative Olin E. Teague (Democrat of Texas), former chairman, Science and Technology Committee, 1973–79.

Former Representative Ray Thornton (Democrat of Arkansas).

Representative Wes Watkins (Democrat of Oklahoma).

James E. Webb, former Administrator, NASA.

Paul J. Weitz, astronaut.

Dr. William G. Wells, Jr., staff director, Subcommittee on Science, Research and Technology, Science and Technology Committee.

Dr. Edward C. Welsh, former executive secretary, National Aeronautics and Space Council.

James E. Wilson, Jr., former staff director, Subcommittee on Space Science and Applications, Science and Technology Committee.

Representative Larry Winn, Jr. (Republican of Kansas).

Theodore W. Wirths, Director, Office of Small Business Research and Development, National Science Foundation.

Representative John W. Wydler (Republican of New York), ranking Republican member, Science and Technology Committee, 1977–

Philip B. Yeager, general counsel, Science and Technology Committee.

References have been included in the text and "Source Notes" are listed on pages 413 - 418, preserving continuity for the general reader. Full access to the voluminous official and personal correspondence files of the committee and its present and former members greatly enhanced the value of the finished product.

To avoid confusing those who are not "insiders," I have, throughout, simplified the dates of fiscal year funding to relate to activities the year they actually took place. This means that the story of what was done in 1979 will actually refer to 1979, instead of "fiscal 1980," for example.

A simple glossary would probably help readers unacquainted with the strange folkways of Congress, but if you assume everyone is a Rip van Winkle the glossary would sound condescending. So I'll just confine myself to the two questions most frequently asked me: What is a conference committee; and what is a "markup"? When the House and Senate pass a bill in different form, a conference committee including the senior House and Senate Members from the committees having jurisdiction over the legislation get together to agree on a compromise version, which is then submitted to the House and Senate for approval. A "markup" is simply a meeting of committee or subcommittee members, usually following public hearings, to amend or mark up a bill in order to move it toward final action by the House or Senate.

The deeper I delved into this subject, the bigger the job grew. It could only be accomplished by compressing several years' effort into one year, giving up evenings, weekends, and holidays, by working longer and harder in order to comprehend the totality of the multi-dimensional edifice being constructed. To measure the full impact of an important congressional committee on public policy, amid the triumphs and the tragedies, the occasional human foibles more than matched by dedicated effort and foresight, was a genuine challenge.

The memory of mankind is short. Congressional committees, with their frequent turnover of personnel, have even shorter institutional memories. To capture and record, and search for the true significance of what an important arm of the Congress has accomplished has been the aim of this work. It will be for others to judge whether that aim has been realized.

I have personally enjoyed the opportunity to reconstruct the events of the past 20 years. Despite the care with which a number of individuals have read the manuscript, I take full responsibility for both the personal judgments which have been made and the errors which have crept in.

KEN HECHLER.

CHAPTER I

In the Beginning, the Select Committee

The House Committee on Science and Technology had its roots in the American public's intense reaction to the Soviet launch of Sputnik, the first satellite to orbit the Earth, on October 4, 1957.

The committee was originally named the "House Committee on Science and Astronautics," which first saw the light of day on January 3, 1959. That was a unique event, because it was the first standing committee of the House to be born since the Legislative Reorganization Act of 1946 had drastically reduced the number of standing committees from 48 to 19. Even more significant, it was the first time since 1892 (when the predecessors of the House and Senate Interior and Insular Affairs Committees had been established) that both the House and Senate had moved to create standing committees on an entirely new subject matter. And, as we shall see, the House committee enjoyed a considerably broader jurisdiction than its counterpart in the Senate.

The 20-year history of the committee traces back to a landmark action of the Congress in 1958.

Scarcely five minutes after the House of Representatives convened on Wednesday, March 5, 1958, and before many Members had reached the floor, Speaker of the House Sam Rayburn crisply but gruffly intoned:

The Chair recognizes the gentleman from Massachusetts.

House Majority Leader John W. McCormack had a real sense of the deep significance of the moment, but his words were simple:

Mr. Speaker, I offer a resolution and ask unanimous consent for its present consideration.

If McCormack sounded matter-of-fact, the actual reading of the resolution was delivered in booming and stentorian tones by House Reading Clerk George Maurer. The Members on the floor stopped their conversations and when Maurer read House Resolution 496, it seemed to take on deeper meaning as each word was emphasized:

Resolved, That there is hereby created a Select Committee on Astronautics and Space Exploration to be composed of 13 Members of the House of Representatives to be appointed by the Speaker, 7 from the majority party and 6 from the minority party, one of whom he shall designate as chairman.

1

The charter of the select committee was simply and directly stated:

> The select committee is authorized and directed to conduct a thorough and complete study and investigation with respect to all aspects and problems relating to the exploration of outer space and the control, development, and use of astronautical resources, personnel, equipment, and facilities.

Once the reading of the brief resolution was finished, it passed unanimously after only one very brief exchange. Republican leader and former Speaker Joseph W. Martin, Jr., who was subsequently appointed vice chairman of the select committee, asked and received an affirmative answer to only one question: Whether the new committee was generally similar in nature to the Senate committee.

The creation of the select committee was the most important single action by the Congress leading to the establishment of the House Committee on Science and Astronautics and the present Committee on Science and Technology.

In designating Majority Leader McCormack as the chairman of the select committee, Speaker Rayburn made one of his wisest decisions. Spurred by McCormack's imaginative leadership, the new committee immediately plunged into a comprehensive series of public hearings which laid the foundation for the Nation's space policy.

ACCOMPLISHMENTS OF THE SELECT COMMITTEE

During the spring and summer of 1958, the select committee worked at a frantic pace. Three important goals were achieved:

(1) Chartering the permanent House Committee on Science and Astronautics, with an expanded jurisdiction covering science as well as space;

(2) The writing of the Space Act, setting up the National Aeronautics and Space Administration; and

(3) Landmark hearings and special committee reports which helped shape the course of the Nation's space program in the crucial year of 1958.

WHAT LED TO THE CREATION OF THE SELECT COMMITTEE?

In 1958, the Congress seized the initiative. The legislative branch, the people's sounding board, responded quickly and decisively. Meanwhile, the executive branch was divided and sounded an uncertain trumpet.

The beep-beep of the Soviet Sputnik I, launched on October 4, 1957, sent shockwaves through the American public. Surprise, fear, humiliation, and anger were intensified less than a month later when

Sputnik II went into orbit with the space dog Laika. How could those ignorant Bolshevik peasants surpass good old American technological know-how? How could they manage to orbit a 184-pound payload, and then follow with the smooth orbiting of a 1,120-pound payload? How did the Russian scientists and engineers overtake us? These questions were on the lips of Congressmen, officials in charge of our missile and satellite programs, and other national leaders. But even more important, the questions were repeated throughout the land by people high and low who were deeply disturbed.

The prairie fire of demands for action swept across the Nation. The clamor rose to a roar. The American people could sense the serious blow to American prestige around the world. The feeling ran deeper than who could put the biggest payload into orbit. There was a widespread uneasiness about our educational system and why we weren't turning out the scientific and engineering talent to meet the Soviet challenge. Fear of Soviet space missiles gripped the Nation.

The House and Senate leadership showed they were much more in tune with public thinking than a seemingly indecisive executive branch. Speaker Rayburn and his majority leader had early and positive reactions. "When Sputnik went up, naturally we discussed it," recalled John McCormack. "And I knew we were not going to meet the challenge that Sputnik presented to us by just talking. We had to act—we had to act ourselves in the field of outer space. * * * Sam realized the importance of it and he said: 'Well, you'll have to be chairman.' I said, 'All right, Sam I will.'"

The second-ranking member of the House Armed Services Committee, Representative Overton Brooks, of Louisiana, was in Paris when the startling news of Sputnik first broke. His staff assistant, Charles Ducander, recollects:

I'll never forget that we were staying at the George Cinq and we came out of the hotel and bought an American language newspaper—I guess it was the old Herald-Tribune, Paris edition—and here on the front page is the headline—Russia had orbited a satellite. Well, Brooks about jumped out of his skin. He could talk of nothing else. As a matter of fact, we came home two days early. He said: "The first thing I'm going to do when Congress goes back in session is to drop in a bill to form a special committee because we have to catch up with them or surpass them."

Brooks, who in 1959 was tapped to be the first chairman of the standing committee (the House Committee on Science and Astronautics), didn't wait for Congress to convene in 1958 before he stuck in his oar on space policy. The New York Times on October 16, 1957—less than two weeks after Sputnik I—quoted Brooks as calling on the President to appoint a czar over America's missile and satellite program.

MEANWHILE AT THE WHITE HOUSE

The reaction of the executive branch was confusing to the general public. President Eisenhower was asked at his October 9 news conference: "Are you saying at this time that with the Russian satellite whirling about the world, you are not overly concerned about our Nation's security?" In his widely quoted response, the President commented: "Now, so far as the satellite itself is concerned, that does not raise my apprehensions, not one iota." Unfortunately, the President had become convinced that the crisis was partially a propaganda and public relations problem. In later years, in his book *Waging Peace*, President Eisenhower wrote that his short-term problem following Sputnik "was to find ways of affording perspective to our people and so relieve the current wave of near hysteria."

At the same time, President Eisenhower renewed the scientific commitment of the United States to the cooperative multinational program called the International Geophysical Year, which included the development of an Earth satellite by the United States. The President called in a number of scientists for personal consultation, delivered two nationwide addresses on science and defense and on November 7, 1957, appointed James R. Killian, Jr. as Special Assistant to the President for Science and Technology. The President's Science Advisory Committee, which had been placed within the Office of Defense Mobilization, was reconstituted and transferred to the White House on December 1, 1957. In a nationwide radio address, President Eisenhower stressed the need for expanding support of science education at all levels of Government.

While Members of Congress were calling for action, and the public was getting frustrated and infuriated by the "Papa-Knows-Best" advice, there were scores of patriotic men of vision and principle who risked their positions within the Government or military hierarchy by speaking out boldly to define the crisis facing the Nation. People like German-born rocket expert Dr. Wernher von Braun, working at the Army's Ballistic Missile Agency in Huntsville, Ala.; Trevor Gardner, Assistant Secretary of the Air Force in charge of Research and Development; and Lt. Gen. James M. Gavin, Deputy Chief of Staff of the Army in charge of Research and Development—and many others were working effectively as well as sounding the alarm bells in the night.

President Eisenhower suffered from some incredible gaffes by his own staff and official family. The press and the Democrats, of course, seized on and magnified these mistakes, all of which helped sharpen the contrast between what appeared to be a timid executive branch and a forcefully articulate Congress which was seizing the initiative. Outgoing Secretary of Defense Charles Wilson ridiculed Sputnik as a "neat scientific trick"; Wilson added that "nobody is going to drop anything down on you from a satellite while you're asleep, so don't worry about it." Another high administration official, Clarence E. Randall, characterized Sputnik as "a silly bauble," and Presidential Chief of Staff Sherman Adams joked about the "outerspace basketball game." When he wrote his memoirs, *Firsthand Report*, Adams conceded that "Eisenhower said he preferred to play down the whole thing. * * * I was only trying to reflect the President's desire for calm poise."

Despite the frenzied flurry of activity at the working levels of Government, the public gained the distinct impression that Congress was the only branch of Government which had the correct sense of urgency. When the Soviets lofted their second Sputnik in November, The New York Times carried an account the tone of which was duplicated throughout the Nation:

> The White House said today that the new Soviet satellite was "no surprise" as it fell "within the pattern of what was anticipated."
>
> Mrs. Ann Wheaton, Assistant Press Secretary at the White House, said the President had received considerable information in advance on expected Soviet achievements.
>
> Members of Congress, however, increased their clamor for an investigation of the U.S. missile and satellite program.

THE JOHNSON COMMITTEE HEARINGS AND VANGUARD

The boldest, most positive reaction on Capitol Hill came from Senate Majority Leader Lyndon B. Johnson. The Preparedness Investigating Subcommittee of the Senate Armed Services Committee on November 25, 1957, started what turned out to be voluminous hearings on the Nation's satellite and missile programs. A future Secretary of State, Cyrus Vance, was assistant counsel of the investigating subcommittee, which helped dramatize the initiative of the Congress to meet the crisis. One of the recommendations of the Johnson subcommittee was to "start work at once on the development of a rocket motor with a million-pound thrust."

While the Johnson hearings were going on, America was plunged into deeper gloom as the eyes of the world were focused on the spectacular failure of the first attempt by the United States to orbit a satellite. The Navy's Vanguard, with a payload of less than 4 pounds, made a pitiful effort to get off the ground, but blew up on the launching pad on December 6, 1957. The December 7 headlines made it awesomely clear that America had suffered another Pearl Harbor for science and technology.

"This administration does not appreciate the urgency of the situation," proclaimed Majority Leader McCormack. Speaker Rayburn stated that we must "put our scientists and engineers to work." At a December White House briefing of congressional leaders, the press reported that "almost to a man, the leaders came from the meeting in a critical mood. The lack of a 'sense of urgency' in the administration was the main complaint."

On the eve of the meeting of the new session of Congress in January 1958, Robert Albright wrote in the Washington Post:

> Sputnik and the battle for survival implicit in the Soviet satellite-missile advances will in all likelihood dominate the second session of the 85th Congress convening Thursday.

EXPLORER I

At the Army Ballistic Missile Agency in Huntsville, Ala., a team of scientists and engineers led by Dr. Wernher von Braun scored a dramatic triumph with the launch of Explorer I on January 31, 1958, from Cape Canaveral, Fla. The Army was jubilant. Despite the fact that the satellite weighed only a little over 30 pounds, von Braun immediately became a hero and a prophet. If anything, the great event underlined the drama of the space race and furnished still another spur toward action by the Congress.

At the same time, the Army's victory helped fan the fires of intense and bitter competition among the three services attempting to get, hold and expand their pieces of the space and missile action. The strong interservice rivalry was not stilled by President Eisenhower's appointment, early in 1958, of Roy Johnson, a General Electric Co. vice president, as head of the newly created Advanced Research Projects Agency in the Department of Defense.

ESTABLISHMENT OF THE SELECT COMMITTEE

Although the Senate voted on February 6 to establish a Special Committee on Space and Astronautics, with Senator Johnson as chairman, the Senate committee was not very active during its early days. Representative William H. Natcher (Democrat of Kentucky) noted in his journal on March 4, 1958, that almost all of the Senate Space Committee members were committee chairmen or their ranking Republican counterparts. "When the announcement was made of the Senate members of the committee, I in turn tried to figure out in my own mind just who the Speaker would appoint on the House committee," he wrote. "It never occurred to me that I would be named as one of the 13 members of the House committee, since my seniority was not comparable."

Speaker Rayburn explained to Natcher, who had only been elected in 1953, that he wanted active members who would represent all sections of the country as well as the different committees with related jurisdictions.

When the House select committee was officially established on March 5, Speaker Rayburn, Majority Leader (and future Speaker) McCormack and Minority Leader (and former Speaker) Martin—who always worked very closely together—all agreed that membership of the new group must include a "blue ribbon" selection of some of the best and most conscientious members from both sides of the aisle.

MEMBERSHIP OF THE SELECT COMMITTEE (1958)

The 13-member committee included the following, in order of seniority:

Democrats	Republicans
John W. McCormack, Massachusetts, *Chairman*	Joseph W. Martin, Jr., Massachusetts
	Leslie C. Arends, Illinois
Overton Brooks, Louisiana	Gordon L. McDonough, California
Brooks Hays, Arkansas	James G. Fulton, Pennsylvania
Leo W. O'Brien, New York	Kenneth B. Keating, New York
Lee Metcalf, Montana	Gerald R. Ford, Jr., Michigan
William H. Natcher, Kentucky	
B. F. Sisk, California	

The personnel of the select committee reflected Speaker Rayburn's decision that this was a top-caliber committee which should include key representatives of major standing committees. Serving in addition to the majority and minority leaders there were also the top-ranking Democratic and Republican members of the Armed Services Committee (Brooks and Arends), the top-ranking Republican member of the Judiciary Committee (Keating), and key members of the Committee on Appropriations (Natcher and Ford), Interstate and Foreign Commerce (O'Brien), Foreign Affairs (Hays and Fulton), Education and Labor (Metcalf), Interior and Insular Affairs (Metcalf, O'Brien, and Sisk), and Banking and Currency and Joint Committee on Defense Production (McDonough).

Fulton, a free-wheeling millionaire lawyer from Pittsburgh, could always be counted on to spark and spice the public hearings with his offbeat manner of presenting startling ideas. Confronting Dr. Wernher von Braun on the opening day of the committee's hearings, Fulton blurted out:

Why do we not try to ask the President to give you that power to take a crack at the Moon even though you do not hit it? Would you like that?

Or, to a witness describing the Navy's Vanguard satellite:

If you took one light year, that would be six trillion miles, and if the nearest star is 26 trillion miles away, the distance our satellite that you put up will go in 200 years is inconsequential, being only 31½ billion miles.

Fulton also had the distinction of being the first Congressman to introduce a resolution calling for a standing space committee. Fulton's interest even predated Sputnik, for on August 29, 1957, Fulton dropped in a resolution "To establish a joint committee on Earth satellites and the problems of outer space." He reintroduced the resolution on January 9, 1958.

THE HIGH HOPES OF OVERTON BROOKS

Brooks, who next to Chairman McCormack proved to be the most active member of the select committee, had high hopes that he rather than McCormack would become chairman of the new committee. As second-ranking member of the House Armed Services Committee, Brooks was constantly in the shadow of the Chairman, Carl Vinson, who ruled his committee with an iron hand. Brooks was an ambitious man who wanted to advance to be Governor of Louisiana or perhaps U.S. Senator, and to do so he wanted and needed publicity in the Shreveport and other Louisiana papers. Brooks was frustrated that Chairman Vinson was the center of attention, always in the spotlight. Vinson, who thoroughly mastered every detail and always did his homework, did not have a very high regard for Brooks and shuddered at the thought that he was in line to be Armed Services chairman after Vinson retired or passed on.

With a nose for publicity, Brooks saw an opportunity to ride the tremendous public interest in space. He dropped in two resolutions— one to create a joint committee on space, and one to set up a House "Special Committee on Astronautics and Space Exploration," thus hoping to be named chairman of the new body. Speaker Rayburn wanted a strong chairman to match Senate Majority Leader Johnson. Therefore, although Brooks had introduced House Resolution 474 on February 10, and McCormack's Resolution 496 did not see the light of day until March 5, it was McCormack's resolution which was taken up in the House. As a matter of fact, the McCormack resolution breezed through the House without so much as touching base with the Committee on Rules, where the Brooks resolution had languished untouched since February 10.

Brooks was miffed, but like a good soldier he told the House:

> It has been decided best not to take up my resolution, but rather to put support behind the present resolution. I am therefore supporting the present resolution with all of the enthusiasm I can command.

Despite his disappointment at failing to be named chairman of the select committee, Brooks showed a clear understanding of the nature and purpose of the committee. He also was one of the earliest to speak out for "the need for scientific education," which he stressed during the debate on the McCormack resolution.

FUTURE PRESIDENT JOINS SELECT COMMITTEE

A 5th-term member from Michigan, Gerald R. Ford, Jr., was named to the select committee as the lowest ranking Republican. Ford blossomed into an active member, both in the hearings and in the executive sessions which involved drafting the new NASA legislation.

As Ford recalls it,

> I had lots of experience in the Defense Appropriations Subcommittee, and at that time virtually all of the funding for space, missiles, etc., came out of the Defense Appropriations.

When it came time to name the conference committee on the Space Act, Ford was included, although lower in seniority than both Martin and Arends.

Chairman McCormack recruited a small but highly competent staff for the select committee. Heading the staff as chief counsel and director was George J. Feldman, a New York attorney and close personal friend of McCormack, who had cut his legislative eyeteeth with Massachusetts Senator David I. Walsh, and later served as Ambassador to Malta and Luxembourg. From the Legislative Reference Service of the Library of Congress came Dr. Charles S. Sheldon II, who served as assistant director, and Spencer Beresford, special counsel. The regular staff was rounded out by Richard P. Hines, Raymond Wilcove, Harney S. Bogan, Jr., and Philip B. Yeager, who served as a special consultant.

HEARINGS BEFORE THE SELECT COMMITTEE

Shortly after he decided to appoint George J. Feldman as chief counsel and director of the select committee, Chairman McCormack sat down in his office one Saturday morning and scratched out on lined, note-pad paper a personal note which he dropped in the mail to Feldman in New York. In addition to advice on people he wanted Feldman to interview, McCormack wound up the handwritten set of instructions with these important words:

> Frankly, I would like to get the jump on the Senate Committee, so our Committee might in the public mind be the leader and not the follower.

Chairman McCormack organized a stellar series of witnesses to appear before the select committee. In addition, he scheduled morning, afternoon, and occasional evening sessions to help meet the challenge of the rapidly developing scientific subject matter necessary for the committee to cover. Some of the Nation's greatest scientists, rocket engineers, military leaders, and administrators were summoned to explore the nature of the major problems faced.

When the committee was not yet three weeks old, the first witness was called on March 20: Allen W. Dulles, Director of the Central Intelligence Agency. The CIA operatives startled the committee by coming up several days in advance of the briefing to "de-bug" the small room in the Capitol where the session was to be held. According to George Feldman, Dulles read a long statement but declined to leave a copy of the statement with the committee because of its classified nature. Although Feldman noted there wasn't anything in the statement that had not already appeared in the newspapers, a partisan argument quickly flared up between the Democrats who insisted they wanted a copy and the Republicans who defended Dulles' refusal. Finally, Chairman McCormack, on the advice of Feldman, quieted the furor by suggesting that any member who wanted to do so could go down to the CIA and look at a copy.

During a discussion before the full committee in 1959, Chairman Brooks mentioned: "We have plans to have the CIA appear in executive session." McCormack countered: "If you get as much out of them as the select committee got out of them, you won't know any more than you do at this time."

In the public hearings through May 12, the hearing record constituted 1,542 pages. Led by Dr. Wernher von Braun, the witnesses included top rocketry and scientific talent in the three military services, on duty and retired, including Rear Adm. Hyman G. Rickover, Maj. Gen. Bernard A. Schriever, Lt. Gen. James M. Gavin, Maj. Gen. John B. Medaris, Rear Adm. John T. Hayward, Brig. Gen. H. A. Boushey, and Dr. Herbert F. York. From the National Advisory Committee for Aeronautics came its Chairman and Director, Drs. James H. Doolittle and Hugh L. Dryden. From the academic world came Dr. Lee A. DuBridge, president of the California Institute of Technology, and Dr. James A. Van Allen of the University of Iowa. Among those testifying from private industry were Kent T. Keller, former president of Chrysler Corp., Dr. Walter S. Dornberger of Bell Aircraft Corp., and Krafft Ehricke of Convair. In addition, representatives and leaders of the National Science Foundation, Weather Bureau, Department of State, and other Federal agencies testified.

All the witnesses discovered that the committee members had done their homework thoroughly, as a result of which they found that most committee meetings extended far beyond their allotted time in probing questions. Aside from the brief flareup during the Dulles hearing, there was a remarkable degree of bipartisanship shown in both the sessions and the reports developed by the committee. A strong partisan Democrat as majority leader, McCormack ran the select committee with a conscious effort to weld a unity of approach. He successfully achieved this aim.

On April 16, Dr. Hugh L. Dryden appeared before the committee, armed with impressive credentials as the Director of the National Advisory Committee for Aeronautics. Dr. Dryden, whose NACA was merged into the new NASA, was everybody's logical choice to head the new space agency up to the time he stumbled before the select committee. In response to a question about manned space flight, Dr. Dryden made the offhand comment that "tossing a man up in the air and letting him come back" had about the same technical value as the circus stunt of shooting a young lady out of a cannon. The remark received wide press coverage, and as a result Chairman McCormack, Representative Fulton, and most members of the committee immediately became disenchanted with Dryden and informed the White House they opposed his appointment as head of the new space agency.

When Dr. Dryden returned to the committee on April 22, Chairman McCormack reminded him of his circus cannon statement and asked:

Do you want to amplify that, or clarify it, or explain it, or anything you want?

In response, Dr. Dryden pretty well stuck to his original position, summarizing:

My statement was not directed in criticism of any specific program, but was intended to illustrate the wide variety of simple experiments, which give you little information, to much more complicated and costly experiments which give you a great deal more information.

Chairman McCormack generally voiced the conclusions of the select committee members when he observed:

Some people thought, assuming an agency were established and you were appointed the Director, the head of it, that it might indicate the state of your mind on your part where you are more wedded to the past activities of your organization than the future activities.

Dr. Dryden, who subsequently served as Deputy Administrator under NASA's first two Administrators, T. Keith Glennan and James E. Webb, also did not endear himself to the select committee by the tone of his testimony on August 1, 1958, during consideration of a construction authorization for NASA. The exchanges developed as follows:

Mr. FULTON. You would say this program is no attempt to leapfrog the Soviets' plans to get ahead of them?

Dr. DRYDEN. In all honesty, I would have to say that the prospective space programs are not such as to leapfrog the Soviets immediately, or very soon.

Mr. FULTON. Thank you.

Mr. BROOKS. Is this an attempt to catch up with the Soviets' program?

Dr. DRYDEN. This is an attempt to establish a national program for the United States. It starts at a beginning which I think is adequate. It most decidedly is not a crash program to catch up with anybody.

Members of the Select Committee on Astronautics and Space Exploration prepare to hear testimony by Dr. Hugh L. Dryden on April 16, 1958. Model of the Echo communications satellite is in background.

Seated at table, left to right: George J. Feldman, chief counsel and director of the select committee; Dr. Dryden (later Deputy Administrator of NASA); Representative John W. McCormack (Democrat of Massachusetts), chairman of the select committee.

Standing, from left: Representatives William H. Natcher (Democrat of Kentucky), James G. Fulton (Republican of Pennsylvania), Lee Metcalf (Democrat of Montana), B. F. Sisk (Democrat of California), Gordon L. McDonough (Republican of California), Leo W. O'Brien (Democrat of New York), Representative Kenneth B. Keating (Republican of New York), and Gerald R. Ford, Jr. (Republican of Michigan). Those select committee members not included in the photo were Representatives Overton Brooks (Democrat of Louisiana), Brooks Hays (Democrat of Arkansas), Joseph W. Martin, Jr. (Republican of Massachusetts), and Leslie C. Arends (Republican of Illinois).

Mr. BROOKS. As I understand it, you mean this is not in any sense competitive with the Soviet program and as you make advances you do not check the advances as against the Soviet program?

Dr. DRYDEN. I would say that this program is not at a level at which we could guarantee to do that; that is correct.

Many other witnesses, including von Braun, Dornberger, General Gavin, Admiral Rickover, and General Boushey, strongly supported the need for manned flight in their testimony before the select committee. General Boushey (Deputy Director of Air Force Research and Development) told the select committee:

Another function which I believe only man can perform effectively is that of interception and mid-space rendezvous. At first, such missions probably would be for the purpose of refueling, thus permitting a manned, maneuvering space vehicle to receive fuel from an uninhabited tanker satellite which might have been circling in orbit for months or years. Eventually the capability to control space would be augmented by the ability of manned military spacecraft to make interception or rendezvous in space.

Chairman McCormack assigned a subcommittee headed by Representative Natcher to examine the issue of unidentified flying objects. At this point, the select committee was not in a position to verify or examine in depth the numerous UFO sightings, but was interested in a briefing of what the Air Force had uncovered in its compilation and analysis of the subject. According to Representative Natcher, "We borrowed Les Arends' minority whip office in the Capitol, hung a little sign on the door reading 'Subcommittee on Upper Atmospheric Phenomena,' and as a result we could conduct our hearing without any outside fuss or interference from anybody."

The select committee hearings were very thorough, developed logically, were well organized, the attendance by members was unusually high, press and public interest was great, and the witnesses were drawn from a wide cross section of knowledgeable leaders in the scientific world, military, the Government, and private industry. Chairman McCormack was clearly the dominant force in leading the questioning in the hearings, and in maintaining the high level of interest which was displayed throughout.

BIRTH OF THE HOUSE COMMITTEE ON SCIENCE AND ASTRONAUTICS

The leadership and jurisdiction of the House Committee on Science and Astronautics developed from a fascinating interplay among congressional personalities and the demands of the times. By the summer of 1958, the Nation had come to the sobering realization that the threat to the United States ran far deeper than a mere space race with the Soviet Union. At stake was a serious challenge to American education, basic research, the training of scientists and engineers, and the entire spectrum of support for the development of science and technology. Soviet mastery of space loomed as a military threat.

The House Committee on Science and Astronautics was authorized by House Resolution 580 on July 21, 1958. The birthday of the Senate Committee on Aeronautical and Space Sciences did not occur until three days later on July 24, 1958. More significant was the fact that the jurisdiction of the House committee was markedly broader than that of the Senate committee.

The jurisdiction of the Senate Committee on Aeronautical and Space Sciences was stipulated as follows:

Aeronautical and space activities, as that term is defined in the National Aeronautics and Space Act of 1958, except those which are peculiar to or primarily associated with the development of weapons systems or military operations.

Matters relating generally to the scientific aspects of such aeronautical and space activities, except those which are peculiar to or primarily associated with the development of weapons systems or military operations.

National Aeronautics and Space Administration.

Meanwhile, the House of Representatives voted to give the new House Committee on Science and Astronautics the following broader charter of jurisdiction:

Astronautical research and development, including resources, personnel, equipment, and facilities.

Bureau of Standards, standardization of weights and measures, and the metric system.

National Aeronautics and Space Administration.

National Aeronautics and Space Council.

National Science Foundation.

Outer space, including exploration and control thereof.

Science scholarships.

Scientific research and development.

The broader jurisdiction of the House committee, as well as the issue of who should chair the new committee, developed primarily because of personality issues. House Majority Whip Carl Albert of Oklahoma, small in stature but brilliant in rhetoric and legislative craftsmanship, was the author and prime mover behind House Resolution 580 which created the new committee.

"They talked about making me chairman of the committee," Albert recalled. "And John McCormack wanted me to be chairman."

McCormack had another interest which had become stronger over the years—that there should be a Department of Science of Cabinet rank. This deep interest manifested itself during the complex maneuvering over the chairmanship and jurisdiction of the new standing committee.

In addition, McCormack observed:

I wanted to create a committee that had strength, because I pictured in my own mind the importance of science in the world of tomorrow. * * * I wanted a committee that had teeth in it, that covered the entire field. * * * I wanted to have a committee that had some power where Members would want to get on, seek the committee assignment because of the challenge it meant to them, as legislators and in connection with the national interest of our country and the world of tomorrow. * * * I considered it one of the most important committees of the Congress.

The catalyst was Carl Vinson, the powerful chairman of the Armed Services Committee. Vinson, as has been noted, looked down his nose at Brooks and was adamant that Brooks, the ranking Democrat on the Armed Services Committee, should never succeed him as chairman of that committee.

According to Carl Albert,

Carl Vinson came over to see Rayburn and they called me over and said: "Listen, we don't want Overton Brooks ever to be chairman of the Armed Services Committee. * * * He's a troublemaker, a griper, and a groucher and (Paul) Kilday is steady and solid and knows the business and he should be the next chairman of the Committee on Armed Services."

Turning to Carl Albert, Speaker Rayburn confided:

We would give you the committee were it not for that fact, but that is an overriding factor. [Ironically, Brooks died in 1961, Vinson stayed on as Armed Services Committee chairman for over three years following Brooks' death, and Vinson celebrated his 96th birthday on November 18, 1979.]

Now came the problem of how to insure that Brooks would be forced to relinquish his post on the Armed Services Committee. It became necessary to broaden the jurisdiction of the House Committee on Science and Astronautics in order to enhance its status, making it a major committee, so that Brooks would be ineligible to remain on Armed Services while chairing a major committee. Here is where McCormack's interest in a Department of Science entered the picture, and McCormack was influential in helping to define the new jurisdiction along broader scientific lines. At the same time, Carl Albert was commissioned as Speaker Rayburn's trouble-shooter to buttonhole Chairman Oren Harris of the Interstate and Foreign Commerce Committee, whose committee would be forced to give up some jurisdiction to the new Science and Astronautics Committee.

"He gave in, but he didn't do it very easily," recalled Albert. "He twitched around a little bit about it, but he had Rayburn and McCormack on his neck so he had to do it."

JOINT COMMITTEE OR SEPARATE HOUSE AND SENATE COMMITTEES?

Early in 1958, both the House and Senate select and special committees were thinking in terms of creating a Joint Committee on Aeronautics and Outer Space. The concept of a joint committee was drawn from the experience of the Joint Committee on Atomic Energy.

Despite the known opposition of Speaker Rayburn to this concept, the joint committee was written into all the early drafts of the proposed Space Act legislation. At executive sessions held in the House Ways and Means Committee room, just off the House floor, on May 13, 14, 19, and 20, the joint committee concept remained in the bill and was unanimously approved by the select committee. When Chairman McCormack introduced a clean bill, H.R. 12575, on May 20, the joint committee not only survived but also received strong support in the committee report (House Report No. 1770), dated May 24:

> The select committee gave serious consideration to the establishment of standing committees (on aeronautics and outer space) in the House and Senate, but decided instead on the establishment of a joint committee. The provisions of title IV of the bill are patterned closely after the provisions of the Atomic Energy Act creating the Joint Committee on Atomic Energy. * * *
>
> Such a committee would provide a number of advantages. In addition to preventing possible conflicts and omissions, as well as unnecessary duplication, it would give Congress the means to oversee executive operations effectively in the highly important and urgent field of space flight.

Something funny happened to the idea of a joint committee on the way to the House floor. Between May 24, when the House select committee had glowingly endorsed a joint committee and June 2, when the bill was taken up in the House of Representatives, the joint committee suddenly fell from favor. Introduced on May 27, Carl Albert's Resolution 580 sailed through the Committee on Rules on May 29, putting the leadership squarely on record in favor of a separate standing Committee on Science and Astronautics for the House. When Majority Leader McCormack was explaining the action in killing the joint committee, he told the House of Representatives on June 2:

> In the bill we provide for a joint committee, but we have eliminated that, and I am going to offer a motion to strike that out, because the Committee on Rules has reported out a resolution within the past few days establishing another standing Committee on Science and Astronautics, which gives it a broad base of legislative action, and in the light of that it will be unnecessary to continue the joint committee in this bill; at least, the members of the select committee feel that way.

Chairman McCormack told the House that on the morning of June 2, just before the bill reached the House floor, the select committee had unanimously agreed to strike the joint committee from the bill. McCormack moved on the floor to excise the joint committee, and his motion received no comment, debate, or objection and was accepted immediately without opposition. When the Senate considered the proposal, there were only passing references to the advantages or disadvantages of a joint committee. Senator Jacob Javits (Republican of New York) asked Senator Johnson whether the joint committee would have oversight jurisdiction over the National Aeronautics and Space

Policy Board. Johnson's response did not seem to indicate any deep and lasting commitment to the joint committee idea:

> We felt that the joint committee, if in its wisdom Congress decides to establish it, will have oversight jurisdiction, and it is given such jurisdiction. We feel that is a very necessary and desirable part of the joint committee's functions.

There was some speculation that the early House support for a joint committee stemmed from the feeling that it might be easier to wrest new jurisdiction away from existing committees toward a joint committee rather than toward a new standing committee. Once the jurisdiction was obtained, it may have been less painful to shift it when the joint committee concept was dropped. Looming larger as a reason was the fact that many House Members feared the Senators on a joint committee might "hog" the limelight.

Although House Members and staff kept pointing at the Senate and charging they were the ones who were trying to "put across" a joint committee, Majority Leader Johnson's commitment seemed to be less substantive and more to regard this item as a bargaining chip which could be used to muscle the House into accepting some other provision which the Senate felt was more important.

The sudden 180-degree reversal in the House position from support of, to opposition to, the joint committee, came while the delicate negotiations to get Overton Brooks off the Armed Services Committee were underway. Speaker Rayburn was at the center of those negotiations. When the House-Senate conference committee convened on July 15, there was a dramatic scene at which the joint committee was buried for good. The smoking pistol of the executioner came clearly into view. Philip B. Yeager of the select committee staff remembers it this way:

> The first hang-up we had was whether there was going to be a joint committee. Johnson, as I remember said:
> "The first thing we are going to do is we are going to have a joint committee. I guess everyone has agreed on that, haven't they?"
> He looked around, and McCormack was just sitting there, shaking his head. Johnson said:
> "We're not going to have a joint committee?"
> "No."
> "Why not?"
> "Mr. Sam says so."
> McCormack added: "If you want to negotiate further, you'll have to settle that at the Texas level."

COMMITTEE ON SCIENCE AND ASTRONAUTICS OFFICIALLY SANCTIONED

Once the joint committee had been shelved, the way was paved for the House of Representatives to act separately to establish the standing House Committee on Science and Astronautics. Two relatively junior Members of the House, both low on the totem pole in the prestigious Committee on Rules, shared honors in reporting Carl Albert's Resolution 580 out to the House of Representatives. The report from the Committee on Rules was written by a huge, St. Bernard-like Irishman from Cambridge, Mass., named Thomas P. "Tip" O'Neill, Jr., later to become Speaker of the House—who in 1958 was only in his sixth year in the House. Floor leader in charge of the debate on Albert's Resolution was a comparatively junior Congressman from Missouri, a protege of Speaker Rayburn named Richard Bolling—who was serving his 10th year in the House.

O'Neill's report included these significant comments:

> The purpose of and reason for this resolution is to set up a committee having full and complete jurisdiction in a broad area that has come to have great significance in recent years. Aside from the spectacular developments which are being made in outer space research and which have both military and civilian importance, mankind has reached that stage in the development of science and the industrial arts where governments must, as a matter of survival, give new emphasis and attention to basic research. Legislative action in those fields is certain to become a matter of greater frequency and greater importance in the near future. We think we have come to the time in which a committee with across-the-board jurisdiction in this area should be established. Our Government is now engaged in considerable research efforts in many fields of pure science, and it is the part of wisdom that these efforts be studied and examined from a legislative angle, and the establishment of this committee emphasizing this field will make a marked contribution in this direction.

When Bolling brought up the Albert resolution on July 21, there was very little debate and no opposition on the House floor. Bolling's statement, as is his custom, was succinct and to the point:

> The standing committee will take over, and continue, the work started by the House Select Committee on Astronautics and Space Exploration. Certain functions of the Committee on Interstate and Foreign Commerce and the Armed Services Committee will be transferred to this committee; namely, legislation relating to the scientific agencies—the Bureau of Standards, the National Advisory Committee for Aeronautics (NASA had not yet officially been launched) and the National Science Foundation. The chairmen of the Interstate and Foreign Commerce Committee and the Armed Services Committee agree with these proposed transfers.

Bolling was perhaps technically accurate when he professed that Oren Harris and Carl Vinson agreed with the proposed transfers. Certainly Vinson, who wanted so badly to remove Overton Brooks from his committee, saw the logic of the jurisdictional transfers. But as time went on, both Harris and Vinson screamed lustily as the long tentacles of the fledgling committee began to reach into areas the old warlords regarded as their private domain. Hell hath no fury like a committee chairman who feels another committee is impinging on his jurisdiction!

During the debate on the Albert resolution, Majority Leader McCormack underlined the new scientific responsibilities which would fall to the Committee on Science and Astronautics:

This is a clear recognition on the part of the House of the importance of basic and applied research and the establishment of this committee as a standing committee to which legislation of that nature will be referred. It includes not only outer space legislation but it takes over other activities, and it is going to be, in my opinion, one of the most important committees of both branches of the Congress.

Republican Leader Martin, in endorsing the resolution, noted:

Mr. Speaker, as one who has been privileged to serve on this special committee, I want to say that I am heartily in favor of making this committee a permanent part of the House legislative system. * * * I want to pay my tribute to the nonpartisan way in which the committee has worked under Congressman McCormack. There was never the slightest semblance of partisanship shown at any of our hearings or in our committee votes.

The Albert resolution passed very quickly and unanimously. Thus the effective date of the new standing Committee on Science and Astronautics was scheduled for the opening of the new Congress on January 3, 1959.

THE WRITING OF THE SPACE ACT AND THE ESTABLISHMENT OF NASA

While the battle was going on to establish the House Committee on Science and Astronautics, the House Select Committee in the spring and early summer of 1958 was hammering out the Space Act which chartered the National Aeronautics and Space Administration.

It would be difficult to duplicate the feats performed by the select committee and its staff; in a few short weeks they accomplished the impossible. Here was a new committee, just established on March 5, with new staff just getting acquainted with each other, headed by the majority leader, minority leader, and minority whip who had other pressing official duties, suddenly plunged into the maelstrom of uncharted seas. The hot glare of publicity shone on their every action. The public was fearful and apprehensive at dramatic blows to America's prestige by the Soviet Union. The staff and members had to perform a triple function simultaneously: to get educated on the complexities of astronautics and space, to exercise oversight over the existing confusion which was the administration's space program, and to draft a new charter for a major regrouping of functions relating to the entire administration program. All these challenges had to be met yesterday, it seemed, and the targets were constantly moving.

The timetable reflected a true sense of urgency in the Congress. President Eisenhower on April 2 sent a message to the Congress along with an administration bill to absorb the National Advisory Committee for Aeronautics into a new National Aeronautics and Space Agency. The message arrived just as Congress left for a brief spring recess. The break-neck speed with which the select committee moved is reflected in the fact that Chairman McCormack opened public hearings on April 15, held 17 sessions through May 12, and charged onto the House floor with the NASA bill on June 2—just two months after the President had first revealed his recommendations on the new space agency. After a difficult and delicate series of negotiations with the Senate, the conference report was approved July 15, passed by the House on July 16, and signed by the President on July 29.

The speed as well as the depth and thoroughness of the committee's work will long stand as some kind of record for dedication and drive under tremendous pressure.

By way of contrast, the Senate moved more slowly, heard fewer witnesses (48 for the House and 20 for the Senate), met fewer days (17 for the House and 6 for the Senate), and called most of its witnesses from within the executive branch. The House, on the contrary, summoned not only military and civilian space experts, but sought the advice of scientists, university professors, and leaders in the aerospace industry.

The value of the hearings by the House select committee was clearly demonstrated by the questioning and probing which revealed the weaknesses and ambiguities in the administration bill. For example, committee members, staff, and witnesses soon discovered that the bill needed to be beefed up to strengthen congressional oversight and control, to cover international cooperation, relations with the Atomic Energy Commission, as well as the overall policy determination and coordination.

At first, administration officials balked at the very idea that staff underlings in Congress could improve on their fine handiwork. William Finan of the Bureau of the Budget, who helped draft the administration bill, lunched at the Congressional Hotel with Committee Staff Director Feldman and Assistant Director Sheldon, and according to Feldman he said:

"Everybody, you know, is getting carried away by this space thing—the Soviet Union beat us into space but we mustn't panic."

And then he handed us a bill that the administration wanted as the basis for the Act and he said:

"We don't want any changes in this." I didn't want to get into any argument with him—I didn't say anything to him—but I did report that back to the committee and the committee paid no attention to it at all, including the Republicans.

Once the administration witnesses and staff discovered the House select committee really meant business, had the facts and the know-how, and wasn't about to be pushed around, attitudes suddenly changed. The administration quickly joined in, acknowledged the need for improvements, and helped draft amendments to try and fill the gaps exposed in the hearings and discussions.

A great amount of time in both open hearings and executive sessions was spent in determining more precisely agency jurisdictions, defining the fine line between military and civilian activities in outer space, and clarifying the machinery for coordination. As time went on, there developed a healthy give-and-take between the administration and the House committee. Because Chairman McCormack was anxious to draft a bill which would be satisfactory to both the committee and President Eisenhower, regular liaison was established between the committee staff and Bryce Harlow, Deputy Assistant for Congressional Affairs, and Edward A. McCabe, Administrative Assistant to the President. In fact, the daily liaison between the White House and the House side of the Capitol proved to be superior to the exchange of information between the House and the Senate. Under the stress of time requirements and pride of authorship, there developed a spirit of competitive one-upmanship between the House and Senate, complete with "confidential" committee prints, and some dog-in-the-manger attitudes toward privileged strategy and tactics.

When the Space Act was voted on in the House of Representatives on June 2, Republican Whip Les Arends made a cogent observation about the investment concept of space spending:

> The original thought of the administration was that the costs of an adequate program under the proposed space agency would be between $100 million and $200 million a year to start with. After going into the matter carefully, however, and in light of the long leadtimes and exploratory activities necessary to the development of astronautical techniques, the committee has concluded that costs may approach $500 million a year for the first several years and perhaps $1 billion a year thereafter.
>
> This is a lot of money. Possibly, on further inquiry, we may find that amount will not be needed. But even if it is, I suggest to you that the probable cost to the Nation of not spending it will be infinitely more. Besides, as other Members have already told you, the peaceful economic benefits and savings to result from the program should begin to more than pay its cost within a few years' time.

In changes which survived the legislative process, the House altered the "National Aeronautics and Space Agency" to a "National Aeronautics and Space Administration," and replaced "Director" with an "Administrator." The House bill as well as the final act added an important freedom-of-information section which, as stated in House Report 1770—

> affirms the intent of Congress to let the people know all the facts, and to promote the spread of scientific knowledge, subject only to necessary security restrictions.

PATENT POLICY

When the administration sent its proposed bill to Capitol Hill, there was no language included concerning patent policy, a subject of intense interest to the aerospace contractors who would play such an important role in the space program. On May 24, when the House select committee made its initial report, a patent provision was included which was based primarily on the language and approach of the Atomic Energy Act, enabling the Government to retain control over patents resulting from research sponsored by NASA. The Senate committee included comparable patent provisions. By the time the bill reached the Senate floor for debate, Senator Johnson was subjected to intense pressure by a number of contractors, patent lawyers, and others who contended that the aerospace industry differed sharply from atomic energy, where most developments were Government controlled. In order to give the Senate a free hand in conference, Senator Johnson withdrew the Senate committee patent provision during the floor debate on the Space Act.

Prior to the meeting of the conference, Chairman McCormack appointed a Patent Subcommittee headed by Representative Natcher, and including Representatives Hays, Metcalf, Arends, McDonough, and Keating. The subcommittee and its staff discussed the problems involved with many interested parties, both Government and private, for several weeks. Although the Natcher subcommittee made some formal recommendations softening the original House patent language, it is interesting that the final version written into the Space Act was based primarily on a new draft produced at the conference committee by the Senate staff. As the conference opened, Select Committee Staff Director George J. Feldman observed:

> Had I seen the Senate version before last night, I would have recommended the adoption of the Senate provision with a few minor changes, instead of the House proposal.

Obviously pleased, Johnson, who was chairing the conference, commented: "Well, there is a sign of a big man." The conference recessed for a few minutes and the House and Senate staffs came up with an agreement which was then incorporated into the Space Act as approved.

The final language written into law essentially gave title to the United States of those inventions made pursuant to NASA contracts, but gave authority to the NASA Administrator to waive title. Controversies over titles were to be referred to the Patent Office, subject to appeal to the Court of Customs and Patent Appeals.

In arguing for the adoption of the conference report, Chairman McCormack made these remarks about the patent provisions during the House debate:

The original patent provision was too closely patterned after the stringent requirements in the Atomic Energy Act which are not fully applicable to the space field. The substitute provision agreed to by the conferees protects both the interests of the Government and affords enough flexibility to the Space Administrator to let him meet needs for preserving the incentives of the individuals and companies whose efforts it is public policy to encourage.

The 1958 Act did not settle the controversy over patent rights, despite the initial attempt by the select committee and the Congress to meet the issue head on. Down through the years, droning through voluminous pages of very legalistic testimony, the issue remained one of the most complex to be tackled by the committee. From the start, contractors felt that the 1958 act went too far in depriving them of the benefit of their inventions which involved heavy investments. As we shall see, the original provisions of the 1958 act on patent policy were softened by administrative rulings in subsequent years.

THE NATIONAL AERONAUTICS AND SPACE COUNCIL

During the debate on the Space Act of 1958, the House had provided for an advisory committee and the Senate for a policy board. To achieve a compromise, Senator Lyndon Johnson persuaded the President that the Space Council would not erode his power if the President were made the statutory Chairman of the Council. Senator Johnson sold the idea to the House and the conference committee.

When Chairman McCormack presented the concept of the Council to the House of Representatives on July 16, he was eloquent:

Like the National Security Council, this new group will bring together a small number of top leaders of Government, and additionally allows the President to recruit leaders in science and administration from private life to advise him on the overall needs for a thoroughgoing national program and how it should be divided and co-ordinated between the Department of Defense and the National Aeronautics and Space Administration. * * * The result is to place the space program at the high level of Government that its great importance deserves.

Subsequently, President Eisenhower never filled the position of executive secretary of the Council. The Council lapsed into innocuous desuetude. James R. Killian, Jr., writing in *Sputnik, Scientists, and Eisenhower*, put it this way: "The Space Council never did very much during the Eisenhower administration, to the relief of the officers of NASA." It was revived under President Kennedy when Vice President Lyndon Johnson was made Chairman of the Council.

ANNUAL AUTHORIZATIONS FOR NASA

The House select committee in a somewhat roundabout way also helped strengthen the power of its successor, the House Committee on Science and Astronautics. History may conclude that it was in spite of, rather than because of, the House select committee. In any event, the issue arose in a somewhat casual way without the kind of preplanning which Speaker Rayburn liked on his taut ship.

A few weeks after the President had signed the Space Act on July 29, the House of Representatives took up a supplemental appropriations conference report on August 20. Congressman Gerald Ford suddenly arose to object vigorously to a conference committee provision which had been inserted in the Senate, reading:

> No appropriation may be made to the National Aeronautics and Space Agency [sic] unless previously authorized by legislation hereinafter enacted by the Congress.

Ford proceeded to denounce a provision which he argued placed an unnecessary burden on NASA. He contended:

> In effect, what you are telling the people of this new agency is that they have to spend about half their time up here first before an authorization committee and then before an appropriations committee to get any money whatsoever for their operations. Instead of * * * spending the maximum amount of time in running their agency and trying to give us the needed impetus to get ahead or stay ahead of the Russians, they are going to be up here justifying every penny they get for operations and construction before four committees of the Congress.

What Ford did not mention, of course, was that the annual authorizations required of NASA were the real tools for legislative oversight needed to give muscle to the House Committee on Science and Astronautics. Looking back in 1978 on his fight against annual authorizations, Ford reminisced:

> I, having been on Appropriations, was always suspicious that annual authorizations would interfere with the appropriations process and I think there was some justification for that concern. I think it is probably less so now than it was at the outset, but the original concept there was a real, legitimate concern * * *. We all suspected and we can't prove it that Lyndon wanted that annual authorization because it gave him a vehicle to keep himself in the spotlight.

Ford's forceful rhetoric quickly won over other members of the House select committee.

"I want to associate myself emphatically with the gentleman from Michigan," proclaimed Representative Kenneth Keating of New York. "This agency is going to be concerned with a great many matters that are vital to the future welfare of this country. To hamstring them this way is a great mistake."

Representative Gordon McDonough of California wanted to know:

> How, for instance, are we going to continue on a program of research on cosmic rays or satellites * * * where it requires research and development for months and

months and perhaps a year, if we come up to a point where we have to come back to a committee and say, "Well, we have gone so far, and we ask for a few more million dollars." This is a ridiculous provision.

The dominoes began to fall. Representative B. F. Sisk (Democrat of California) joined the concert:

I, too, was concerned about this language when I found it in the Senate discussion of the matter. Does the gentleman agree with me that this is in direct contradiction to the language we placed in the original authorization bill for the agency?

Ford decisively responded: "It seems to me it is about 95 percent in opposition to the basic legislation for the space agency."

Everybody rushed to get into the act. Representative Walter Judd (Republican of Minnesota) eloquently declared:

The gentleman from Arkansas (Mr. Hays) and I were in England in the late summer of 1944 when Hitler sent across his first V–2's. That was just two months or so after our Expeditionary Force had left England to land in Normandy. It was said in all quarters that had Hitler been able to launch his V–2 just 4 months earlier, he might have won the war. Are we here today to take chances on four, six, eight months, or a year of delay in this most important field?

Ford warmed to his task:

If this language which is in disagreement is included, before the Space Agency can hire one clerk, one single clerk to do some typing, they have to come to Congress and get an authorization by the Congress on an annual basis.

Against the flood of oratory, there was only one Congressman (not on the House select committee) who dared to stand up and fight for the annual authorization. He was Representative Albert Thomas (Democrat of Texas) chairman of the Independent Offices Subcommittee which handles NASA appropriations. Thomas lifted his glasses high on top of his forehead and twanged away:

What is wrong with them coming over and letting the Congress determine? After all, we do the legislating. Is there anything wrong in the Congress legislating? Whose duty is it to legislate, the executive's or that of the legislative branch?

But Thomas was at this point like King Canute trying to sweep back the ocean. Ford and his allies won the day to knock out the provision requiring annual NASA authorizations, by a decisive vote of 236 to 126.

The aftermath of the House action was sudden.

Dr. Sheldon reached Chairman McCormack in New Hampshire where he was vacationing. According to Sheldon:

It's the only time ever that McCormack really chewed me out. He was furious. It turns out that he had made an agreement, in private, with Lyndon Johnson to accept the change and had failed to tell any of us. Lyndon Johnson then called Sam Rayburn, and said, "This has to be undone."

Speaker Rayburn moved his big guns out to reverse the August 20 decision in the House. The Senate voted 86–0 to stand firm in its position and the conferees reassembled. A compromise was reached, and the provision was reworded to require authorizing legislation only for the period until June 30, 1960. Ford thundered:

It is my opinion that this one-year trial run will prove the unsoundness of following the Senate position.

Keating chimed in: "I am sure it will." The chairman of the Committee on Appropriations, Representative Clarence Cannon (Democrat of Missouri) concluded, "The effect of it is to leave the whole matter for the next Congress."

With very little fanfare, it might be noted, Congress in 1959 quickly reintroduced the requirement for annual NASA authorizations and the measure went through with no objection. But just as Lyndon Johnson consistently and effectively supported the space program from the very start, and furnished strong executive leadership as President, so it must be noted he furnished the House Committee on Science and Astronautics with one of its most powerful weapons for effective oversight.

REPORTS OF THE HOUSE SELECT COMMITTEE

By year's end in 1958, the House select committee had turned out an impressive series of detailed studies and reports which received wide and favorable recognition. Perhaps the most popular was the space primer, entitled "Space Handbook: Astronautics and Its Applications," prepared by the Rand Corp. in accordance with policy guidance and with the editorial assistance of the committee staff. The Space Handbook with illustrations was a 252-page product which covered space environment, trajectories and orbits, rocket vehicles, propulsion systems, propellants, internal power sources, structures and materials, flight path and orientation control, guidance, communication, observation and tracking, atmospheric flight, landing and recovery, environment of manned systems, space stations and extraterrestrial bases, nuclear weapons effects in space, cost factors and ground facilities, observation satellites, meteorological and navigation satellites, balloon satellites, bombing from satellites, scientific space exploration, and astronautics in the U.S.S.R. and other countries. The handbook was so popular it was reprinted in paperback.

One of the earliest publications of the select committee was "The National Space Program," a 236-page document in layman's language which was published May 21, 1958, and was written primarily by Frank B. Gibney, a committee consultant.

Among other studies and reports completed by the select committee were the following:

International Cooperation in the Exploration of Space.
Survey of Space Law.
The International Geophysical Year and Space Research.

Of interest also is an unofficial "Legislative History of the Space Law," totaling 1,346 pages of text and copies of numerous drafts of the 1958 Space Act, prepared by Raymond Wilcove of the staff of the House Select Committee but never formally approved for release. A copy of Mr. Wilcove's study is deposited in the Library of Congress.

A very significant study compiled by the House select committee, entitled "The Next Ten Years in Space 1959–1969" was not officially approved for release until early 1959 and therefore will be discussed in the next chapter on the beginnings of the House Committee on Science and Astronautics.

The House Select Committee on Astronautics and Space Exploration came to an end on January 3, 1959. The select committee provided a smooth launching pad for the House Committee on Science and Astronautics and its successor, the present House Committee on Science and Technology.

Membership and staff of select committee: Front row, from left, Representatives William H. Natcher (Democrat of Kentucky), Lee Metcalf (Democrat of Montana), Overton Brooks (Democrat of Louisiana), John W. McCormack (Democrat of Massachusetts), Joseph W. Martin, Jr. (Republican of Massachusetts), Gordon L. McDonough (Republican of California), and James G. Fulton (Republican of Pennsylvania). Second row, from left, Mary Myron, Richard P. Hines, William Coblenz (on loan from LRS), Raymond Wilcove, Dr. Charles S. Sheldon II, George J. Feldman, Spencer M. Beresford, Philip B. Yeager, Joseph Moran, and Jean Cameron. Other members of the select committee not included in the photograph were Representatives Hays, O'Brien, Sisk, Arends, Keating, and Ford.

Representatives R. Walter Riehlman (Republican of New York) and Emilio Q. Daddario (Democrat of Connecticut) converse with President Dwight D. Eisenhower during inspection of George C. Marshall Space Flight Center at Huntsville, Ala., September 8, 1960.

The committee in its first year:

From left: Representative David S. King (Democrat of Utah), Charles F. Ducander, (executive director and chief counsel), Representatives J. Edward Roush (Democrat of Indiana), Ken Hechler (Democrat of West Virginia), Dr. Wernher von Braun, Representatives Walter H. Moeller (Democrat of Ohio), James M. Quigley (Democrat of Pennsylvania), Victor L. Anfuso (Democrat of New York), James G. Fulton (Republican of Pennsylvania), Dr. Charles S. Sheldon II (technical director).

In front: Representatives Overton Brooks (Democrat of Louisiana) and B. F. Sisk (Democrat of California).

CHAPTER II

The Overton Brooks Years, 1959–61

At the stroke of noon on January 3, 1959, the House Committee on Science and Astronautics officially came into being.

A grim reminder of the challenge facing the new committee was contained in screaming black headlines announcing that on January 2 the Soviet Union had launched another heavy rocket with an instrumented payload of 796 pounds, headed toward the Moon. Even though "Lunik" missed the Moon and eventually orbited the Sun, it was front-page news during the first week of the committee's existence.

The new committee was authorized to have 25 members—16 Democrats and 9 Republicans. The increased number of Democrats reflected the larger Democratic majority in the 86th Congress. It took until January 19 to complete the delicate process of tapping new members, and on that date the following were officially certified as charter members of the new committee:

Democrats	*Republicans*
Overton Brooks, Louisiana, *Chairman*	Joseph W. Martin, Jr., Massachusetts
John W. McCormack, Massachusetts	James G. Fulton, Pennsylvania
George P. Miller, California	Gordon L. McDonough, California
Olin E. Teague, Texas	J. Edgar Chenoweth, Colorado
Victor L. Anfuso, New York	Frank C. Osmers, Jr., New Jersey
B. F. Sisk, California	William K. Van Pelt, Wisconsin
Erwin Mitchell, Georgia	A. D. Baumhart, Jr., Ohio
James M. Quigley, Pennsylvania	Perkins Bass, New Hampshire
David M. Hall, North Carolina	R. Walter Riehlman, New York [1]
Leonard G. Wolf, Iowa	
Joseph E. Karth, Minnesota	
Ken Hechler, West Virginia	
Emilio Q. Daddario, Connecticut	
Walter H. Moeller, Ohio	
David S. King, Utah	
J. Edward Roush, Indiana	

OVERTON BROOKS AS CHAIRMAN

Every congressional committee carries the imprint of its chairman in its mode of operation, areas of activity, and effectiveness. Overton Brooks clearly set the tone of his new committee which plunged into wide-ranging investigations, studies, and hearings covering space

---

[1] Riehlman was appointed on Jan. 29, 1959.

propulsion, scientific manpower and education, missile development, chemical warfare, agriculture, space law, communications satellites, inventions, weather, and biomedical experiments—to mention just a few of the areas covered.

The 61-year-old chairman of the new Committee on Science and Astronautics was a tall and courtly gentleman, hard-driving, highly ambitious, proud, demanding, controversial, and determined to leave a record of activity and accomplishments for the committee. A 6-footer, Brooks was not flashy in appearance, dressed conservatively, and although he had a ready laugh he was serious minded. His graying blond hair was combed in a semipompadour. His left eye was slightly out of focus, and when he peered out from behind his horn-rimmed glasses it sometimes seemed he wasn't looking directly at you.

Born into a family of public servants, Brooks was the nephew of U.S. Senator John H. Overton, of Louisiana, and another uncle, Winston Overton, served on the Louisiana Supreme Court. Brooks was born in Baton Rouge in 1897 and served overseas as an artilleryman in the 1st Division in World War I. After only 30 days of training, he was thrust into combat in France with the 1st Division, astride a horse pulling a caisson. Never having ridden a horse before, Brooks recalled: "I fell off three times." He earned a law degree at Louisiana State University, practiced law in Shreveport, and served as U.S. Commissioner for a 10-year period. In the Roosevelt landslide of 1936, Brooks was first elected to the House of Representatives from the Shreveport district in northwestern Louisiana.

Assigned to the old Military Affairs Committee which merged into the Armed Services Committee, by 1949 Brooks had risen to No. 2 in seniority. He chafed at the fact he could not get the attention and publicity which Chairman Vinson was receiving. He dreamed of the day when he could run the Armed Services Committee with the same awesome power exercised by Chairman Vinson.

The story goes that while Lyndon Johnson was a member of the old House Naval Affairs Committee when Vinson was chairman, Johnson wanted to ask a question about the Corpus Christi Naval Base during testimony by a Navy admiral. His glasses perched on the end of his nose, Chairman Vinson peered down toward the end of the rostrum and growled:

"And how long have you been a member of this committee?"

"Six years, Mr. Chairman," Johnson answered.

"Well, then, if you've been here for 6 years, I guess you're entitled to one question," Chairman Vinson barked.

As the new chairman of the House Committee on Science and Astronautics, Brooks at first tried to emulate Chairman Vinson. According to Executive Director Ducander:

> He tried to copy him, word for word, sentence for sentence. There was only one Mr. Vinson, and because of the way Congress has changed, there'll never be another one like him because he was a benevolent dictator. Well, Mr. Brooks tried to be the same thing, but he failed in one respect, he was not benevolent and he was not like Mr. Vinson so he couldn't emulate him.

During the first year of the new committee's operation, Brooks centralized power in his own hands, and he also declined to set up any subcommittees. This created rebellion among the senior members of the committee, some of whom had been persuaded by the leadership that their seniority on the new committee would enable them to have new responsibilities and rise within the congressional hierarchy. Strong pressure from the senior members finally persuaded Chairman Brooks to appoint subcommittees. But even then, he was reluctant to assign subject matter titles to the subcommittees, and preferred to number the subcommittees 1, 2, 3, and 4 in conformity with the practice of the Armed Services Committee.

For the eight freshman Democrats on the committee, Brooks was a good leader even though he was peripatetic. To be sure, it took a long time for a freshman legislator to be reached down the line in order to question a witness during a hearing. By 11:30 a.m., when all the whipped cream had been skimmed off the really important issues, and the press table was deserted as newsmen peeled off to file their afternoon copy, the freshman at last had his chance. At this juncture, of course, he was prodded along by the chairman's warning that the House would convene at noon, and there were eight eager freshmen who had to divide up the remaining time. But the subject matter was so fascinating, and Brooks was so enthusiastic about getting into new topics that his interest was contagious. On some very rare occasions, Brooks might even flatter the freshman members by announcing: "Today, we'll start the questioning at the bottom of the committee instead of by seniority."

Chairman Brooks was very sensitive and eager to score a good record for his new committee. A tireless worker, he frequently remained in his office until 8 or 9 p.m., was always in the office on Saturdays and often on Sundays. He made a special effort to get a favorable press, called frequent news conferences, urged radio and television stations that he was available for interviews, and arranged to be taped for rebroadcast over "Voice of America." On the negative side, he was always fearful lest his committee or any of its members might do something which possibly could be portrayed unfavorably in the press. For example, he made a big public issue of the fact that he was personally turning down the request of Representative Victor Anfuso (Democrat of New York) to take his subcommittee to the Soviet Union to meet with Khrushchev. In a play for press and editorial attention—which he received—Brooks publicly expressed his doubt that the Anfuso subcommittee—

no matter how talented, sincere, and devoted, could add much to our international cooperation by a visit to Khrushchev, the Butcher of Hungary, and by the action of personally eating caviar and drinking vodka with him.

COMMITTEE MEMBERSHIP

Six veterans of the select committee helped form the nucleus of the new committee—Brooks, McCormack, Sisk, Martin, Fulton, and McDonough. McCormack seemed reluctant to play an active role on the new committee, lest he upstage the new chairman. McCormack's duties as majority leader became more burdensome during Speaker Rayburn's last term, and McCormack left the committee after the close of the 86th Congress in 1961. Martin was minority leader when he was chosen for the committee, but on January 6, 1959, Representative Charles Halleck of Indiana scored a stunning upset by wresting the leadership position from Martin, 74–70. Although Martin remained on the committee, he rarely attended hearings and was inactive. Martin encouraged Fulton to take the lead as the next highest ranking Republican, and be the spokesman for the minority. McCormack rejoined the committee, but he did not stay long.

With McCormack fading out of the picture, the effective ranking Democratic member of the committee was Representative George P. Miller of California. For Miller, vaulting from 14th ranked member of the Armed Services Committee up to No. 2, the Committee on Science and Astronautics furnished a great new opportunity for responsibility and leadership. For Olin E. "Tiger" Teague of Texas, membership on the new committee resulted from several conversations with Speaker Rayburn. As chairman of the Veterans' Affairs Committee—a committee Teague enjoyed and worked hard on—Teague nevertheless asked Speaker Rayburn to be assigned to another committee. "I was working on the past * * *. I also wanted to work on something that pertained to the future of the country," Mr. Teague told Speaker Rayburn, and Rayburn then asked him to go onto the Committee on Science and Astronautics.

Representative Anfuso attracted a great deal of publicity, not only because of his planned trip to the Soviet Union—which he eventually took by himself—but through other activities and statements. A colorful, popular Congressman, Anfuso held the first congressional hearings on women astronauts. Meanwhile, Representative B. F. Sisk of California did a quiet and workmanlike job, as he had on the select committee.

Although eight freshman Democrats were assigned to the new committee, the Republican members were all veterans of several terms of service. Congressman Chenoweth of Colorado was serving his ninth term; Riehlman of New York and Osmers of New Jersey were in their seventh terms; Baumhart was in his fourth term; and Bass, in his third term.

RECRUITMENT OF STAFF

One of the first tasks which faced Chairman Brooks was the recruitment of staff. He wanted to have his own people with whom he had worked, who knew his methods of operation, and could sense how he would react while he was not available personally—chairing hearings, visiting his district, touring installations, or occupied with other pressing business. This was perhaps even more important because Brooks himself tended to be unpredictable in some of his actions and reactions. At the same time, Brooks sincerely attempted to assemble a staff which was both professional in its competence, and technically proficient.

Chairman McCormack had developed a firm and healthy relationship with the select committee staff, headed by George Feldman. The members of the staff overflowed with affection and respect in their deeply appreciative letters to Chairman McCormack and all members of the select committee on July 21, 1958, which included these words of praise:

> The whole record of hearings makes clear the active and intelligent participation of the members of the committee. This kind of interest and support has made our work for you more meaningful. We also appreciate the opportunity we have been afforded in executive sessions to express our views on the reports and the draft legislation before the committee. * * * It was the inspiration of leadership shown by the members and the chairman which made it easy for us to devote the long hours we did to our efforts, and which turned a burden into a rich and satisfying experience. * * * We have been treated with unfailing courtesy and friendship without partisanship ever influencing the treatment we have received, any more than party lines influenced our warmth of feeling for all the members.

Representative Ford in a response to Feldman noted: "This joint letter reaffirms my belief that a real 'team-work' job was done."

It was against this backdrop that Brooks began considering how he should organize his staff. At a meeting in Majority Leader McCormack's office, Brooks asked Feldman to stay on, but Feldman had already decided to leave. "Then I made a real hard pitch for Dr.

(Charles) Sheldon,'' said Feldman. ''I just went all out, only for one reason, that was because Sheldon was not only dedicated but he knew more about this than anybody else and he was far and above anybody that they could have, or get.''

Brooks decided to keep Sheldon, Spencer Beresford, Richard P. Hines, Raymond Wilcove, Harney S. Bogan, Jr., and Philip B. Yeager—all members of the loyal and dedicated select committee staff. They were all present at the creation of the standing committee, contributed a great deal toward launching the committee, and all had euphoric memories of the idyllic days when working for John McCormack had been such an inspiration. This fact alone caused unfortunate comparisons which affected staff morale. Also, Sheldon and Beresford had high hopes of moving up to the two top posts on the staff. They were rudely disappointed.

During the period when Rayburn, McCormack, and Vinson were negotiating to move Brooks from ranking Democrat on the Armed Services Committee to chairman of the new Science and Astronautics Committee, Brooks did a lot of soul-searching with Charles Ducander, his Shreveport staff counsel on the Armed Services Committee. Ducander had been with Brooks since 1949. ''Duke'' advised Brooks that if the latter were ever going to realize his ambitions for the governorship or the Senate, ''he would have to get out from behind Mr. Vinson.'' Finally, Brooks told Ducander: ''Well, I've made up my mind. I'll go over and take that chairmanship if you come with me.''

Ducander balked. He was getting no advance in salary, and was moving from a happy situation into an unknown jungle of tangled and uncertain relationships, led by a chairman who could not hold a candle to Carl Vinson in power, prestige, and respect. But there was no slipping out. According to Ducander:

> Mr. Vinson called me in and said: ''Duke, you've got to go. Now if anything goes wrong, you can always come back.''

With Vinson, that was not a request or a suggestion; it was an order.

So it was that early in January 1959, Charles Ducander became executive director and chief counsel of the new standing committee. Brooks named Sheldon as technical director, Beresford became special counsel, Yeager was called special consultant, and the other staff members were given titles ranging through various degrees of ''consultant'' or ''counsel.'' John Carstarphen, a Shreveport lawyer, was brought in to serve as counsel (he later became chief clerk and counsel) and several other Louisiana residents were recruited in relatively minor positions. Brooks also borrowed on reimbursable detail a series of officers from the Army, Navy, and Air Force who came over, one at a time, to assist the committee staff. These officers were generally of high caliber, and the committee gained substantial support through their technically competent staff work.

In terms of arranging hearings, producing a monumental number of professional staff reports, and keeping the 25 committee members briefed during a very fast-moving situation in a technically complex field, the staff performed remarkably well during the Brooks chairmanship, 1959–61. Ducander assumed the dual role of briefing the chairman and directing the staff. Because of the difference in backgrounds of various members of the staff, there were serious and voluble disagreements over countless points of jurisdiction, leadership, direction, and quality of performance. Occasionally, these disagreements erupted into public print, to the horror of Chairman Brooks.

Perhaps the most serious attack publicly made on the committee was printed in the widely read editorials of Robert Hotz in Aviation Week and Space Technology. Hotz had printed some highly complimentary remarks in his editorial columns about the select committee during 1958, and in 1959 he began to compare the new committee, its staff, and its leadership very unfavorably with the select committee. Then on February 1, 1960, Hotz blasted the committee and Chairman Brooks in particular. On the staff, he leveled these charges at Brooks:

> He has failed to appoint a technically qualified professional staff, without which the committee cannot hope to be taken seriously, and has apparently used residency in his home district of Shreveport, La., as the sole qualification for what staff appointments have been made. This failure to provide the committee with a professionally qualified staff and the curious practice of Chairman Brooks forbidding staff members to provide questions to other committee members has turned the current hearings into a series of petty squabbles and allowed them to drift into bayous of technical stagnancy rather than keeping sharply in the mainstream of current space problems.

Chairman Brooks was stung by the editorial. He telephoned Hotz and invited him to have lunch with him at the Capitol, where he discussed at length these and other accusations. He wrote a lengthy rebuttal to the editorial which was a masterful response covering every point which Hotz raised (the answer was printed in the February 22, 1960, issue). Among Brooks' comments on the staff were the following:

> Knowing as we do that Members of Congress are, generally speaking, not experts in science and in space technology and exploration, this committee has tried to gather together a competent and experienced staff. We feel that we have done so. It is headed by a career congressional employee who is regarded as one of the foremost professional staff experts on Capitol Hill, with more than 11 years' experience.

Representative Overton Brooks (Democrat of Louisiana), right, the first chairman of the Committee on Science and Astronautics, receives a model of the Saturn launch vehicle from Dr. T. Keith Glennan, the first Administrator of NASA.

Another member of the staff has had 16 years of congressional experience. Most of the staff is composed of veteran members of the staff of the Select Committee on Astronautics and Space Exploration which preceded this committee and worked with House Majority Leader John McCormack in drawing up the Space Act of 1958 which created NASA. The staff is a highly professional, competent, and nonpolitical group which won the praise of Mr. McCormack and the other members of the select committee, both Republicans and Democrats.

It is true that the staff is not composed of scientists and technicians. This is a legislative committee and not a scientific body. However, there are several highly competent men on the staff with broad technical knowledge and the committee implements the work of these technical experts by employing scientists and engineers as special consultants on a per diem basis. No effort is overlooked to supply the committee with the best technical advice possible.

Nevertheless, the editorial had a more dramatic effect than Brooks' written response. The chairman called in Mr. Ducander, and laid down the law: no more Louisiana staff appointments. When a young lawyer named Frank R. Hammill, Jr. (not from Louisiana) filed an application for a staff appointment in mid-February 1960, he was hired with breakneck speed. Suddenly, staff questions for members other than the chairman began to be circulated. The net effect of the scorching editorial was generally salutary within both the staff and the entire committee operation.

GETTING THE COMMITTEE ORGANIZED

The early days of January 1959 were bedlam for the new committee. "The 86th Congress, bursting its buttons with ideas and Democrats, may be in for a historic run," predicted the Washington Post. Catapulted into the space age, the new committee's biggest problem was finding adequate space. While the staff was scurrying around to borrow hearing rooms from the Veterans' Affairs Committee, the Armed Services Committee, and making arrangements to use the caucus room in the Cannon Office Building, the architects and carpenters were frantically hammering away in room 214–B of the Longworth Office Building, across the hall from the basement cafeteria. Not until mid-March were the makeshift rostrum and other arrangements completed so the committee could have its own space. But the new room was terribly cramped for both members, staff, and most of all for the many spectators who crowded in, or tried to stand in the back.

While the staff was rushing around to arrange for the parade of witnesses, and getting the subject matter background lined up, they also had to double as purchasing agents for the new drapes, arrange to push out the typewriter repair shop which occupied part of the space, and rush to get everything ready for the grand opening.

FIRST MEETING OF THE COMMITTEE

On January 19, 1959, the Democratic and Republican caucuses had completed their work and the new membership of the Science and Astronautics Committee became official. Wasting no time, Chairman Brooks immediately called an executive session of the new committee for 10 a.m. on January 20, in the cavernous caucus room of the Cannon Building.

On that historic day when the House Committee on Science and Astronautics first met, the outside world paid no attention at all to their deliberations. In fact, the news was pretty dull and routine that day, and the headlines might have applied to nearly any year before or since. "President Asks Action To Curb Rising Prices" blared a 4-column headline in the Washington Evening Star. The Washington Post dutifully reported a recurring, predictable situation: "Iced-Up Roads Snarl Traffic." The story reiterated the obvious: "The American Automobile Association received hundreds of distress calls, and found difficulty dispatching aid because of the same conditions that caused the trouble."

To be sure, it was an executive session with the press barred. But Chairman Brooks, an eager seeker after good publicity, was strangely silent in his public comments. The acoustics were atrocious as the members eagerly leaned forward to interpret the stream of resolutions and 19-paragraph committee rules which were read; 23 out of 24 committee members (Representative Riehlman was not appointed until January 29) showed up for the first meeting, which proceeded smoothly with passage of resolutions to organize the staff and adopt the committee rules.

Chairman Brooks welcomed the new members, and also spent a few minutes stressing how important it was to have Majority Leader McCormack and the former Speaker and former Minority Leader Martin serving with the committee. He stated there was no room for narrow or partisan considerations in the future operation of the committee. He noted that it would be his policy to conduct a maximum amount of the committee's business in open session, and he added a caution to all the members to guard classified information which would be brought out in executive session or documentary materials made available to the committee.

Perhaps the biggest accomplishment of that first meeting was to enable the members of this new committee to mill around and get to know each other a little better. But it was dramatically different from the organization meeting of any other committee. There was a sense of destiny, a tingle of realization that every member was embarking on a voyage of discovery, to learn about the unknown, to point powerful telescopes toward the cosmos and unlock secrets of the universe, and to take part in a great experiment. To be a charter member of a new committee was exciting enough. But to take part in deliberations which held such a great promise for the benefit of all mankind was a challenge which stirred the blood of all the members.

THE FIRST PUBLIC HEARING

Chairman Brooks drove his staff hard to get off to a fast start, and to hold a series of public hearings which would focus attention on the space program, missile development, and space sciences. Ducander bluntly observed, after the fact, that "Brooks was insane to have committee hearings as soon as he was confirmed as committee chairman." But Brooks was determined, and the chairman's will prevailed.

In publicly announcing on January 31 that hearings would open on February 2, Brooks stated: "The purpose of these hearings is to present to the members of this committee a picture of the situation as it exists today in the fields of science and astronautics." He added:

How does the United States stand in these areas, so vital to the continued existence of the free world?

Is Russia really ahead in science? In astronautics? In space exploration? In missiles and rockets?

Conflicting claims have been made as to the relative positions of the United States and the Soviet Union. We hope that the testimony presented to the committee at these hearings will clarify the picture and bring it into sharper focus.

Ours is a new committee and one which will, in my estimation, grow increasingly important as time passes. These hearings represent a start in the task which this committee has set itself—to help advance science and astronautics in the interests of national defense and the security of the free world.

Brooks scheduled Dr. T. Keith Glennan, NASA Administrator, as the leadoff witness, and announced four days of hearings including the Army, Navy, and Air Force. Care was taken to specifically designate that one of the Army witnesses would be the renowned Dr. Wernher von Braun.

Finally, the magic day arrived—Monday, February 2, 1959—the first official hearing of the new committee. As he pounded his gavel to open the hearing in the big Cannon caucus room, Chairman Brooks expressed his personal feelings about the historic event:

Gentlemen of the committee, this is the first public activity of the newly constituted Committee on Science and Astronautics. * * * Although perhaps the principal focus of the hearings for the next several days will be on astronautics, it is important to recognize that this committee is concerned with scientific research across the board.

Brooks expressed concern, not so much about the known lag behind the Soviet Union, but in the fact that different governmental authorities were furnishing different appraisals about where America stood. "The public is confused," Brooks declared. "These hearings, if they do anything, should clear up this confusion among authorities. * * * This is no time for kid-glove conversation, but it is a good time to present to the public the plain and unvarnished truth."

Brooks then lapsed into a discussion of strictly military matters, which soon became a bitter bone of contention with Carl Vinson and his Armed Services Committee over issues of jurisdiction:

> We are definitely behind Russia in the development of the intercontinental ballistic missile, so important to our survival. We must overtake and surpass Russia in this respect, and I am sure this committee is resolved to do everything within its power to encourage and stimulate our leaders to reach the goal of overtaking and surpassing Russia in this part of our national defense.

Speaking for the Republican side, Fulton abandoned his customary stance as a proponent of far-out concepts and soberly declared:

> The field is much broader than a race with Russia, and we in this committee, I hope, on the Republican side, will see that the implementation is given for broad scientific advances, not only for our security in a race with Russia but for the benefit of all mankind.

> Lastly, I believe we on this side want to see these scientific advances made available for the whole world—all the scientists—so that every people, that is, our allies as well as the people behind the Iron Curtain, can move ahead, raise their standards of living, and arrive at a peaceful world.

In the course of the first group of hearings, originally planned to last 4 days but which actually stretched out over 11 days, morning and afternoon, the committee scored some telling points. The overwhelming thrust of the committee questions and observations added up to stressing a sense of urgency on the witnesses and the agencies they represented. At the same time, the opening hearings provided a wealth of informative material to help publicize the entire program, educate the public as well as the members of the committee, and awaken the Nation.

"THE NEXT TEN YEARS IN SPACE, 1959–1969"

The first publication officially sanctioned by the House Committee on Science and Astronautics was a prophetic report entitled "The Next Ten Years in Space, 1959–1969." At its executive session on February 2, 1959, the committee authorized the publication of this provocative study which actually had been completed under the aegis of the select committee under the direction of George Feldman and Dr. Charles Sheldon.

"This report is one of the most fascinating studies ever prepared for the Congress," lyrically states the introduction. It all started when the thinking of the leading scientists, engineers, industrialists, military officials and public servants was solicited to give their prophecies under the pretentious title of "Whither the Space Age in the Next Decade." Naturally, when eminent authorities in the United States, Great Britain, Germany, Italy, and the Far East were asked by the majority leader of the House of Representatives to give their considered opinions on the world of the future, they responded quickly in a remarkable series of analyses.

It is interesting to measure the predictions against what actually happened during the decade. So far as manned flight to the Moon and return, the most optimistic was Dr. Herbert York, Director of Defense Research and Engineering, who prophesied that man could first set foot upon the lunar dust in "just about 10 years (perhaps in as little as 7, if a high priority were placed on this goal)." Donald W. Douglas felt that—

certainly within 10 years manned flights around the Moon and return can be accomplished, and possibly during that time manned landings on the Moon and return will be possible.

Dr. Wernher von Braun correctly noted that neither Soviet nor American technology would be far enough advanced in the next decade for manned flights to Mars or Venus, but that instrumented probes to those planets "are a certainty." Arthur C. Clarke, English scientific author, very correctly foresaw the day when stationary satellites would make television available to everyone on Earth. The predictors were perhaps too optimistic in their assessment of the precision of weather predictions which might result from weather satellites.

The dunce cap for the worst prediction perhaps should go to the unnamed expert who was sure that mail delivery in the space age would be considerably speeded up.

One of the most significant aspects of the committee publication was the focus it concentrated on the goal of reaching the Moon within a decade, which later became the most dramatic aspect of the space program.

THE COMMITTEE JURISDICTION

One of the outstanding contributions which Chairman Brooks made toward the development of the Committee on Science and Astronautics was his incessant effort to both preserve and broaden the jurisdiction of the committee. Lacking the clout of a Carl Vinson, without the great personal and official power of a John McCormack, and absent the finesse of many other committee chairmen, Overton Brooks kept up such a whirlwind of activity in so many different fields that he was a difficult target to contain. His technique was to continue to "test the jurisdiction" of the new committee through a manifold series of hearings, reports, speeches, and activities which frequently remained unchallenged because of the rising popularity of and interest in space by the American public.

Brooks fully appreciated the fact that the House committee was chartered with a broader scientific jurisdiction than the Senate committee. He encouraged a number of hearings and reports in this area to demonstrate the committee's responsibility for the National Science Foundation, the dissemination of scientific information, basic research, the Bureau of Standards, scientific manpower and education, weather modification, and a host of other scientific subjects. Brooks realized that each of these areas had constituencies of varying public interest and support, but he also appreciated the fact that the real glamor subject which excited the most press and public attention was space and the issue of whether America would overtake Russia.

With a good background of long years of service on the Armed Services Committee, and with a veteran staff director who had served on the committee, Chairman Brooks felt very much at home with military issues. He did not hesitate to test and push the new committee's jurisdiction to the point which incited frequent and bitter challenges by Vinson. At first, Vinson was inclined to laugh and snort at "Ole Overton" and the committee which had been created from one of Vinson's ribs, so to speak. But when Brooks began to hold hearings on why the Army wasn't given the green light on the Nike-Zeus anti-ICBM system, Vinson vented his fury at Brooks for clearly violating the jurisdiction of the Armed Services Committee.

The battle between Brooks and Vinson raged on throughout 1959. Taking the position that the best defense was a good offense, Brooks personally dictated a curt letter to Vinson on May 9:

> I note from a number of sources that the Special Investigations Subcommittee of the House Armed Services Committee has been holding hearings on the Vega vehicle and on contracts relating thereto and on other phases of space.

> Of course, these matters are clearly within the jurisdiction of the Science and Astronautics Committee. This fact shows how easy it is to transcend the jurisdictional lines of committees. I think, however, that I should call this to your attention.

Chairman Vinson was outraged at the charge, and on May 11 he fired back an angry letter intended to put Mr. Brooks in his place with withering words like these:

> Obviously, I am unaware of the "sources" to which you refer and upon which you seem to have relied for your sole information. Had either you or your "sources" made an effort to determine the facts before you wrote your letter, I am confident this is one letter that would not have been written.

Vinson went on to document through copies of the Hébert sub-committee transcript that his subcommittee was only inquiring into management of the Atlas booster, and not the NASA-controlled Vega; then unleashed a final swipe at Brooks:

> If, in the future, you should feel it necessary, or desirable, to raise this same sub-ject or other related matters, I trust you will extend to me the courtesy of first making inquiry as to the facts before assuming, on the basis of rumor, that this committee has transcended its jurisdiction.

Brooks couldn't resist getting in the last word, and wrote Vinson on May 12 to express his appreciation for the "attitude" of Vinson and Hébert. Then he confessed that one of his sources was Aviation Daily.

The running fight between Brooks and Vinson and their respective committee jurisdictions erupted with greater fury at the end of July 1959, when Vinson blasted Brooks with a three-page letter charging again that the Science and Astronautics Committee was invading the jurisdiction of the Armed Services Committee. Vinson pointed to a July 22 Brooks press release announcing that the Brooks committee was inquiring into various aspects of the Atlas and Polaris missiles. Vinson told Brooks that "I fail to find anything within the rules of the House which grant the jurisdiction which you have announced you intend to exercise." Vinson concluded:

> I trust that you will find it advisable to reconsider your decision to assume juris-diction over a subject matter which is clearly within the jurisdiction of this com-mittee. In the event of your unwillingness to accede to this request, it is my further judgment that the matter should be submitted to the Speaker for resolution.

In a masterpiece of understatement, Brooks responded on July 28 that "I believe that a misunderstanding has arisen between us." But he stood his ground. Brooks told his committee, and repeated his statement in response to Vinson, that "this committee was not at-tempting to poach on the jurisdiction of any other committee." He defended the inquiry into Atlas because it was a booster for the Mercury man-in-space program, and explained that the committee was interested in Polaris because it was concerned with research and development on solid propellants. Brooks then related a little piece of personal history:

> You will recall that before leaving the House Armed Services Committee and before accepting the place as Chairman of the House Committee on Science and As-tronautics, I came to your office to discuss the jurisdiction of the two commit-tees. * * * At that time you, in substance, assured me that we would have no trouble in establishing jurisdictional lines since your committee with its $40 billion juris-diction (the largest in the House) had more jurisdiction than it needed and could use. It has, therefore, given me pleasure to consult with you repeatedly on various matters affecting our committees.

It is a tribute to Chairman Brooks that he barged ahead and refused to be embarrassed in the face of a storm of criticism from many different committee chairmen. Among the most upset was Representative Oren Harris (Democrat of Arkansas), chairman of the Committee on Interstate and Foreign Commerce, who raised the issue on the House floor. Harris objected to hearings by the Brooks committee on communications satellites and their operation. Brooks merely fended off the challenge by insisting that his committee was only inquiring into R. & D., and was not interested in operation.

In some other areas in the first few years, the Science and Astronautics Committee held hearings which seemed to stretch its jurisdiction pretty far. The committee was barely a month old when Brooks asked Ducander to start some new hearings on space food by calling the Department of Agriculture over to testify. The hearings were a disaster in their lack of planning, and almost total lack of any useful information elicited. The Agricultural Research Service gave an extended dissertation on their administrative operations, but had little to offer about space food. The hearing would have completely collapsed had it not been for Congressman Fulton's determination "to spur you on to new ideas and new approaches. * * * We are trying to get you to raise your sights." After getting nothing but wooden responses to his questions, Fulton finally erupted with a question which literally stunned the witness and was long and fondly remembered as the greatest Fultonism of all time:

> Possibly in space the approach to vegetables might be different. Did that ever strike you—because we are thinking of three-dimensional vegetables, maybe in space, where you have a lot of sunlight, you might get a two-dimensional tomato. It might be 1 million miles long and as thin as a sheet of paper, aimed toward the sun—a tomato.

There was a long silence, as the Department of Agriculture witness blinked, and finally blurted out softly: "It is an interesting thought." He was completely flabbergasted.

In addition to his jurisdictional fights with other committees, Chairman Brooks had one serious jurisdictional fight which arose within his own committee. Brooks was eager to expand his jurisdiction to cover oceanography, and he fashioned a bill for the development of teaching facilities and aiding graduate students which he managed to get referred to the Committee on Science and Astronautics. But Brooks received an angry reaction from the ranking member of his committee, George P. Miller of California, who also served as chairman of the Oceanography Subcommittee of the Merchant Marine and Fisheries Committee. Miller insisted that jurisdiction over oceanography really belonged to the Merchant Marine Committee. Brooks' next approach was to set up a Special Subcommittee on Earth Sciences, of which he made himself chairman, and in August issued a special invitation to Miller to sit with the committee.

Rather than acknowledge the jurisdiction of the Science Committee, Miller did not sit as a member of the committee, but appeared instead as a witness at the Brooks hearing on August 25, 1959. There he sparred gently with Brooks, pointing out the work in oceanography which was already underway in the Merchant Marine Subcommittee which he chaired. In turn, Brooks asked for Miller's printed hearings and very courteously stated:

> We want to study those so there will be as little overlapping as possible ***. There is no need for duplication, because we will develop the whole program of Earth sciences in this particular committee.

On September 1, Brooks tried to persuade Miller with a letter which stated:

> It is late in the session to discuss the subject of jurisdiction but I want to assure you that your Committee on Science and Astronautics is going to proceed with the bills before it and I am satisfied no conflict will arise between the Committee on Merchant Marine and Fisheries and the Science and Astronautics Committee. Our viewpoint is scientific, while the Merchant Marine and Fisheries is that of operating the merchant marine and supporting the fish and wildlife of the ocean.

Miller was not mollified. The following year Brooks again scheduled hearings on his bill, this time before the full committee on April 28 and 29. When the Merchant Marine Committee got wind of the hearings, its staff made a vigorous protest to Ducander and indicated that Miller would be upset also at the news. On April 18, Ducander in a memo alerted Brooks that a big storm was brewing, and that he had dispatched Dr. Sheldon to brief Miller, just returned from Geneva:

> Accordingly, Dr. Sheldon went to Miller's office the following morning and explained that we were merely setting up a two-day briefing on oceanography, after which Miller became quite angry and said we had no authority to do this. He further told Sheldon that if we persisted in going into the field of oceanography, considering the fact that the Merchant Marine and Fisheries Committee had a Special Subcommittee on Oceanography of which he was chairman, he intended to take the matter up with the Speaker and the House leadership. He further stated to Sheldon that if the hearings went on as scheduled, he would be present at the hearings, and publicly protest this committee's unwarranted usurpation of the Merchant Marine and Fisheries Committee's jurisdiction.

The hearings went ahead, as scheduled, and Miller was an active participant. The hearings were published under the title of "Frontiers of Oceanic Research." But Brooks did not venture again into the deep and turbulent waters of oceanography. The Subcommittee on Earth Sciences became moribund and held no more hearings during Brooks' tenure. The sequel to the story is that when George Miller moved up to the chairmanship of the Science and Astronautics Committee, he did not sanction any activity in the field of oceanography either.

ESTABLISHMENT OF SUBCOMMITTEES

On January 26, 1959, a few days after the organization of the committee, Chairman Brooks sent out memos to all members asking their personal preferences on which of four subcommittees they would like to serve:

Scientific Training and Facilities, No. 1.

Scientific Research and Development, No. 2.

International Cooperation and Security, No. 3.

Space Problems and Life Sciences, No. 4.

The replies of the members had barely started to come in when Chairman Brooks started an intensive series of morning and afternoon hearings of the full committee which summoned NASA, National Science Foundation, the military services, Department of Agriculture, National Bureau of Standards, private industry, and many other witnesses. These full committee hearings made it impossible for the subcommittees to operate.

Brooks tapped the four senior committee members to head the respective subcommittees, as follows:

No. 1, Representative George P. Miller, California.

No. 2, Representative Olin E. Teague, Texas.

No. 3, Representative Victor L. Anfuso, New York.

No. 4, Representative B. F. Sisk, California.

Chairman Brooks also set up a Special Investigations Subcommittee, making himself chairman. George Miller was made chairman of another subcommittee to make recommendations on and exercise oversight over the National Bureau of Standards. Toward the close of the summer, Chairman Brooks also made himself chairman of a Subcommittee on Earth Sciences. Finally, he formed a Special Subcommittee on Patents and Scientific Inventions, which did not begin work until August 1959, and was chaired by Representative Erwin Mitchell of Georgia.

The four chairmen of the numbered, permanent subcommittees soon discovered that it was impossible for them to organize and operate with any independence and responsibility. For 2 months, they protested the fact that Brooks was arrogating to himself all the power, and delegating none of the responsibility. As the protests mounted, Brooks would hand out new assignments—as, for example, the special subcommittee to investigate whether the Soviet Lunik was a hoax. Congressman Anfuso was assigned to chair a special ad hoc subcommittee on the subject.

NASA AUTHORIZATION IN 1959

When NASA sent up its budget, Brooks again went into full committee hearings. Then he very quickly divided up the NASA authorization into four parts, assigning construction of facilities to the Miller subcommittee, splitting research and development down the middle for the Teague and Sisk subcommittees, and giving salaries and expenses to the Anfuso subcommittee. With a grand rush, he gave the four subcommittees a week from April 24 to comb over the NASA budget and make their recommendations to the full committee. It was murderous work, but it certainly kept Chairman Brooks insulated from criticism from the senior members, temporarily at least.

The Miller subcommittee was the only one which acted to change an item during the race to analyze the authorization in 1959. The Miller subcommittee knocked out a $4,750,000 NASA-requested item for a research facility for high energy solid and liquid rocket propellants. The subcommittee quite correctly argued that NASA didn't have the foggiest notion where the site was to be located. "It doesn't hurt to serve notice that we are going to be very vigilant in watching what they are doing," Miller reported to the full committee in executive session. The committee approved the cut, the cut was sustained by the House, but NASA subsequently made a special appeal to the Senate and got the cut restored in conference.

Teague's role in the early subcommittee hearings was to pound some clarity and simplicity into NASA's high-blown, abstruse language. At the very first meeting of his subcommittee on April 24, 1959, Teague opened an executive session with NASA officials by pointing out that he had spent until 1 a.m. the night before poring over the backup books, and he just wished NASA would try to present their program in everyday language:

> You know a lot of people come before Congress and if they can word things in a way that nobody can understand, they think maybe it will be better. In NASA's case the simpler the language and the more explanatory it can be, instead of using words that 99 percent of the Members won't know—I think you will be a lot better off.

Dr. Hugh Dryden, the veteran and brilliant Deputy Administrator of NASA, who had graduated from college as a boy genius at the age of 17, protested: "This is a difficulty with any highly technical subject, to state it in terms that the ordinary person understands." Teague shot back:

> But I do think, Dr. Dryden, that the language could be simplified a whole lot, so that somebody who takes this and reads it will know more about it than I knew when I got through reading it. There are just so many technical terms. * * * Many of them aren't in the dictionary.

Again in 1960, Teague told the witnesses that he wanted them to present their programs in a way that could easily be explained to other Congressmen on the floor:

> These are complicated programs, and we need their explanation in layman's language. I would like you to prepare what you would say to somebody on the floor who has heard none of the technical hearings, and knows little about it. Otherwise there is a good chance before this is over that you will get cut in appropriations.

It is instructive to read the record of the growing sophistication of the subcommittees and their members, as each year progressed. When the first budget hearings were held in 1959, almost everything had to be accepted on faith. But as the years went by, fortified by intimate knowledge drawn from field inspections, excellent staff investigation, and private conversations with field officials or private industry representatives out on the firing line, the subcommittees took more initiative in reviewing the budgets of agencies under their jurisdiction.

Late in 1959, Chairman Brooks decided to make another move to change the general jurisdiction of the four permanent subcommittees which in January had been given subject-matter titles. On December 15, he wrote to every member of the committee, and used a very persuasive argument to support his desire to do away with the subject-matter titles and simply give numbers to the subcommittees, retaining the same chairmen. His December 15 letter explained it by arguing this way:

> By now it has become apparent that the jurisdiction of the committee far exceeds the functional titles which have been allotted to the subcommittees. As a result, it has become necessary to handle different matters because the subject did not properly come within the jurisdiction of the subcommittees originally appointed. If we would merely use numerical subcommittees, each subcommittee could be expected to have considerable additional jurisdiction without being bound by its present functionalized title and jurisdiction.
>
> As a matter of fact, I have checked into this matter, and I find that other committees, such as the Armed Services Committee, have also found it best to use numerical subcommittees rather than limiting the jurisdiction of subcommittees to their titles.

Chairman Brooks solicited the responses of all members of the committee, slipping in the phrase at the end of the letter: "In the absence of serious objection I plan to put this program into effect in the near future." Perhaps it was the spirit of good will of the Christmas season. Perhaps it was the large number of "it makes little difference to me" responses the chairman received from many members. But when the committee convened in January 1960, Chairman Brooks could report that an overwhelming majority of his committee had endorsed the idea, so it became a fait accompli.

ANNUAL AUTHORIZATION

As noted in chapter I, Senator Lyndon Johnson in 1958 inserted an amendment requiring annual authorizations for NASA. However, the powerful opposition of Representative Gerald R. Ford and others had forced a compromise limiting the requirement to one year. Now Chairman Brooks took the leadership to extend the annual authorization requirement—thus insuring that the Committee on Science and Astronautics would have a powerful oversight weapon which by precedent became permanent. In a dramatic presentation to a May 6, 1959, executive session of the committee, Brooks termed his annual authorization amendment—

the crux of the whole thing, the important thing that we have to battle for. * * * this is a vital section * * * you are blazing a path and you want to look ahead at the type of work they do in development. * * * what we want to do is to bring them back year by year for the next few years until they become an established agency and we have fashioned a program.

Teague commented immediately: "Mr. Chairman, I am going to be for your amendment." After a little discussion, the amendment passed.

It was perhaps lucky for the Science and Astronautics Committee that when the authorization bill hit the floor on May 19, 1959, it came up under suspension of the rules which barred any amendment. To many of the members of the Science and Astronautics Committee, it was outrageous to bring out their first authorization bill with only 40 minutes of debate allotted under suspension of the rules. To be sure, they were only authorizing $480,550,000 that first year—less than 10 percent of the high watermark of funding for NASA through most of the 1960's. But here was a great chance to educate all the Members of the House and the time limitation was cruelly constricting.

Under the circumstances, however, the leadership had to reckon with the mood of the House which in 1958 had soundly rejected the annual authorization concept by a whopping 236–126 margin. The same cast of characters, led by Congressman Ford, were sharpening their weapons to try and remove the requirement for annual authorization.

When the bill came onto the floor on May 19, the Brooks forces caught their opposition napping. The opponents were entitled to control 20 minutes of the 40-minute debate if they had been alert enough to "demand a second" when the Speaker asked at the outset: "Is a second demanded?" But they were really asleep at the switch. Congressman Gordon McDonough of California, a member of the Science and Astronautics Committee and strong supporter of the bill, grabbed the microphone and claimed the time before the opposition realized it was being stolen from under their noses. But although McDonough controlled the time, he graciously yielded to Ford to make his pitch.

Once again, Ford spoke eloquently against annual authorizations. He pointed to his work on the select committee, and how he had worked to draft the Space Act setting up NASA. He charged that the authorization process would slow down the space program since—

time is of the essence if the United States is to move forward in space competition with the Soviet Union. * * * I doubt if it is necessary to have a complete and total authorization each year plus a review by the House and Senate Committees on Appropriation.

Ford was joined by a new ally, Representative Albert Thomas of Texas, who in 1958 had argued on the other side in favor of the annual authorizing power. Thomas told Brooks that "I am sure that my able, congenial, and distinguished friend realizes that he is departing from the normal procedure in the House." The atmosphere became tense. Needing a two-thirds vote to pass the bill, the leadership wasn't quite sure it had the horses. When the debate finished, the bill was pulled from the floor and a 24-hour delay imposed while votes could be rounded up.

It wasn't an easy task to get a two-thirds majority against the powerful opposition of the entire Appropriations Committee, led by opponents like Clarence Cannon, George Mahon, Gerald R. Ford, and a coalition of conservatives such as John Rhodes, William Colmer, and Howard Smith and liberals like Wright Patman and Hale Boggs. But the Science and Astronautics Committee won the day by a vote of 294–128.

The principle of true oversight and annual authorization was now embedded in the power of the new committee.

PANEL ON SCIENCE AND TECHNOLOGY

The long parade of witnesses from in and out of the Government who appeared before the Science and Astronautics Committee during 1959 came in response to a specific summons from Chairman Brooks, and their testimony covered a wide variety of issues which the committee identified as important, timely, and useful. Special consultants were available to address specific problems when needed. But it soon became apparent to thoughtful members, the staff, as well as to the chairman, that there were many immediate and long-range problems which were recognized by farsighted scientists and engineers but which were not reaching the Congress soon enough for constructive action.

In February 1959 Chairman Brooks first mentioned to Dr. James A. Van Allen the concept of a panel of scientists and engineers, plus other objective individuals in the field, who could advise or meet with the committee. After a number of exploratory talks, Chairman Brooks wrote to the heads of the three key agencies over which the committee had jurisdiction—NASA, the National Bureau of Standards, and the National Science Foundation. Identical letters went out on September 25, 1959, to Dr. T. Keith Glennan, Dr. Allen V. Astin, and Dr. Alan T. Waterman. In the letters, Brooks mentioned that 1959 had been spent on the regular authorization bills and oversight, spot investigations plus "special attention to the problems of clarifying the jurisdiction of the committee within the congressional structure." He added:

> Now I would like to move into our second year with more attention to an orderly program of work, keyed to the most urgent scientific needs of the country where public policy questions are involved.

Brooks asked the three agency heads to nominate about a dozen outstanding scientists and engineers with some attention both to specialties and experience, as well as geographic and institutional distribution.

The objective and mode of operation were stated in this way:

> Such a panel might meet once or twice a year to offer its suggestions on programs particularly in need of congressional study, and might also serve on an ad hoc basis to give other recommendations if this committee has questions for them.
>
> I would like to see a closer tie develop between the Congress and the scientific community, and it seems to me that this might be a good starting place.

Based upon the list of names submitted, the first batch of letters inviting these distinguished individuals went out on December 15, 1959. On January 12, 1960, Chairman Brooks aired the concept more formally in an executive session of the committee, soliciting additional suggestions from the members. The committee was enthusiastic in its support of the idea, and made a number of recommendations on those to be invited.

On January 17, Chairman Brooks publicly announced the members of the panel:

Dr. Edward J. Baldes—biophysics—senior consultant in biophysics, Mayo Clinic, Rochester, Minn.

Dr. Clifford C. Furnas—chemical engineering—chancellor, University of Buffalo, Buffalo, N.Y.

Martin Goland—applied mechanics—Southwest Research Institute, San Antonio, Tex.

Prof. W. Albert Noyes, Jr.—general chemistry—University of Rochester, Rochester, N.Y.

Dr. Clarence P. Oliver—genetics and zoology—University of Texas, Austin, Tex.

Dr. Sverre Petterssen—meteorology—professor of meteorology, University of Chicago, Chicago, Ill.

Dr. Roger Revelle—geophysics and oceanography—director, Scripps Institution of Oceanography, University of California, La Jolla, Calif.

Prof. Richard J. Russell—geology—Louisiana State University, Baton Rouge, La.

Dr. H. Guyford Stever—aeronautical engineering—Massachusetts Institute of Technology, Cambridge, Mass.

Prof. James A. Van Allen—nuclear physics, cosmic rays—University of Iowa, Iowa City, Iowa.

Prof. Fred L. Whipple—astronomy—director, Astrophysical Observatory, Smithsonian Institution, Cambridge, Mass.

Prof. Maurice J. Zucrow—jet propulsion—professor of engineering, Purdue University, Lafayette, Ind.

Dr. Lee A. DuBridge—physics—president, California Institute of Technology, Pasadena, Calif.

Dr Thomas F. Malone—meteorology—director of research, Travelers Insurance Companies, Hartford, Conn.

On March 25, the Panel on Science and Technology held its first meeting in the Whittall Pavilion of the Library of Congress. The main reason for holding the meeting in the Library was to get away from the atmosphere of the committee room which would put the panel members in the position of being witnesses rather than advisers.

At that meeting, Chairman Brooks sketched in the history of the select committee and the standing committee and their accomplishments. He noted that the committee, in examining ways to "increase the effectiveness of the committee commensurate with the challenge of the times," had decided to call on the panel and "set a pattern of cooperation between the scientific and technical community and the people's Representatives in the Congress of the United States."

In announcing the establishment of the new panel, Brooks stated.

We shall provide for these men of science a forum in which they can speak out to the world on the problems that face it in basic and applied science, in space technology, and in space exploration.

It is unfortunately true that too many times scientists with important ideas that would help advance the interests of the United States and mankind in general have been unable to find anyone to listen to them. Theirs have been, on too many occasions, voices in the wilderness.

Now, through this panel, we shall make available to them a public forum in which they can be heard.

At the first meeting of the panel, Martin Goland, president of the Southwest Research Institute in San Antonio, Tex., presented a paper on energy conversion and also prospects for the marginal or secondary recovery of petroleum. Dr. Sverre Petterssen, professor of meteorology at the University of Chicago, delivered a paper on "Expected Developments in Meteorology During the Coming 10-Year Period." Both papers helped sketch in future developments with which the committee later became involved, in the areas of energy research and the development of weather satellites.

In his remarks at the first panel meeting, Dr. James A. Van Allen, of the University of Iowa, provided some of the stimulus for committee support of NASA's sustaining university program, as well as the efforts of the National Science Foundation. Dr. Van Allen told the committee:

> Industry, for the most part, delivers what we might call tangible products. Universities, on the other hand, deliver a product which is much less tangible. I think I might say that our product, in idealized form, consists of competent, enthusiastic, and tough-minded young men and women who are devoted to a life of study and a life of inquiry. I may say that our product is in very great demand. It is not clear to me that the Federal Government recognizes the value of our product in a way which I think it deserves. * * * Consideration might well be given to the idea of what one might call lump subsidies to general scientific areas within universities.

The concepts outlined helped furnish some of the ammunition which the committee effectively used to support NASA and NSF programs of university support for research and the training of scientists and engineers.

The first panel meeting was so successful that the committee hastened to schedule another meeting in June of 1960. This time the panel met for two days, on June 2 and June 3, enabling fuller discussion of the issues presented. Although the meetings were held in the regular committee room in 214–B of the Longworth Building, Chairman Brooks introduced the meeting by observing:

> We are sitting here today in this special meeting situation, not in any sense a formal committee hearing. For this reason I have interspersed members of both the committee and the panel in these seats, not in accordance with any concept of seniority, but just as is convenient as people arrive. This is to help preserve the atmosphere of free exchange of ideas on both sides.

At the second panel meeting, a number of additional papers were presented on the need for extended geologic research, studies of the lunar surface, radio astronomy, micrometeorites, the need for national research planning, the interrelation between the Earth sciences and space sciences, scientific education in the Soviet Union, world population growth, desalinization of water, and minerals research.

Representatives Miller and Fulton, who were deeply interested in pushing forward America's progress toward the metric system, raised the issue during the second panel meeting. Dr. Richard J. Russell, acting dean of the Graduate School at Louisiana State University, in boosting the metric system, noted that one of the reasons arithmetic was difficult for so many school pupils was that "our whole system has this millstone around its neck, of the obsolete system of weights and measures." Considerable encouragement and support was given to the committee by the panel members, stimulating action on legislation introduced by Chairman Miller to move toward establishment of the metric system in the United States. At the fourth meeting of the panel on March 21–22, 1962, the panel formally endorsed the establishment of a metric system of measurements in this country.

Three more sessions of the panel were held during 1960 and 1961 under Brooks' chairmanship, and 10 sessions were held in the 11 years from 1962 through 1972 under Miller's chairmanship. Down through the years, the following topics were discussed at the panel sessions:

Mapping and Geodetic Satellite Programs,
Advanced Propulsion for Space,
International Scientific Activities,
European Organizations for Space Research,
Availability of Scientific Advice to Congress,
Aeronautics,
Application of Science and Technology to Economic Growth,
Government, Science and International Policy,
Data Processing,
Transportation and Communication,
Science and Technology in Latin America,
Modern Evolution of Science and Technology in Japan,
Applied Science and World Economy,
Utilization of Scientific and Technical Resources in Canada,
Science, Technology and the Cities,
Management of Information and Knowledge,
Forces for Change in the Seventies and Eighties,
Education in Post-Industrial America, and
Earth Resources Satellites.

The meetings of the panel were always accompanied by informal receptions at which members of the panel, invited guests, members of the committee, and other Members of Congress had an excellent opportunity to exchange ideas in a relaxed atmosphere. The invited guests included not only governmental officials, but also representatives from industry, labor, universities and research organizations, and specialists interested in the topics under consideration. The committee "flower fund," built up through assessing the members $10 or more apiece, was insufficient to cover the cost of receptions and luncheons for the visiting dignitaries. Aerospace contractors occasionally hosted luncheons of the National Space Club which coincided with the panel meetings, and helped cover the cost of social affairs which included the panelists at Capitol Hill gatherings.

Chairman Brooks originally envisioned that about 16 panelists would be appointed, with about 5 or 6 new members of the panel to come in each year, on a rotational basis. Brooks' death intervened following the third panel meeting, and his concept of panel rotation was not carried into effect until the late 1960's.

Starting with the March 21–22, 1962 panel meeting—the first under Miller's chairmanship—the concept of panel moderators was started. Chairman Miller designated as guest moderators, Dr. George B. Kistiakowsky, former Science Adviser to President Eisenhower, for the March 21 meeting and Dr. Harrison S. Brown, professor of geology at the University of California, for the March 22 meeting. Also taking part as a guest participant was Sir Bernard Lovell, director of the Jodrell Experimental Station in England.

Chairman Miller introduced a new organizational idea for the panel, starting in 1965, by devoting the panel discussions to a central theme. The topic of the 1965 meeting was "Aeronautics." At the same time, Chairman Miller persuaded Prof. Luigi Broglio, chairman of the Italian Space Commission for the National Council of Research, to serve as guest panelist. The guest moderators for the January 26–27, 1965 panel meeting were Prof. René H. Miller, Slater professor of flight transportation at the Massachusetts Institute of Technology, and Dr. Edward C. Welsh, executive secretary of the National Aeronautics and Space Council.

Most participants favored the "central theme" concept introduced by Chairman Miller in 1965. The multiple-subject agendas used for the earlier panel meetings provided a fascinating and free-wheeling opportunity for everybody to pitch in and sound off on any subject, resulting in great freedom but little continuity. The new structure allowed more time for discussions in depth, and the subject of the discussions unfolded more logically with the skillful guidance of the panel moderators.

Another interesting twist was introduced by Chairman Miller at the 1965 panel meeting: over 150 prominent persons representing Government, industry, and the scientific and academic communities attended at Miller's special invitation. For the first time, audience participation was encouraged. This served to broaden the discussion, but of course made the interchanges between committee and panel members somewhat less intimate in nature. However, the social opportunities for panel and congressional personnel to get together in between these formal sessions obviated this difficulty.

As noted in chapter V, Representative Daddario, as chairman of the Subcommittee on Science, Research and Development, helped organize and shared top billing at several of the panel meetings, starting with the 1965 session. The 1965 panel meeting was also unique in that it was the first use of the new committee quarters in the spacious room 2318 of the Rayburn Building.

Under Chairman Miller's leadership the keynote speakers, guest moderators, and guest panelists constituted a stellar array of talent, including:

Vice President Hubert H. Humphrey.
Lord Snow, British Ministry of Technology.
Gerard Piel, publisher, Scientific American.

Hon. Dean Rusk, Secretary of State.

Robert Major, director, Royal Norwegian Council for Scientific and Industrial Research, Norway.

Dr. Kankuro Kaneshige, member, Council for Science and Technology, Japan.

Dr. H. W. Julius, chairman, Central Organization for Applied Scientific Research, the Netherlands.

Dr. Donald F. Hornig, Director, U.S. Office of Science and Technology.

Dr. S. Husain Zaheer, chairman, National Research Development Corporation of India.

Dr. C. Chagas, president, Brazilian Academy of Sciences.

Lady Jackson (Barbara Ward), Foreign Affairs Editor, "The Economist," London.

Dr. O. M. Solandt, chairman, Science Council of Canada.

Dr. Jorge A. Sabato, National Commission for Atomic Energy, Buenos Aires, Argentina.

Hon. George D. Woods, president, International Bank for Reconstruction and Development.

André de Blonay, secretary general, Interparliamentary Union, Switzerland.

Dr. Philip Handler, Chairman, National Science Board.

Dr. Alexander King, Director for Scientific Affairs, Organization for Economic Cooperation and Development, France.

Hon. John W. Gardner, former Secretary of Health, Education, and Welfare, and chairman of the Urban Coalition.

Constantinos A. Doxiadis, president, Athens Center of Ekistics, Athens, Greece.

Richard Llewelyn-Davies, professor of architecture, University of London.

Zivorad Kovacevic, secretary general, League of Yugoslav Cities, Belgrade, Yugoslavia.

McGeorge Bundy, president, Ford Foundation.

Hon. Earl Warren, retired Chief Justice of the United States.

Daniel J. Boorstin, Director, National Museum of History and Technology, Smithsonian Institution.

Don K. Price, dean, Graduate School of Public Administration, Harvard University.

Osmo A. Wiio, Helsinki University, Finland.

Ioan D. Stancescu, counselor, National Council of Scientific Research, Rumania.

Hon. William P. Rogers, Secretary of State.

Dr. Adriano Buzzati-Traverso (Italy), UNESCO.

Prof. Abdus Salam (Pakistan), International Centre for Theoretical Physics.

Dr. Viktor A. Ambartsumian (U.S.S.R.), International Council of Scientific Unions.

Prof. Thomas Odhiambo, International Centre of Insect Physiology and Ecology, Nairobi, Kenya.

Hon. Staffan Burenstam Linder, Member of Parliament, Stockholm, Sweden.

Capt. Jacques-Yves Cousteau, Centre d'Études Marines Avancées, Marseilles, France.

Dr. Edward E. David, Jr., science adviser to President Nixon.

Dr. James C. Fletcher, Administrator, NASA.

Dr. Robert M. White, Administrator, National Oceanic and Atmospheric Administration.

Dr. Fernando de Mendonça, general director, Instituto de Pesquisas Espaciais, Brazil.

Dr. Franco Fiorio, chairman, United Nations Working Group on Remote Sensing of the Earth by Satellites, Italy.

Dr. Norman Fisher, chairman, Australian Committee on Earth Resources Satellites, Australia.

Armin Spaeth, Ministry of Science and Education, Germany.

Among the members of the committee's advisory Panel on Science and Technology were the following: Front row, from left, Dr. Athelstan Spilhaus, Dr. Harrison S. Brown, Dr. Lee A. DuBridge, Dr. Clifford C. Furnas, and Dr. Fred L. Whipple. Rear row, Martin Goland, Dr. Richard J. Russell, Dr. Maurice J. Zucrow, Dr. W. Albert Noyes, Jr., Dr. James A. Van Allen, and Dr. H. Guyford Stever.

Among those taking part in sessions of the Panel on Science and Technology were Representative George P. Miller (Democrat of California), left, Dr. Harrison S. Brown, Dr. Roger Revelle, and Dr. Robert C. Seamans, Jr., of NASA.

During the early sixties, several personnel changes occurred in the panel's permanent roster. Dr. Sverre Petterssen resigned in 1963, and Dr. Harrison S. Brown of the California Institute of Technology and Dr. Walter J. Hesse of Ling-Temco-Vought Corp. were added, bringing the total membership to 15. Dr. Clay P. Bedford of the Kaiser Aerospace and Electronics Corp. was appointed in 1966. In 1967, Dr. Clarence P. Oliver resigned and was replaced by Dr. Athelstan Spilhaus, president of the Franklin Institute of Philadelphia, Pa.

By 1968, it was felt that some new blood should be infused into the panel. It was concluded that perhaps some of the charter members of the panel, although ideally oriented toward the needs and challenges of the early sixties, did not in some instances perform as effectively in the disciplines and areas most needed by the committee in the seventies. It was also felt that there was some repetition in the views advanced by some panelists. Chairman Miller, supported by Representative Daddario, became convinced that the panel process would be enriched by the rotation of some of its members. As a result, the full membership of the permanent panel by 1972 included the following:

Dr. Ivan L. Bennett, Jr.—medicine—New York University.
Dr. Harrison S. Brown—geochemistry—California Institute of Technology.
Dr. A. Hunter Dupree—history—Brown University.
Dr. David M. Gates—ecology—University of Michigan.
Mr. Martin Goland—applied mechanics—Southwest Research Institute.
Dr. Walter J. Hesse—aircraft and missile systems—LTV Aerospace Corp.
Dr. Herbert E. Longenecker—biochemistry—Tulane University.
Dr. Thomas F. Malone—meteorology—University of Connecticut.
Dr. Roger Porter—microbiology—University of Iowa.
Dr. William F. Pounds—management—Massachusetts Institute of Technology.
Dr. Roger Revelle—geophysics—Harvard University.
Dr. Athelstan Spilhaus—oceanography—Woodrow Wilson International Center for Scholars.
Dr. H. Guyford Stever—aerospace engineering—Carnegie-Mellon University.
Dr. James A. Van Allen—physics—University of Iowa.
Dr. Fred L. Whipple—astronomy—Smithsonian Astrophysical Observatory.
Dr. John T. Wilson—psychology—University of Chicago.

The panel members and guest panelists were reimbursed for their transportation and subsistence expenses, plus a consultant fee of $50 per day for the period of the panel meetings.

Months of advance planning preceded the meetings of the panel. In addition, the activities of the committee on the subject matter of the panel never ended with the rap of the gavel which marked the formal termination of the sessions. For example, following the 1969 panel meeting on "Government, Science and International Policy," the committee undertook follow-on studies on U.S. policy regarding scientific relationships with other countries. Panelists Harrison S. Brown and Roger Revelle were joined by Dr. Philip Handler, Chairman of the National Science Board, as a steering committee. This steering committee assisted the committee in assembling and assessing information from 11 Federal departments and agencies, analyzing their international science programs, the limitations and potentialities of each, the funding, problems, and possibilities. Similar activities preceded and followed most of the panel sessions on a wide variety of topics of concern to the committee.

In summary, the panel during its operation helped develop a background of scientific, technical, and policy information for the committee which was authoritative, timely and candid. One of the useful byproducts of the interchanges was the improved understanding by scientists, both American and from other nations, of the legislative process and the manner in which Congress and the Science Committee operated with respect to science and technology. The panel sessions helped to identify spheres of scientific and technological research which offered exceptional promise for the welfare and security of the Nation, and which needed legislative attention. The committee was exposed to updated methods of conducting research, and the assembling and analysis of data by modern means. Also, the committee through the panel meetings received updates on issues such as the availability of scientific manpower and educational or training needs; international cooperation and organizations concerned with science and technology; and a general appraisal and assessment of the priorities being followed in the committee's work.

Naturally, the information and inspiration provided through the panel sessions had differing influences on the various members. Certainly the ease of the dialogue enhanced the mutual respect between Congressmen and the scientific community.

The final meeting of the Panel on Science and Technology—the 13th—took place in a three-day span, January 25–27, 1972. The number 13 proved unlucky because the panel did not meet after that date. Various reasons have been advanced for the abandonment of the panel meetings after 1972. Under Chairman Miller, the panel concept developed to its fullest flower, and when Miller left the House of Representatives in 1972 the panel lost its greatest champion. In addition, Representative Daddario shouldered a vast amount of the burden of arranging, moderating, and providing leadership for the panel, and with his departure in 1971 another strong supporter and active worker was lost. There is no question that the amount of staff time devoted to arranging and following up on the panel meetings constituted a heavy drain away from other duties. Some unfavorable publicity was generated through the contributions of aerospace contractors toward the lunches and receptions for panelists and their guests.

The record indicates that prior to becoming chairman, Representative Teague was not a frequent participant in the panel discussions. Chairman Teague felt that the ongoing work of the committee should not be in any way delegated to those on the outside. Although a vast amount of work and effort went into general receptions enabling the committee to meet with scientists, engineers, astronauts, and those associated with the space program and other activities of the committee, the formal sessions of the Panel on Science and Technology were not revived after 1972.

THE PASSING OF THE SCEPTER

Although not a completely well man, Overton Brooks kept up a supercharged schedule during the 2½ years that he chaired the House Committee on Science and Astronautics. Under his chairmanship, the new committee became firmly established, staffed, tested, and expanded to the limits of its jurisdiction, produced a plethora of studies and reports of high quality, earned the respect of the scientific community, served as a sounding board for the public in new and challenging areas, fully established the principle of annual authorizations by law which was an essential tool for oversight, and helped educate all Members of Congress in the complex fields of science and space.

In 1960, the McNaught Syndicate presented to the House Committee on Science and Astronautics the Holmes Alexander Award as the "House Committee of 1960." The award noted that:

> This committee has distinguished itself by its inspiring work in the sciences, in space exploration and in astronautical research. * * * The committee, established in January 1959, is the only committee with jurisdiction over science in general in addition to space. The devotion of the Members to the activities of this committee has been inspiring to behold.

Although Chairman Brooks met some angry opposition from other committee chairmen who felt the fledgling committee was infringing on their jurisdiction, he had a powerful ally in the leadership in Majority Leader John W. McCormack, the chairman of the original select committee who remained on the standing committee through most of Brooks' tenure. When Brooks wrote McCormack in mid-December 1959 to ask his opinion about changing the subcommittees, McCormack, in a typical response, said simply: "I will follow your leadership." Other committee chairmen and executive branch officials who threatened to go over Brooks' head to Speaker Rayburn soon found out that John McCormack was in the doorway.

After the very strenuous sessions of 1959, 1960, and 1961, Brooks finally entered Bethesda Naval Hospital for a long-dreaded operation in August 1961. His gall bladder was removed, and it was decided to allow him to recover his strength before a further operation. His staff director, Charles Ducander, visited the hospital every morning and, at Brooks' insistence, brought him sheafs of committee papers. Several days later, Brooks called Ducander and asked him to visit him at his home. Surprised that Brooks was out of the hospital that fast, Ducander was even more surprised when he found Brooks at home, fully dressed and lying down on a sofa downstairs. "He had big piles of papers, correspondence he was going through—congressional work, some committee matters I had brought out for him to read. And just about 5 or 6 days from then, he had a heart attack and he died."

Chairman Brooks died on September 16, 1961.

The changing of the guard occurred smoothly. Congressman George P. Miller of California moved up to become chairman of the House Committee on Science and Astronautics on September 21, a position he held until Congressman Olin E. "Tiger" Teague became chairman in January 1973.

White House meeting of space leaders. From left: James E. Webb, NASA Administrator Senator Robert S. Kerr (Democrat of Oklahoma), chairman of Senate Committee on Aeronautical and Space Sciences; President John F. Kennedy, Vice President Lyndon B. Johnson, Chairman Overton Brooks, and Edward C. Welsh, Chairman of National Aeronautics and Space Council.

While chairman of the Manned Space Flight Subcommittee, Representative Olin E. Teague (Democrat of Texas), right, took his subcommittee on frequent trips to oversee the work at NASA installations and their contractors. Along with Representative Edward J. Patten (Democrat of New Jersey), left, Teague is conferring with James S. McDonnell, head of McDonnell Aircraft Corp. in St. Louis, Mo., producers of the Mercury and Gemini space capsules.

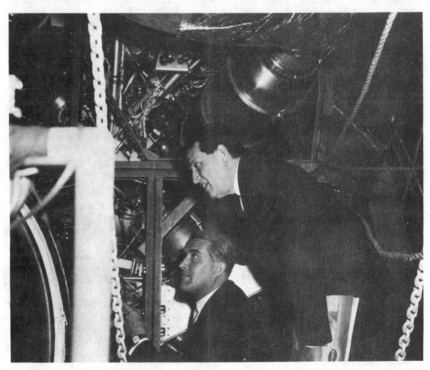

Dr. Wernher von Braun and Representative James G. Fulton (Republican of Pennsylvania), right, inspecting a Saturn engine used in the Apollo program.

CHAPTER III

Racing for the Moon

The members of the Committee on Science and Astronautics were easy to spot, even though they were scattered throughout the crowded chamber of the House of Representatives early on the afternoon of May 25, 1961. They applauded long and loudly when President Kennedy suddenly announced to a joint session of Congress the bold commitment "to achieving the goal, before this decade is out, of landing a man on the Moon and returning him safely to Earth."

The ranking Republican on the Science Committee, Representative James G. Fulton of Pennsylvania, was applauding so vigorously that a press gallery occupant pointed down at him and remarked to a fellow newsman: "They must make a lot of space vehicles in Pittsburgh." For Senator Kenneth B. Keating (Republican of New York), who had served on the select committee, President Kennedy's address was "an alarm clock to awaken the Nation."

Up until the moment the dramatic announcement was made, there had been considerable argument over the feasibility of the goal both within the executive branch and in Congress. In the committee publication "The Next Ten Years in Space, 1959–1969" written by the select committee, but approved and released by the standing committee in 1959, many scientists, engineers, and military men had focused on the timetable for a manned flight to the Moon.

The committee itself was clear and specific in its recommendation contained in its July 5, 1960 report entitled "Space, Missiles, and the Nation":

A high priority program should be undertaken to place a manned expedition on the Moon this decade. A firm plan with this goal in view should be drawn up and submitted to the Congress by NASA.

There was a tug-of-war going on within NASA and also among scientists generally. When the seven newly chosen Mercury astronauts first appeared before the Science Committee on May 28, 1959, Representative J. Edward Roush (Democrat of Indiana) asked Gus Grissom whether he was thinking beyond the preliminary suborbital and Earth-orbiting flights. Grissom immediately answered:

Surely. We have thought in terms of extending this on out further—to the Moon and other planets—but there has to be a first step and we feel this is the first step.

If there were one persistent note which the committee repeatedly sounded in 1959 and 1960, it was the need for a greater sense of urgency. Officially, NASA drew back from an early and firm commitment. When Richard E. Horner, Associate Administrator of NASA, appeared before the Science and Astronautics Committee on January 28, 1960, his official view of the manned Moon flight timetable was very modest:

> It appears to be clear, from a careful analysis of launch vehicle requirements as we now understand them, and recognizing the need for information yet to be developed, that a manned landing on the Moon will fall in the time period beyond 1970.

This timetable was not good enough for the Science and Astronautics Committee. On December 30 and 31, 1959, two of the most articulate committee supporters of the manned space flight program, Representatives Olin E. Teague of Texas and Emilio Q. Daddario of Connecticut and staff visited several space industries, including Chance Vought Corp. On that occasion and subsequently, Vought Astronautics, a division of Chance Vought, made a presentation to the committee and staff which they contended "could place a manned expedition upon the Moon in 8 years, by 1968, if the effort were begun immediately."

KEITH GLENNAN

Dr. T. Keith Glennan, the first NASA Administrator, was the sparkplug in pushing the first manned flight program—Project Mercury—and toward that end he had the fullest support of President Eisenhower. "It would be no exaggeration to say that the immediate focus of the U.S. space program is upon this project," Glennan told Congress early in 1960. But Glennan had a strikingly unemotional attitude toward the lunar program which contrasted sharply with his successor, James E. Webb, and repeatedly caused clashes with the Committee on Science and Astronautics. Although the committee respected Glennan's professional knowledge and general administrative abilities, they felt impelled to prod, push, and occasionally berate Glennan for his somewhat casual attitude toward the speed of the space program.

A clue as to Glennan's inner feelings is contained in a private memoir he wrote for his family, quoted by Presidential Science Adviser James R. Killian, Jr., which confesses:

> I had taken no more than casual interest in the efforts of this Nation to develop a space program following the successful orbiting of Sputnik I by the Russians on October 4, 1957.

Killian himself was somewhat blunter in his own attitude: "I would be less than candid about the role I played if I did not make clear my lack of enthusiasm for some of our man-in-space projects and for the manned lunar program." Like a good soldier, Glennan was reflecting the feeling of his administrative superiors, especially President Eisenhower.

Repeatedly, the committee tried to get Glennan to admit that if he had more money for different aspects of the space program, then NASA could perform its various missions faster. Chairman Brooks, in opening "space posture" hearings on January 20, 1960, made a sharp and challenging assertion:

> Those of us on this committee would be indulging in fanciful thinking if we did not admit to ourselves that the U.S. space effort has reached neither the pace nor the proportions we had hoped for when we passed the National Aeronautics and Space Act in July 1958. Perhaps we expected too much. But there are definite indications—these have existed some time—that a true sense of urgency has not constantly attended the American space program.

THE SPACE RACE

To focus attention on the importance of speedier and more significant forward progress, the committee called Livingston T. Merchant, Under Secretary of State for Political Affairs, and George V. Allen, Director of the U.S. Information Agency. Allen testified that "It is hardly an overstatement to say that space has become for many people the primary symbol of world leadership in all areas of science and technology." Merchant underlined the obvious point that "the performance of the United States and the Soviet Union in outer space will inevitably be compared by the rest of the world." The committee made no bones of the fact that the forum of the committee hearing, well covered by the news media, was being utilized as a sounding board to spur a higher level of activity by both the administration and NASA.

Individual members of the committee hammered home the theme that America must wake up to realize that we were in a true space race. Addressing a conference on "Electrical Engineering in Space Technology" in Dallas on April 13, 1960, Subcommittee Chairman Teague noted that "when Russia first put her Sputnik into orbit, we lost an important battle in the eyes of the uncommitted. Also, when Russia—through Lunik 2—implanted the hammer and sickle on the Moon—we lost still another important battle. We cannot afford to lose many more such battles." On March 17, 1960, Representative Daddario bluntly accused NASA of lacking "foresight and urgency" by failing to develop a plan to land a man on the Moon prior to 1970. Referring to rumors that there might be a Russian manned flight to the Moon, Daddario, in an address in Baltimore, claimed that the future of the free world might "depend on whether or not a U.S. mission is already on the Moon when that event occurs."

Glennan maintained his serenity and aplomb in his appearances before the House committee. But a small anecdote reveals the fact that the committee was getting its point across. Glennan was waiting to testify one morning, following an official committee photograph which had been arranged.

"We want the Doctor up here too. C'mon up and get a picture with the committee," suggested one of the members.

Glennan quipped: "Do you want me up there in my usual position, on my knees?"

The committee persisted in its efforts to get NASA and the administration to raise their sights, and the net effect was to stimulate greater support both in Congress and in the Nation. When some Congressmen felt inclined to cut funds from the program, it was easy to produce a huge majority in favor of the proposition that "rather than being cut, the program should be increased."

EXECUTIVE PRIVILEGE

Early in 1960, the committee had a bruising battle with Glennan over documents needed by the committee to review contract awards to the Rocketdyne Division of North American Aviation, Inc., for development of the 1½ million-pound-thrust engine, and with McDonnell Aircraft Corp. for the development and manufacture of Mercury capsules. In examining NASA's contracting procedures for the $102 million Rocketdyne contract and the $28 million McDonnell contract, the committee asked for certain documents and NASA refused to furnish the committee with reports of its Source Selection Board. Prodded by subcommittee Chairman B. F. Sisk (Democrat of California) and Chairman Brooks, NASA repeatedly refused on grounds of "executive privilege." Glennan's position was expressed in this way:

> This document contains the personal evaluations and recommendations of certain officials of NASA whom I consulted to aid me in reaching my decision on the selection of a prospective contractor. Since this document discloses the personal judgments of subordinates made in the course of preparing recommendations to me, I am sure you will agree with me that it would not serve the interests of efficient and effective administration of this agency for such a document to be reviewed by anyone outside of NASA.

Both contracts were negotiated contracts, and neither of the final awards were given to the lowest bidders. Private meetings with NASA, Chairman Brooks and the staff failed to provide a solution, so Brooks called an executive session for January 12, 1960, to obtain the advice of the committee. Teague pointed out: "This is not a matter peculiar to this committee, the policy comes right out of the White House. We are wasting time unless you," turning to Brooks "and the chairman of Foreign Affairs and others get together on the fundamental right of Congress to know." Fulton urged: "Do the dramatic thing. Call him right in here." Sisk added: "We wanted to find out the criteria, the procedure used, because this agency is going to spend billions in the future, and we felt right now was the time to establish once and for all that this committee should have the right." Miller stated: "If we are going to keep scandal away from NASA, and the rest of them spending this money, they certainly can stand the scrutiny of this committee and the scrutiny of the Comptroller General's office, and I for one think they flout the will of Congress."

The Comptroller General, asked by the committee to assist, issued a devastating report following a further denial of the documents to the General Accounting Office. GAO contended that refusal of the documents was "an interference in our statutory responsibilities" and failed to "promote confidence in the conduct of public business."

Sisk concluded that "I feel it is absolutely imperative if we, as Members of Congress, are to fulfill our responsibility to our constituents as taxpayers of this country, that we must have some information on negotiated contracts." Glennan, however, disagreed and added: "I discussed this matter with the President personally and with his staff. The position I take has his approval."

Fulton, as was his frequent custom, shifted his ground once the hearing was under way and defended the practice of executive privilege in the withholding of the documents in question. Most members of the committee became angry, frustrated, and aghast at the belligerent refusal of Glennan to budge as much as a millimeter from his position.

Yet after many hours of emotional confrontation in a morning and afternoon session on January 29, 1960, Chairman Brooks closed the book on the hearings with this conciliatory statement:

> I want to assure Dr. Glennan and his staff that this committee is going to continue to work in cooperation with NASA, difficult as it might be under the circumstances, in the interests of speeding up our program in space and in the further interests of our country.

In another age, it seems probable that other congressional committees or members thereof would have exacted some form of retribution in slashed funding, legislative restrictions, or highly critical oversight. But the Committee on Science and Astronautics was firmly dedicated to the proposition that the space program must succeed. There were no recriminations. The committee felt a deep obligation to point out how wrong NASA was, and the committee discovered that the same problems occurred when James E. Webb became NASA Administrator. Yet all memory of the unpleasantly harsh words was quickly washed away overnight as the committee went on to tackle the more important issue of how best to reach the Moon quickly.

TRANSFER OF THE VON BRAUN TEAM TO NASA

Early in 1959, members of the Science and Astronautics Committee visited Cape Canaveral, Fla., witnessed the launches of several missiles, and spent a considerable amount of time at the Army Ballistic Missile Agency in Huntsville, Ala. The boom of a Bomarc missile shattering the predawn darkness at Cape Canaveral took second place to the powerful influence of a personal visit with Dr. Wernher von Braun.

It was easy for the committee members to see and appreciate at Huntsville why von Braun was such a towering figure in the space program. Beyond his stellar technical ability, von Braun demonstrated the inspirational leadership around which thousands of determined scientists and engineers rallied. Not only the repatriated group of newly naturalized associates of von Braun who had been with him at Peenemünde, but countless other experts in the developing new field of rocketry and space found von Braun a leader whom they trusted and admired. To the committee members, von Braun was a symbol of success. His predictions always seemed to come true, he spoke in graphic terms which carried beautifully etched imagery, and he demonstrated to the committee and the world that he practiced what he preached. He was also a popular figure with whom Congressmen and the public quickly identified.

The Science Committee early on recognized and took steps to protect the integrity of one of von Braun's greatest assets—his team. Even before the standing committee was formally organized in 1959, the select committee recognized the team concept which von Braun was stressing. In response to a question from Representative Gerald Ford, von Braun told the committee: "To build up a good team takes years, to wreck it takes a few moments. And yet, these experienced development teams are our greatest single national asset in the race for leadership in missiles and space exploration."

The startling success of von Braun's Explorer I, and his continued success as a supersalesman for space, made the committee even more determined to preserve the "team." NASA made a number of attempts to negotiate the transfer from the Army of those experts who had experience with developing large boosters needed to launch sizable spacecraft.

All three military services stepped up high-powered propaganda campaigns designed to gain public and congressional support for expanding their own programs in space. Secretary of the Army Wilber Brucker threatened to resign if von Braun's group were taken from the Army; the Air Force endured some gentle kidding from the Science Committee by redefining all space as "aerospace;" and the Navy argued before the committee that mobile sea launches like Polaris were the best path to a workable space program. The Science Committee aired, monitored, and probed into this power struggle.

Throughout the heated fights of 1959, the Science Committee underlined the basic fact, which was also argued strongly by the Eisenhower administration, that NASA must be the dominant agency for the use of space for peaceful purposes. Beyond this, the committee insisted that the von Braun team must remain intact.

The transition was not an easy one. Von Braun's boss was the fearlessly outspoken Maj. Gen. John B. Medaris, commander of the Army Ballistic Missile Agency at Huntsville, Ala., a salesman in his own right. General Medaris so impressed the committee that Chairman Brooks tried to persuade him to join the committee staff as a consultant when Medaris eventually retired in 1960. According to James R. Killian's account, "Medaris and von Braun campaigned with fierce religious zeal to obtain a central role in space for the Army." At stake was not only von Braun's 4,000-man team at Huntsville, but also the prestigious Jet Propulsion Laboratory at Pasadena, Calif. "JPL" was operated by the California Institute of Technology under Army contract, and its team was headed by Dr. William H. Pickering. NASA desperately needed the in-house capability possessed by both the von Braun and Pickering organizations, and here is the point where the committee played an extremely helpful role.

Because of strong opposition from the Army and some individual Senate and House Members—not primarily on the Science Committee—President Eisenhower initially decided on December 3, 1958, to move only JPL from the Army to NASA, and to allow von Braun's team to work in Huntsville and accept assignments from NASA while technically remaining with the Army. The work already underway at Huntsville on the Saturn program—absolutely essential for the flights to the Moon—prompted NASA to keep pressing for a full transfer of the von Braun team until the Department of Defense and President Eisenhower finally gave their full support in October, 1959. General Medaris and von Braun supported the transfer to NASA when it appeared that a plan was afoot to give the Saturn program to the Air Force, which the Huntsville group feared might reduce its long-range priority.

Although the actual decision to make the transfer was clearly made by President Eisenhower, the Science and Astronautics Committee made two important contributions. First, the committee from the start indicated its confidence in and strong support for von Braun and the team he represented. Second, the committee held hearings on February 3, 1960, to demonstrate its support for House Resolution 567, sponsored by Representative B. F. Sisk of California.

Dr. Wernher von Braun, Director of the Marshall Space Flight Center at Huntsville, Ala., greets members of the Science Committee during one of their many field trips to that installation. From left, Representatives Richard L. Roudebush (Republican of Indiana), R. Walter Riehlman (Republican of New York), James G. Fulton (Republican of Pennsylvania), Col. Earl G. Peacock, committee staff; Col. Harold F. Dyer, committee staff; Representative Olin E. Teague (Democrat of Texas), K. K. Dannenberg of MSFC, Dr. von Braun, Harry H. Gorman of MSFC, Representatives Ken Hechler (Democrat of West Virginia), Joe Waggonner (Democrat of Louisiana), John W. Davis (Democrat of Georgia), and Erich Neubert, MSFC.

Representatives J. Edgar Chenoweth (Republican of Colorado), B. F. Sisk (Democrat of California), Joseph W. Martin, Jr. (Republican of Massachusetts) and Joseph E. Karth (Democrat of Minnesota), far right, with Dr. James A. Van Allen (second from right), after whom the "Van Allen Radiation Belts" were named.

The Sisk resolution called for immediate approval of the transfer, rather than waiting the customary 60 days allowed under statutory reorganization procedures.

During the formal hearing by the committee on the von Braun team transfer, NASA commended the committee for its support of the manner in which the transfer was being arranged. "NASA at this point is confident that the plans are realistic and that, with the support of this committee and the Congress, the proposed transfer can be accomplished in a manner which will greatly strengthen this Nation's space program, both civilian and military," Albert Siepert of NASA told the committee. Siepert added that early passage of the resolution would smooth the transfer and remove employment uncertainties at Huntsville.

In presenting his resolution to the House of Representatives on February 8, 1960, Representative Sisk noted that:

This joint resolution expresses the intent of the Congress that the von Braun team will remain essentially intact, and that our programs for space science and exploration will thereby be materially expedited and advanced.

The major opposition to the resolution came on military grounds and was expressed by Representative Sam Stratton (Democrat of New York) of the House Armed Services Committee. Stratton urged the House to disapprove the resolution because—

I find it difficult to see why the program of an agency that has already demonstrated its ability to get the job done should be switched to an agency which in my judgment has not yet demonstrated its ability successfully to manage a program so vital to our national security.

House Majority Leader John McCormack indicated that "Congress is showing leadership in accelerating the approval of this transfer." Republican support was voiced by Representative R. Walter Riehlman (Republican of New York), who stated:

Dr. von Braun and Secretary Brucker, both of whom appeared before the committee to discuss this matter, left no doubt in the minds of the members that it was entirely satisfactory to them that this transfer be made. They were both in favor of this resolution being passed immediately because of the psychological effect this will have on the von Braun team.

When Representative Stratton demanded a "division" on the adoption of the Sisk resolution, there was an overwhelmingly favorable vote of 92 to 2.

The Senate later held hearings on the joint resolution, but because of a civil rights filibuster, no action was taken prior to the effective date of the transfer on July 1, 1960. Nevertheless, the huge majority by which the House supported the resolution gave the Committee on Science and Astronautics the chance to reaffirm its support of von

Braun's work on the Saturn project. Without that support, confusion and uncertainty would have resulted. The transfer did occur, and the committee helped smooth the way to insure the successful operation of the von Braun team working under new leadership. On July 1, the Marshall Space Flight Center was officially designated by NASA, with von Braun as its first Director.

LIFE SCIENCES

In preparation for eventual Moon flights, as well as the preliminary Mercury and Gemini flights, NASA was prodded by the House Committee on Science and Astronautics to focus on life support systems as well as the more striking priorities such as propulsion. In the early years of the committee, despite the multiplicity of subjects dealt with in hearings and reports, press and popular interest centered on the space race with the Russians and who would get to the Moon first. Members of the committee devoted a great amount of their efforts to educating the public to think more in terms of the need for American preeminence in space, for which the race to the Moon was only one symbol.

During his freshman year in Congress and at the beginning of his service on the committee, a Connecticut lawyer named Emilio Q. "Mim" Daddario gained early renown by developing as a specialist in the life sciences. Some 17 years after leaving NASA, Dr. T. Keith Glennan still vividly recalls Daddario's 1960 questions on life sciences during House hearings. Daddario boned up on everything that was being done by the Army, Navy, Air Force, Federal Aviation Agency, and other Federal agencies on the stress effects of space flight on the human organism. He then performed a very useful function in sending NASA officials scurrying to get themselves briefed on the most up-to-date information available in other agencies on the human factors in space flight.

Every agency and every bureau possesses a prideful desire to save the world in its own way. At a time when there was fierce competition among the military services, and between the military and NASA, for who should control space projects having both civilian and military significance, there was a tendency to build and control duplicating tasks. One of Daddario's early contributions was to point out forcefully the importance of coordination in the life sciences. The results were salutary. Not only did NASA avoid the expense of building competing installations, but also recruited knowledgeable military personnel who had gained their expertise in life sciences—outstanding people like Dr. Charles Berry. Dr. Berry was trained as a flight surgeon in the Air Force, and later went on to become Chief of Medical Operations at the Manned Spacecraft Center, and the personal physician of the astronauts.

PROJECT MERCURY

The major groundwork for Project Mercury, the first manned space flights, was laid by the Eisenhower administration. To the dismay of the Air Force, and to some extent the Army, Mercury was transferred out of the military and assigned to NASA upon the establishment of that agency in October 1958. President Eisenhower, while insisting that the program be administered under civilian supervision, nevertheless directed that the original Mercury astronauts be drawn from test pilots serving in the Armed Forces.

Even though the committee did not materially shape policy with respect to Project Mercury, the committee members were intensely interested in both the funding and progress of the program and the astronauts themselves. From the day the first seven Mercury astronauts appeared on May 28, 1959, in executive session before the committee, the members developed a close and personalized relationship with the first men in space.

"Do you feel you are being prepared for this flight with as much precaution as the Wright brothers took when they jumped off in their first plane?" asked Representative Gordon McDonough (Republican of California).

John Glenn, in answering affirmatively, also added a rare look into his own future as he replied: "Perhaps the dangers in your profession are more than they are in this."

On numerous occasions prior to the first Mercury suborbital flight of Alan Shepard in 1961, the committee met with the astronauts during their preflight training at Cape Canaveral, Fla. There, the committee had rare opportunities to talk with the astronauts about their training, the safety measures being designed for their protection, the configuration and status of the equipment being developed for their flights, and their own personal suggestions concerning the dramatic experience they faced.

In its first interim report on Project Mercury, the committee on January 27, 1960, underlined the high priority which was placed on the flights. But the committee in its report raised the question of "whether the national interest is best served by a single approach to this problem * * *. If there is an element of criticism in this report, it is not of what is being done or of the people involved, but rather that we are not doing more with other programs dedicated to the broader end of attaining a useful man-in-space capability."

The committee continued in 1960 and 1961 to vote full funding for Project Mercury. While the committee was urging accelerated action on the schedule for Moon flights, NASA personnel were quietly working behind the scenes on the future programs to send instrumented and then manned flights to the Moon. NASA remained confident that they would retain strong support in the Congress for these ventures.

On July 29, 1960, an unmanned Atlas-Mercury booster exploded one minute after launch at Cape Canaveral. But NASA announced the same day that planning had commenced on an entirely new manned space flight program called "Apollo," a project to carry three men in sustained orbital or circumlunar flight. The committee moved fast to support the new program, and also to apply pressure to speed up Mercury.

A NEW ADMINISTRATOR FOR NASA

With the inauguration of President Kennedy, Dr. T. Keith Glennan resigned as NASA Administrator on January 20, 1961. A struggle ensued over whether the new administrator should be a scientific or technical expert, or whether he should be an individual with proven administrative experience. There is no evidence that the House committee influenced the decision, but it is clear that Capitol Hill was the dominant force in directing the final choice. In meetings with President-elect Kennedy and Vice President-elect Johnson in Palm Beach during December 1960, Senator Robert Kerr (Democrat of Oklahoma) was thoroughly briefed on his prospective role as the new chairman of the Senate Aeronautical and Space Sciences Committee which Johnson had chaired. Chairman Brooks joined Kennedy, Johnson, and a special group of scientists headed by Jerome Wiesner, for a January 10, 1961, confab in Johnson's Capitol Hill office to discuss the future of space. Wiesner headed up an "Ad Hoc Committee on Space" for the President-elect, and later was named as President Kennedy's special assistant for science and technology. Wiesner and Johnson clashed on several issues, including what kind of person should be Administrator of NASA; Johnson wanted a man with political savvy and administrative ability, and Wiesner leaned toward an individual who had more scientific and engineering background.

In addition, Wiesner's committee issued a report labeling Project Mercury as "marginal", expressed the fear that an astronaut might be killed or not recovered from orbit, and urged a deemphasis of manned space flight.

Johnson, winning the power struggle with Wiesner, proceeded to interview a large number of possible appointees. In his book, *The Vantage Point*, Johnson relates that President Kennedy wanted to offer the job to retired Gen. James M. Gavin, who had headed Army research and development. Johnson persuaded the President that it would be a serious mistake to put a military man at the head of NASA in light of the strong feeling expressed by Congress in the Space Act, committing the United States to develop the use of space for peaceful goals.

Senator Kerr was a key factor in the eventual selection of James E. Webb. After a long delay during which Kerr attempted to obtain for Webb appointment as Secretary of the Treasury, the pieces began to fall into place. Johnson felt strongly that the appointment should not be determined by scientific knowledge as much as administrative ability, and he also resisted the pressure of the powerful groups lobbying to turn the space program over to the Air Force. Kerr knew Webb intimately, not only as a fellow Oklahoman but as a director and officer of Kerr-McGee Oil Industries, Inc. Far more important, everyone concerned appreciated that Webb had precisely the qualities necessary to lead, inspire and manage a massively expanding organization like NASA.

From a congressional standpoint, Webb was a perfect choice. He had actually served on the Hill as Administrative Assistant to the Chairman of the House Committee on Rules, Representative Edward Pou (Democrat of North Carolina), in the early Roosevelt years. As Director of the Bureau of the Budget and Under Secretary of State, he had cultivated excellent congressional relations both institutionally and personally. A lawyer, Marine aviator in World War II, associated with several nonprofit educational foundations, Webb had the breadth of experience to handle problems and issues across the board. He also had many scientific ties and a good personal friendship with Wiesner, even though the latter did not at the time consider Webb to be his first choice.

"Senator Kerr originally called me in Oklahoma City toward the latter part of January, and said that Mr. Johnson would be calling me, or the White House would be calling me to ask me to come to Washington to talk about the job. He hoped very much that I would take it, and he knew then that he was going to be chairman of the Senate Committee on Aeronautical and Space Sciences," Webb recalls. As it turns out, Webb was called from the dais to take a Washington call from Wiesner, while attending a luncheon in Oklahoma City honoring Senator Kerr. At President Kennedy's direction, Wiesner phoned Webb and asked him to come to Washington on January 30 to discuss the job.

Webb, after lunching with NASA Deputy Administrator Dr. Hugh Dryden, called at the White House. President Kennedy told Webb he wanted as NASA Administrator not a scientist but, "someone who understands policy. This program involves great issues of national and international policy." In accepting the appointment, Webb asked that Dr. Dryden be renominated as Deputy Administrator. He also announced his intention to retain Dr. Robert C. Seamans, Jr., as Associate Administrator; Seamans had come aboard at NASA in September 1960. Webb was sworn in as NASA Administrator on February 14, 1961.

Webb made an immediate hit with the Committee on Science and Astronautics. His enthusiasm for the space program was contagious. His outgoing personality and unbounded optimism were in marked contrast to his more reserved predecessor, Dr. Glennan.

Webb's assets included an ability to win the confidence of Republicans. On February 27, 1961, he honored the House Science Committee with his maiden appearance and this introduction:

This is not only my first appearance before this committee, but the first appearance in public of any kind or description since I took the oath of office.

On the occasion, former Speaker Martin welcomed Mr. Webb with these words:

I first want to congratulate the country in getting Mr. Webb as the head of NASA. It has been my privilege to know Mr. Webb almost longer than memory would permit accurate recollection. When Ed Pou was chairman of the old Rules Committee of the House, Mr. Webb did wonderful service with him. I have come to know him through the years in all aspects of his career. NASA is to be congratulated upon getting a dedicated public servant such as Mr. Webb.

The members of the House committee admired Webb's effectiveness as a salesman. Some members probably disliked his tendency to give longwinded answers to pointed questions, but this was a trait which was also very familiar among congressional colleagues, hence accepted in a bemused fashion.

Perhaps Webb's strongest asset in his relations with the Science Committee was his accessibility and close working relationship with the successive chairmen of the committee. Brooks lived for less than a year after Webb assumed office, and their relationship was never close. But both Miller and Teague quickly developed a personal rapport with Webb which enabled frequent, frank, behind-the-scenes conversations to anticipate problems in advance, and to tackle issues which otherwise might have ballooned into controversies.

The issue of whether to go to the Moon and how soon was resolved in a somewhat different fashion, and with somewhat less input from the Science Committee than in other cases where the committee influenced NASA policy. This was partially due to the fact that Webb was feeling his way as a new Administrator in 1961 and was treating the committee in the more traditional, formal fashion used by those downtown who respect the separation of powers. In addition, Brooks was exerting leadership to fashion the committee into an independent force, rather than an appendage of the agencies over which it was exercising oversight. Nevertheless, the committee continued to press for an early decision to commit the Nation to a manned series of flights to the Moon. In the early months of the Kennedy administration, the committee demonstrated that it was far more hawkish on manned space flight than even Mr. Webb would publicly admit at that time.

THE AIR FORCE CHALLENGE TO NASA

The Committee on Science and Astronautics and its predecessor, the select committee, had entrenched NASA as the custodian of the Nation's space program and endowed it with a distinctly civilian flavor. The Army, Navy and Air Force, each of which had a clear military interest in the development of space weaponry, struggled insistently to wrest more of the space budget away from NASA. The Science Committee listened intently to a parade of military witnesses advocating more power for the military in space in order to protect America's national security. With the exception of a minority of the committee on both sides of the political aisle, generally the committee wound up on the side of NASA and the peaceful uses of space.

The Air Force launched the most vigorous campaign to support its own role in space, and for several reasons that campaign reached a fever pitch late in 1960 and in the early months of 1961. The 1960 Presidential campaign had debated the so-called "missile gap," and many of John F. Kennedy's statements stressed the national security aspects of space. When President Kennedy first assumed office and before NASA Administrator Webb began to assert himself, there was a vacuum in leadership on space matters which was not filled until some clear budgetary decisions were made later in the spring of 1961. Finally, the Wiesner report contained some criticisms of NASA which the Air Force picked up to strengthen its case. The House Committee on Science and Astronautics printed a very revealing December 1, 1960, memorandum from the Office of the Secretary of the Air Force to all Air Force commanders and contractors urging a larger role in space for the Air Force and its contractors.

When the Wiesner report was unveiled in Chairman Brooks' presence on January 10, 1961, he became disturbed by some of its implications in threatening the role of NASA in space. In February, the Science Committee held its annual round of hearings on defense interests in space, and heard from the Department of Defense as well as the Army, Navy and Air Force. Following these hearings, Chairman Brooks was impelled to write to President Kennedy on March 9, 1961:

I am seriously disturbed by the persistency and strength of implications reaching me to the effect that a radical change in our national space policy is contemplated within some areas of the executive branch. In essence, it is implied that United States policy should be revised to accentuate the military uses of space at the expense of civilian and peaceful uses.

Of course, I am aware that no official statement to this effect has been forthcoming; but the voluminous rash of such reports appearing in the press, and particularly in the military and trade journals, is, it seems to me, indicative that more than mere rumor is involved.

Moreover, I cannot fail to take cognizance of the fact that emphasis on the military uses of space is being promoted in a quasi-public fashion within the defense establishment. Nor can I ignore the suggestion, implicit in the unabridged version of the Wiesner report, that the National Aeronautics and Space Administration role in space is purely one of scientific research and that the military role in the development of space systems will be predominant. Such an assertion not only seems to disregard the spirit of the law but minimizes the values of peaceful space exploration and exploitation.

Brooks stressed in his letter how important it was to support the civilian space program as a means of "preserving the peaceful image of the United States." He wrote the President that he did not want to see "the military tail undertake to wag the space dog" and that "if NASA's role is in any way diminished in favor of a space research program conducted by a single military service, it seems unlikely to me that we shall ever overtake our Soviet competition which, by the way, has been peculiarly effective because of its public emphasis on scientific and peaceful uses of space."

The letter from the chairman of the House Committee on Science and Astronautics to President Kennedy had a very healthy effect in strengthening the determination of the President to protect NASA's turf. The President responded to Chairman Brooks on March 23, 1961:

It is not now, nor has it ever been my intention to subordinate the activities in space of the National Aeronautics and Space Administration to those of the Department of Defense. * * * Furthermore, I have been assured by Dr. Wiesner that it was not the intention of his space task force to recommend the restriction of NASA to the area of scientific research in space.

Even though the letter left unanswered issues like the possible future interest of the Air Force in developing large space boosters, or manned flights moonward, the interchange cleared the air and helped strengthen NASA's position against the forces eager to get a bigger cut of the space budget pie.

MANNED FLIGHT AND THE KENNEDY BUDGET

As the House committee started its 1961 hearings, one by one every NASA official informed the committee that they favored speeding up the timetable for a manned flight to the Moon. Webb informed the House committee that his first task was to make a thorough review of the Eisenhower space budget, and the implication was clear that the conclusions would justify stepping up the total effort. Dr. Dryden, who had shocked the select committee in 1958 by downgrading manned flight by comparing it with the circus stunt of shooting a woman out of a cannon, exuded a new spirit of buoyant optimism on March 14 as he exulted about the successful suborbital flight of the chimpanzee Ham in a capsule atop a Redstone missile. Dr. Dryden added:

> You will recall that in the budget submitted by Mr. Eisenhower there was a statement that he could see no reason for proceeding beyond Project Mercury. This I think you know is not in accordance with my own recommendations and ideas.

At the same March 14 House committee hearing, Dr. Seamans made his own position crystal clear: "As an individual, I'm irrevocably committed to pushing the man-in-space program at a maximum speed consistent with budgetary matters and things of that sort." Seamans could sense that "budgetary matters and things of that sort" would soon give a big push to plans which were already being formulated for a speeded-up Moon flight. Dr. Robert Gilruth's Space Task Group, working out of Langley, and George M. Low, at NASA's headquarters, had underway detailed studies which were far more optimistic than NASA's old 10-year plan which pegged the manned Moon landing as sometime after 1970.

On April 10, Webb made another appearance before the House Science Committee, telling the House Members that the President was asking $125.6 million more for NASA—most of the stepped-up funds to go toward development of the Saturn booster.

Webb was followed on April 11 by George Low, who brought the committee up to date on Project Mercury. Low clashed with Miller on the issue of whether the January 31, 1961, flight of the chimpanzee "Ham" had been a success, as listed by Low on a huge chart he showed the committee. Miller challenged Low in sharp terms:

> Mr. MILLER. In the case of "Ham," that was the January 31 flight that you showed as successful there?
> Mr. Low. Yes, sir.
> Mr. MILLER. Wasn't there a condition there where you had planned on one orbit and something went wrong and you kicked him out 120 miles further?
> Mr. Low. We went about 120 miles farther than planned.
> Mr. MILLER. Shouldn't this be put down as a failure for the booster?
> Mr. Low. Perhaps I should add another column to this chart for the booster, to indicate its performance.
> Mr. MILLER. I think you should. I think you are trying to fool us.
> Mr. Low. I am not ——
> Mr. MILLER. I suggest you correct the chart before it goes into the record.
> [The chart printed in the hearing record still listed the flight as a success, with the notation: "Booster difficulties resulted in longer range than planned."]

As the noon hour approached, Low announced that he had a movie of the Ham flight, just produced the day before, which he wanted to show the committee. Chairman Brooks responded: "We better recess at this point, so you will have a prelude to your movie when you present it to the committee."

Low had scarcely finished his testimony on April 11 when an event as shocking as Sputnik occurred, which spurred a radical change in attitudes and timetables from the top to the bottom of the space program.

EFFECT OF GAGARIN FLIGHT

On March 9 and 25, the Russians had successfully orbited and recovered dogs in their spaceships. Suddenly on April 12 came the electrifying news that Maj. Yuri Alekseyevich Gagarin in a 5-ton Vostok spacecraft had orbited the Earth in 89 minutes, returning safely to Earth without any problems caused by weightlessness or reentry.

In the early morning hours, many telephones rang to alert committee members and NASA officials with the skimpy details of the Gagarin flight. Newsmen awakened John A. "Shorty" Powers, "the Voice of Mercury Control" and public affairs officer for the Mercury program at Cape Canaveral, to ask for a public statement. Powers responded candidly: "We're all asleep down here."

The House committee had scheduled Dr. Edward C. Welsh, Executive Secretary of the National Aeronautics and Space Council, for April 12. Welsh appeared on behalf of legislation to make the Vice President Chairman of the Space Council. He was battered with questions about the Gagarin flight. Initially, he told Chairman Brooks:

> You said before this hearing that you were woke up about 3 o'clock this morning to receive some information, and so was I. So each of us missed that much sleep.

Welsh realized what every NASA official soon discovered also, that the Science Committee was expressing the insistent, demanding, sentiment of most Americans that it's about time we start doing something to demonstrate our capabilities in space. The blows to national pride caused angry reactions. There had been a brief honeymoon after President Kennedy took office, but it was obviously now over. NASA officials, in turn, were expected to demonstrate that they shared the sense of urgency being strongly expressed through the committee.

Representative James G. Fulton (Republican of Pennsylvania), who had become the laughing stock of many Members by his repeated announcements he wanted to make a flight himself, startled Welsh by suggesting:

> I think we are getting to the point where if they are afraid, let more of us go who aren't afraid of the risk. If it is good enough for Ham, and a chimpanzee can do it, why couldn't a man do it?

I am always laughed at, but I would go in a minute.

Suave, poised, and experienced in appearances before congressional committees, Welsh was stunned into speechlessness. Chairman Brooks broke the tension by observing dryly:

We will get you a one-way ticket there sometime.

A delighted chimpanzee "Ham" reaches out from his couch to take an apple from a crewman of the U.S.S. *Donner*. This was the first food for "Ham" after a 420-mile ride in a Mercury capsule prior to the first manned flight in space.

Fulton added:

The thing to do is to get some space enthusiasts, as they had when they were developing airplanes, who are willing to take certain risks. I am getting awfully tired of the Mother Hubbard approach of "Tie your apron up after the Russians do it." All you have to do is put a little overtime on and go around the clock on some of these programs instead of knocking off at 5 o'clock. I think we in the United States should stand the expense of it and put some overtime in on this and pay for it.

Chairman Brooks, King, Karth, Anfuso, and other committee members also suggested that the United States should be making an all-out drive to overtake the Soviet Union. Miller entered a cautionary note, stating:

I think we are justified in proceeding with celerity but not trying to get into competition on this thing. * * * We are more than justified in taking our time and doing a thorough job rather than just trying to be in competition.

Fulton was insistent, and he told the committee:

Spend the money that is necessary and let's be preeminent in science. I think you can win with all the international conferences in the world and you lose one on man-in-space such as we did yesterday, you are rated a second-rate power whether you like it or not.

With one eye on the press table, Fulton also added a one-liner which eventually reverberated all the way up Pennsylvania Avenue to the White House: "I am tired of coming in second best all the time," he told Welsh.

REACTION OF PRESIDENT KENNEDY

A few hours later when President Kennedy held his news conference, a reporter asked:

Mr. President, a Member of Congress said today that he was tired of seeing the United States second to Russia in the space field. I suppose he speaks for a lot of others. Now, you have asked Congress for more money to speed up our space program. What is the prospect that we will catch up with Russia and perhaps surpass Russia in this field?

The President answered at some length, and among his remarks were the following:

Well, the Soviet Union gained an important advantage by securing these large boosters which were able to put up greater weights, and that advantage is going to be with them for some time. However tired anybody may be, and no one is more tired than I am, it is a fact that it is going to take some time and I think we have to recognize it * * *.

So that in answer to your question, as I said in my State of the Union message, the news will be worse before it is better, and it will be some time before we catch up. We are, I hope, going to go in other areas where we can be first and which will bring perhaps more long-range benefits to mankind. But here we are behind.

The April 12 hearing with Welsh spent precious little time on the more mundane issues of how to hurry along the legislation to make the Vice President head of the Space Council. Typical of the observations was the reaction of Representative Victor L. Anfuso (Democrat of New York), fourth-ranked member of the committee and a subcommittee chairman:

We can't just wait on this 10-year period program because if we carry out the things that they say in 10 years the Lord knows where the Russians will be by then and whether America will still be in existence.

Finally, Anfuso added: "I think we have debated this bill long enough. We have had an interesting session here. I move that we report out H.R. 6169." The committee quickly went into executive session, reported out the bill, 20 to 0 and got ready to send it to the House floor. The bill cleared the House and Senate with remarkable speed, and was signed by the President on April 25. Once the committee had completed its executive session, the members got ready to sharpen their knives for the top officials of NASA who appeared on April 13.

From the perspective of many years later, NASA Associate Administrator Seamans described the atmosphere after the Gagarin flight:

> The day after Gagarin went into orbit was one of the more hectic days in NASA's existence * * *. Jim Webb and Hugh Dryden testified and people were pounding the desks, and why aren't we going faster, and why aren't we working triple time, and we can't let the Russians do this and keep doing this to us.

THE COMMITTEE SEIZES THE INITIATIVE

At the pressrooms and radio and television gallery of the Capitol, demands from news editors were pouring in to obtain congressional reactions to the Gagarin flight. Recognizing the tremendous popular interest in the issue, Chairman Brooks convened the April 13 hearing with Webb and Dryden in the huge Cannon caucus room. In opening the hearing, Brooks observed that "Because of the events of the last few days, we expect a large audience, and we thought it would be more comfortable for some of our friends, especially the press, radio, and television people, to meet here."

Webb related in a memo to President Kennedy's assistant, Kenneth O'Donnell, what happened on April 13 "in the atmosphere of great excitement and focusing of public interest in the hearings held in the caucus room." Webb added:

> The members of the committee, almost without exception, were in a mood to try to find someone responsible for losing the race to the Russians and also to let it be known publicly that they were not responsible and that they were demanding urgent action so that we would not be behind. Pursuing this further in the days that followed, the committee steadily bored in on every phase, trying to get every bit of detailed information that would focus public interest on the committee, and the role it had chosen for itself as the goad to force a large increase in the program.

It was one of the few occasions in Webb's experience when the enthusiasm of the committee far exceeded his own. "The committee is clearly in a runaway mood," Webb warned, adding that "I believe I can assure you that NASA personnel have not so conducted themselves as to cause the type of hearing now being conducted." Webb was cast in the unusual role of the calm, cool, and collected defender of a program who refused to be affected by the supercharged effort of the enthusiasts to get him to speedup his program. The former Director of the Budget reported:

> On every point in the budget which the committee has covered, they have specifically pressed to ask what our presentation to the Bureau of the Budget included

and have asked the question as to why it was denied * * *. I based my position on the fact that the U.S. effort was "a solidly based, step-by-step program, based on a long period of effort * * *."

One after another, members of the committee communicated their intense, sometimes emotional, concern over whether NASA was moving fast enough. Representative Joseph E. Karth (Democrat of Minnesota) stressed the need to forget the 40-hour week, and pay overtime where the critical bottlenecks were occurring. Representative Jessica McC. Weis (Republican of New York) remarked to Mr. Webb: "It must be very refreshing to be before a committee that is anxious to give you more money than you seem to want," to which Mr. Webb quite diplomatically responded: We "certainly appreciate this committee." The exchanges were tense, the atmosphere highly charged, and once again the committee successfully transmitted the overpowering sentiment of the people that they wished President Kennedy would fulfill his campaign promise in space to "get the country moving again."

In the small, cramped room 214–B of the Longworth Building, Dr. Seamans and George Low returned to testify on April 14. Low had planned to show the film of the flight of the chimpanzee Ham, originally scheduled for his presentation on April 11. He later confessed:

> We thought it would not be in our best interest to show how we had flown a monkey on a suborbital flight when the Soviets had orbited Gagarin. The chairman did say, "Well, we thought we were going to start with the movie." We looked around and the projectionist wasn't there, and we fumbled and said, "We don't have it with us today."

Dr. Seamans' testimony demonstrates how a congressional committee can frequently affect both the timing and substance of a Presidential decision, even though the decision itself is made in the White House and is exclusively a Presidential responsibility.

Under questioning by Karth, Dr. Seamans indicated that additional funds for Saturn enabled NASA to step up the schedule for a manned landing on the Moon. He stated that as a result of the NASA review of the Eisenhower budget, $308 million had been requested above that budget, and the President had approved an increase of $125.6 million of that amount. Asked for his personal opinion as to what NASA could do with additional money for the Apollo program, Seamans responded: "My own opinion is that the country is capable of more effort in this area than it is now expending."

SPEEDING UP THE LUNAR LANDING TIMETABLE

Among the welter of questions which bombarded Dr. Seamans on April 14, the one most vividly etched in his memory was posed by freshman Democrat David S. King. Son of a U.S. Senator, representing a very shaky district in Salt Lake City, Utah, King was one of the prime critics on the committee who was advocating use of solids instead of liquids in space boosters. On this particular day, he recited the biblical parable of the king who found himself in the unenviable position of confronting 50,000 enemy troops while he possessed only 10,000. He noted: "The point of the parable being that before engaging in contest one must very carefully evaluate and appraise the strength of the adversary." When Dr. Seamans could not answer pointedly the obvious question whether the United States would get to the Moon before the Russians, King then posed this question:

> The Russians have indicated at various times that their goal is to get a man on the Moon and return safely by 1967, the 50th anniversary of the Bolshevik Revolution. Now specifically I would like to know, yes or no, are we making that a specific target date to try to equal or surpass their achievement?

Seamans answered:

> As I indicated in earlier testimony this morning, our dates are for a circumlunar flight in 1967, and a target date for the manned lunar landing in 1969 or 1970.

King then asked whether, through a fuller marshaling of manpower and resources it might be possible to meet a target date of 1967. Dr. Seamans responded that "to compress the program by 3 years means that greatly increased funding would be required for the interval of time between now and 1967. I cannot state that this is an impossible objective * * * my estimate at this moment is that the goal may very well be achievable."

Pressed by Representative J. Edgar Chenoweth (Republican of Colorado), Seamans indicated that to speed up the lunar landing goal would cost many billions of dollars. Chenoweth raised the question of "whether our economy can stand perhaps double or treble the present funding or even go higher than that, by putting up money to achieve this lunar shot, say in 1967, or even before. It is a question of whether such an accomplishment has that much national and international significance and importance. Do you agree with that?"

> Dr. SEAMANS. Yes——
> Mr. CHENOWETH. You say the United States can do it if we increase the money?
> Dr. SEAMANS. I did not say we could do it. I said we would review our plans and advise whether it was possible. I think it may be possible.

Chenoweth became even more disturbed as Seamans remained optimistic. With rare vision, he virtually predicted what would happen as a result of the discussion:

> Mr. CHENOWETH. I think you have to be very careful of what you tell this committee because there will be those who will say, "All right, let's boost up our appropriation, double it, treble it. The most important thing is to put this man on the Moon."

I don't know that it is. I doubt it. But some feel that way. I think it is a high policy decision to be made and to be made shortly. I think it is important you word your answers carefully here, because the wrong interpretations may be placed upon them not only by this committee but by those who will read the news stories that will go out.

Dr. SEAMANS [continuing]. I feel this committee is a most important forum for discussion of this issue. I believe there are other important forums. I agree this is a most important national issue.

Mr. CHENOWETH. The question is whether it is of such great importance that we can afford to neglect other programs that perhaps may involve a change of our whole fiscal program in order to accomplish this one objective. Is it that important, in your opinion?

Dr. SEAMANS. Obviously I cannot answer that question.

Mr. CHENOWETH. It is a decision to be made at a higher level.

Dr. SEAMANS. I think it is a decision to be made by the people of the United States.

Mr. CHENOWETH. How will they make it?

Dr. SEAMANS. Through the Congress and through the President. It is a matter of national importance to have specific objectives for our space effort.

Mr. CHENOWETH. I disagree. The people of this country do not have the technical knowledge on this subject that you have. When you talk about placing a man on the Moon, they don't know what you are talking about. They don't know what expenditure is involved, nor the scientific and research work that has to be done. We can't expect them to make that decision.

Mr. MILLER. Is this not our responsibility as the representatives of the people?

Mr. CHENOWETH. We can make the decision. But I think when it comes to affecting the economy and the fiscal policies of this country and the tremendous amounts of money that are involved, I think perhaps this will have to be made at a higher level of the administration.

A battery of television cameras, tape recorders, microphones, and pushing reporters with notebooks greeted Dr. Seamans when he emerged from the hearing room shortly before noon. As the committee remained behind for an hour-long executive session, beads of sweat slowly gathered on Seamans' forehead from the hot glare of the lights in the narrow corridor of the Longworth Building. There was cold sweat when he picked up The New York Times and the Washington Post the following morning. And in the Oval Office at the other end of Pennsylvania Avenue, there was an explosive reaction from President Kennedy.

The Washington Post headlined the testimony: "Reaching Moon First Would Cost Billions, Expert Tells House Unit." The lead began: "A multibillion-dollar crash space program might put an American on the Moon by 1967—perhaps ahead of the Russians—a top Government official said yesterday." The New York Times correctly interpreted Seamans' testimony in the following terms: "Pointing to the large expense involved in a 'crash' effort to land a man on the Moon, Mr. Seamans repeatedly emphasized that such a venture presented 'a most important national issue,' and that the American people, Congress, and the President would have to decide whether it was in the national interest * * * Mr. Seamans confirmed reports that the administration had refused funds requested by his agency for development and fabrication of the Apollo capsule. The space agency, he said, asked for an overall budget of $1.4 billion but had its request cut by $182 million by the administration.''

Seamans relates that the President was very upset "that some minion of his that he didn't know was talking about going to the Moon, and I thought it might be the end of my existence at NASA." As a matter of fact, Seamans' job was in real jeopardy as a result of the incident. But the Gagarin flight, the testimony before the House Science Committee and the issue of what must be done to restore America's badly bruised prestige spurred the President to focus on the space program and lunar landing in the next few weeks.

With this backdrop, President Kennedy called Vice President Johnson to his office April 19. The next day, the same day Congress approved the legislation to make Johnson Chairman of the Space Council, the President wrote a memorandum to his Vice President commencing: "In accordance with our conversation, I would like for you as Chairman of the Space Council to be in charge of making an overall survey of where we stand in space." By April 19, the abortive Bay of Pigs invasion had degenerated into a total failure, and historians will argue how much this defeat may have related to the lunar landing decision. On April 21, the President in a press conference stated bluntly: "If we can get to the Moon before the Russians, then we should."

While Johnson was holding almost nonstop conferences to assemble the best advice from sources both within and outside the Government, the House Science Committee continued to hold daily hearings on the NASA budget. In addition to fine-tooth combing that budget, the House committee continued to press very hard for a general speedup in the entire space program. This pressure certainly was not lost on NASA officials making their frequent appearances by day and relaying their assessments by night. By April 28, Johnson had a preliminary memorandum ready for the President, recommending a manned lunar landing as the centerpiece of the space program.

Early in May, activity intensified on both ends of Pennsylvania Avenue. Johnson telephoned many Congressmen to learn whether they would support a greatly accelerated step-up in the space program; he was encouraged to receive enthusiastic bipartisan support. In response to Johnson's request, Chairman Brooks submitted a 10-page memorandum entitled "Recommendations re the National Space Program," dated May 4, starting: "We cannot concede the Moon to the Soviets, for it is conceivable that the nation which controls the Moon may well control the Earth." Brooks noted that he and his committee believed that "the United States must do whatever is necessary to gain unequivocal leadership in space exploration." The committee recommended an immediate acceleration of programs for communications, television, weather, and navigation satellites. Also suggested was an orbiting astronomical observatory aimed at discovering "the origin, evolution, and nature of the universe." The memorandum also argued the economics of a larger space program, pointing out that the Soviets were devoting 2 percent of their gross national product to space, and "a $5 billion a year space program represents only about 1 percent of our gross national product, even half of which offers returns crucial to the leadership, the prestige, and perhaps even the survival of the United States."

In a series of executive sessions between May 1 and May 4, the committee broke open the budget. The committee voted to restore every penny of the nearly $200 million cut from NASA's requests by the President's Bureau of the Budget. In addition to a $15 million add-on authorization for solid fuel propulsion, the committee voted to fund $50.2 million above the Budget-approved figure for Project Apollo. The committee voted every penny that NASA had requested, and had been cut by the Budget Bureau, and added $7.6 million for additional Apollo tracking facilities and staffing the Apollo program.

By these actions, the Science Committee sent a clear message to the President that he could and should raise his own sights on the future of the space program.

At a final executive session on May 4, 1961, the committee in morning and afternoon sessions worked feverishly to hammer home the final details of the authorization bill. Knowledge of the Johnson study was piecemealed to the committee, and had its effect in the bullish attitude of the committee, as the following typical colloquy indicates:

Mr. Mosher. Do you have information that the administration will come in with recommendations that will completely differ from what we are accomplishing?

Chairman Brooks. All I know is a study is being made of this at this time. The public is pretty well shaken up that the Russians did orbit the Earth with a man and we haven't——

Mr. Mosher. I would assume that this shakeup in public opinion would be reflected in the administration's recommendations——

Chairman Brooks [continuing]. What we know is that they are making a study. I think their study is very appropriate in the light of the fact that we have just put through a measure implementing the Space Council and the Space Council is at work.

Mr. MOELLER. Would it not seem that the public would then expect and Congress here in particular would expect that if this shakeup is having any kind of effect at all, it ought to be reflected by the activities of this committee, to push this thing even a little harder and with more money?

Chairman BROOKS. The committee certainly got very favorable comment from its recent actions in apparently pushing the program.

Mr. MOELLER. I don't think we should wait for the President's recommendations.

Chairman BROOKS. We are not——

All along the line, the committee took an aggressive position in support of almost every aspect of NASA's program. This was especially true on Apollo-related activities. With the leadership of Teague and Daddario, the Teague subcommittee held intensive hearings on the life sciences program. As Teague reported to the full committee: "I think the fact that Mr. Daddario brought this up caused them to search their minds for a more aggressive program than they had in mind."

SHEPARD'S SUBORBITAL FLIGHT

It was almost 5 p.m. when the committee finished its markup session on May 4. Members barely had time to grab their overnight bags, rush to Andrews Air Force Base, and enplane for the flight to Patrick Air Force Base, Fla. There they motored to Cocoa Beach to prepare for an early-morning bus ride to Cape Canaveral, where thousands of spectators awaited the first attempt to put a man in space. Alan Shepard's 15-minute suborbital flight seems rather puny today, but to the committee all of the prestige of the Nation and the future of the space program rested on the absolute necessity for its success.

Meanwhile, back in Washington, Vice President Johnson learned that he would be departing May 8 for a 2-week tour of Southeast Asia. So he ordered NASA and the Department of Defense to have their reports on the future of space in his hands before his departure, necessitating round-the-clock work over the weekend even by those officials who made the trip to witness the Shepard flight.

The members of the committee knew Shepard and the other six Mercury astronauts, had followed their training, voted funds for their support, and had a personal as well as official stake in the success of the flight. The committee strongly opposed the views expressed by Senators John Williams (Republican of Delaware) and J. William Fulbright (Democrat of Arkansas), who had urged President Kennedy either to postpone the flight or close it to the press. The committee supported the view that the open media coverage of every detail of the Shepard and subsequent flights was a real plus in contrast to the Soviet practice of hiding failures and never announcing a space success until after it had been achieved.

At 1:30 a.m., May 5, Alan Shepard was awakened and started the long series of preparations necessary prior to his historic flight after daybreak. By that time, most of the committee members had bedded down, awaiting the predawn phone call alerting them for their bus trip out to the Cape. Not all members of the committee could sleep, however.

At 2:30 a.m., the telephone jangle awoke one of the members of the committee. He was deeply chagrined and embarrassed when the voice on the phone gruffly proclaimed:

> This is Tiger Teague. The whole crew is getting on the bus, and we've been waiting for you for 10 minutes. Get your tail down to the lobby right away or you'll be left behind!

Three minutes later, an unshaven, dishevelled Congressman breathlessly asked a sleepy-eyed desk clerk: "Where is everybody?"—only to learn that the wake-up calls would not be made until 5 a.m.

Out at the Cape, the committee members watched closely as final checks were made at Mercury Control. NASA Administrator Webb was visibly nervous. Teague recalls Webb had three statements ready; one if the flight succeeded, one if Shepard had to be ejected in case of malfunction, and a third in case the astronaut was killed. One of the members, standing next to Bill Hines of the Washington Evening Star, heard him report over a live telephone line to his paper:

> Two, one, zero, ignition!
> There it goes!!
> This is the moment, the first time an American has entrusted his life to one of these things. I am covering this story, but God help this man.

Shepard's flight, viewed by millions live on television, was an unqualified success. His wife, Louise, remarked: "This is just a baby step, I guess, for what we will see."

Back in Washington, many officials were frantically putting together the final report which Vice President Johnson had ordered three days after the successful Shepard flight. It was their responsibility to project that baby step into a giant stride.

After Shepard addressed a joint meeting of Congress on May 8, Johnson was handed the final memorandum which he took to President Kennedy without change. The memorandum began: "It is man, not merely machines, in space that captures the imagination of the world." In the ensuing time frame, President Kennedy polished the historic declaration which he was to make to the Congress on May 25.

The day before President Kennedy appeared before the joint session of Congress to announce the goal of a lunar landing, the Science and

Astronautics Committee brought its authorization bill to the floor. The timing was perfect. Six weeks after the shocker delivered by Gagarin's orbital flight, and less than three weeks after America swelled with pride at Shepard's achievement, the committee presented to Congress a bill with a price tag of $1.37 billion—some $142 million beyond what had been budgeted in the March revision made by the Kennedy administration. Chairman Brooks was not breaking anything "top secret" when he told the House that "Tomorrow there will be recommendations by the President * * * for a considerably larger sum." Majority Leader John McCormack, in supporting the huge bill, also mentioned the President's impending appearance on behalf of increased funding and added:

That clearly shows the judgment of the chairman and the members of the committee is sound and that they were looking to the future * * *. Read the reports of this committee. They are ahead of the Executive * * *. This committee stands not for catching up—but for surpassing.

Former Speaker Martin in supporting the 40-percent increase in funding over what had been voted in 1960, also underlined the bipartisan character of the Science Committee's operations:

I have been here for 37 years, and I have never seen more dedicated service than the members have devoted to this subject. Let me say, too, that the subject is one that is very difficult and very technical. It requires great study. That it has had. Above all, what impressed me was the fact that there was no partisanship displayed in this committee in any instance. We all had, on both sides, but one purpose, and that was to do what was best for America and for the development of science.

When President Kennedy appeared to announce his recommendation that a manned lunar landing within the decade be set as a national goal, he told the Congress that "no single space project in this period will be more impressive to mankind, or more important for the long-range exploration of space; and none will be so difficult or expensive to accomplish."

In asking the Congress for over half a billion dollars in additional funds for NASA and the Department of Defense, President Kennedy also helped to pinpoint the responsibility, not only of the Nation, but the immediate tasks facing the committees in the Congress:

Now this is a choice which this country must make, and I am confident that under the leadership of the Space Committees of the Congress, and the Appropriations Committee, that you will consider the matter carefully.

The House Committee on Science and Astronautics went to work again after President Kennedy's address. In subsequent discussions, both the committee and the House considered and in effect ratified the new goal of a manned lunar flight within the decade. Although both the committee and the House had force-fed NASA with $142 million more than President Kennedy's figure unveiled in March, the President leapfrogged Congress with a startling new request for some half a billion dollars beyond the House-passed figure at the time he made his dramatic May 25 announcement.

The Senate docilely voted every penny of the President's request on June 28. The House held 3 days of hearings starting July 11. Members wanted to know why NASA in March had sworn that extra money would be uneconomic, wasteful, and not speed up the program, and now they were coming in to do a 180-degree turn. Dr. Dryden explained it as a policy decision: "Shall the recommendation of President Eisenhower be accepted that the manned space flight program be confined to research and development beyond Project Mercury or should steps be taken to move the follow-on vehicle development?" The issue was joined in this colloquy between Dr. Dryden and Representative J. Edgar Chenoweth (Republican of Colorado):

Mr. CHENOWETH. We have great respect for you. You have changed your attitude a little bit here in the last few months.

Dr. DRYDEN. In what way?

Mr. CHENOWETH. You didn't present such a program when you were here before. What has caused the change in your thinking?

Dr. DRYDEN. Two or three months ago you had a document before you from President Eisenhower which said that he saw no reason for going beyond Project Mercury with manned flight. I could not submit this kind of budget under the rules as you know.

Mr. CHENOWETH. I don't think it makes much difference who is President of the United States.

Dr. DRYDEN. I think it makes a lot.

A few days after the new House committee hearings had concluded, the conference committee met on July 19 and accepted an increase of $408 million beyond the authorization bill the House had passed on May 24.

Representative Perkins Bass (Republican of New Hampshire), fifth-ranked Republican on the House Science Committee, led the fight against the conference report. Representative Charles A. Mosher (Republican of Ohio) did not speak against the report, but he joined Bass in voting against it. Bass lost an important ally in his efforts to defeat the test endorsement of the manned lunar goal. Representative George P. Miller (Democrat of California), soon to become chairman of the House Science Committee in September, had teamed up with Bass to denounce the May 24 committee authorization bill when it reached the floor. In supplemental views and on the floor in May, Miller advised that we should move with "celerity" rather than with "haste and hysteria," adding: "We can ruin this program, we can ruin our position in the world if we seem to think that we can buy our way with money through these programs." But in July, following the address of President Kennedy, Miller supported the committee and the President, and Bass was hard put to round up many allies for his opposition.

THE COMMITTEE AND THE LUNAR LANDING GOAL

On a rollcall vote, which can be interpreted as an endorsement of the manned lunar landing goal, the House of Representatives on July 20, 1961, voted 354–59 to authorize $1,784,300,000 for NASA.

President Kennedy's bold stroke of leadership not only had a profound effect on the Nation, in mobilizing vast scientific, technical, and engineering resources toward the goal of a manned lunar landing; the decision itself immediately impacted on Congress and more specifically the House Committee on Science and Astronautics. The more glamorous aspects of manned space flight, which immediately attracted public attention, vastly expanded the work of the committee. The less publicized aspects of the space program—instrumented planetary probes, basic research, the tracking network, astronomy, and other fields—did not suffer from the light of the Moon, but received greater emphasis because of the increased public and congressional support for NASA. Likewise, the related scientific agencies like the National Science Foundation prospered rather than being squeezed out by the emphasis on Project Apollo. A major scientific revolution, including a surge of interest in education, was sparked by the decision to go to the Moon.

Once the committee had matured beyond the adolescent thrills of such glorious experiences as meeting astronauts and their families— which never ever ceased to be a thrill—the committee dug in to exercise genuine oversight over the agencies under its jurisdiction. The May 25, 1961, decision simply made the job bigger, more important, more exciting and more exacting. The job, to be done right, required travel and firsthand, on-the-spot investigation, a practice encouraged and stimulated by the examples set by successive chairmen—Miller and Teague.

The man who served the longest as NASA Administrator, James E. Webb, has referred to the 1961 decision as a goal rather than a commitment. Because of the difference in these two terms, the Committee on Science and Astronautics assumed greater importance each year in forging the congressional and public support toward that goal. Each year the battle had to be won over again, while at the same time carefully maintaining the necessary oversight to insure the maximum efficiency and economy in the program.

President John F. Kennedy, Representative Albert Thomas (Democrat of Texas), chairman of the appropriations subcommittee handling NASA and NSF funding; Representative George P. Miller, chairman of the House Committee on Science and Astronautics and James E. Webb, NASA Administrator, at Rice University, Houston, Tex., September 12, 1962.

Chairman Miller inspects astronaut training at Manned Spacecraft Center, Houston, Tex.

CHAPTER IV

The Early Miller Years

As one of the few Members of Congress with a civil engineering background, Representative George P. Miller (Democrat of California) used to like to needle his lawyer colleagues in the House of Representatives this way:

> You guys think in circles. I am one of the few guys around here who has been trained to think in straight lines.

Upon the death of Overton Brooks, George Miller officially was named chairman of the House Committee on Science and Astronautics on September 21, 1961. In the dozen years he served until January 3, 1973, Miller strengthened the committee's internal structure, broadened its activity in the scientific area, created the influential Subcommittee on Science, Research and Development first chaired by Representative Daddario, presided over the highly successful Gemini and Apollo programs under the jurisdiction of Representative Teague's Manned Space Flight Subcommittee, and also helped stimulate the growth of the National Science Foundation, international cooperation, weather and communications satellites, and the many unmanned space ventures handled by Representative Karth's subcommittee.

Born in San Francisco in 1891, Miller represented the East Bay area from the southern edge of Oakland south and eastward through Alameda, and towns like San Leandro and Castro Valley. Like Overton Brooks, he had served in World War I, graduating from the School of Fire for Field Artillery, Fort Sill, Okla., and was a lieutenant in the field artillery from 1917 to 1919. He was a practicing civil engineer both before and after the war, having studied engineering at St. Mary's College near San Francisco.

After running a travel agency that failed during the Great Depression, for a brief period he helped sweep the streets of Alameda, Calif., to qualify for relief allotments. This proved to be a good entrance into politics, and Miller served two terms in the California State Assembly from 1937 to 1941, and in 1941 became executive secretary of the California Division of Fish and Game. At the age of 53 he was elected to the House of Representatives in 1944, the same year Franklin D. Roosevelt was elected to his fourth term as President.

Miller gained some renown as the freshman Congressman who blew the whistle on Elliott Roosevelt for bumping one of Miller's serviceman-constituents from an airplane in order to ship Roosevelt's huge dog Blaze. On the Merchant Marine and Fisheries Committee, he became chairman of the Subcommittee on Oceanography, which he helped to create. His progress on the Armed Services Committee was much slower. "I sat on the Armed Services Committee for about 8 years," Miller told his committee in a frank executive session shortly after he became chairman. He confessed: "I was never quite taken into the confidence of the people to the extent you are. There was never an opportunity to serve on a subcommittee such as this, to bring this stuff right home to you." His experience on the Armed Services Committee, where he looked up toward Chairman Vinson and saw he was only the 14th in seniority, influenced Miller's decision to switch to the Science and Astronautics Committee. Reacting against Vinson's practice, Miller was liberal in delegating authority to subcommittee chairmen on his new committee.

CONTRASTS BETWEEN BROOKS AND MILLER

Early in Miller's chairmanship, a staff member remarked: "Under Brooks, I turned out three press releases a week. Now under Miller, there haven't been three in six months." At the organization meeting of the committee on January 17, 1962, Miller quickly organized standing subcommittees, gave them specific names and jurisdictions and encouraged the subcommittees to exercise full responsibility.

James R. Kerr, in a Ph. D. dissertation written at Stanford University, recorded in 1962 an interesting series of interviews with committee members and staff which were very frank because of their anonymity. "Brooks was more inclined to emphasize publicity for the committee, and put this ahead of the work of the committee. We covered a very broad area, but never got to the specifics of the program. * * * We have better cooperation and working together under Miller—there was a feeling of resentment that was there under Brooks," said one member.

A junior Democrat made this observation:

Under Brooks we had all full committee hearings—a parade of scientists, military men, civilian experts. But nothing was done about the specifics of the program. We didn't know where the money went. But subcommittees are different. You can get a close look at what needs looking at. Generally, you are confined to a small portion of the budget. Miller is a fine chairman, but Brooks served a valuable purpose although he epitomized the layman's point of view. He asked that kind of question. He sought publicity, educating the public.

Another committee member put it this way:

There were Brooks and Fulton, two prima donnas. It really became quite impossible with Brooks and Fulton acting like prima donnas baiting each other in patronizing but insulting terms. Brooks kept all authority right in his hands. He never passed around opportunities to participate in floor debates on the authorization bills. * * * Miller is a more practical, down-to-earth chairman, and we are all grateful for the change.

Another member of the committee indicated that there had been a change of attitude when Miller acceded to the chairmanship:

Since he has become chairman, he has become much more conservative, loath to move. His attitude has now become don't rock the boat, keep relations with NASA smooth and unruffled. * * * He was a much more vigorous questioning committee member when he was the ranking majority member. Perhaps becoming chairman makes one become more fatherly and protective. He is, without doubt, a great improvement over Brooks in every sense.

In stressing the independence and responsibility of the subcommittees from the start, Chairman Miller built up respect and a high morale among both members and staff. Republican members of the committee were particularly strong in their praise of the Miller regime, and the bipartisan approach to issues which arose. Some members stated that Chairman Miller had a "short fuse," but they all commended his fairness and genuine prestige that developed as the committee delved into new areas.

Of medium height, gray haired, bespectacled, possessed of a good sense of humor, a good storyteller and wonderful traveling companion, Miller was inclined to deal arbitrarily with those who disagreed with him on the committee. But he never held grudges. There were times when staff members and members of the committee felt that his wide-ranging anecdotes, reminiscences and philosophical observations, although interesting and stimulating, were time-consuming. Yet Miller earned and won the respect of his committee and colleagues, and through his service raised the prestige of the committee.

On May 2, 1962, the following exchange took place between Miller and Teague, who at the time was also chairman of the Veterans' Affairs Committee:

Mr. TEAGUE. The Speaker of the House recently called a meeting of all the committee chairmen. I was the only committee chairman that had a chance to comment on another chairman. I told him I had a complaint, that I had a chairman that was working the hell out of me, and that was the chairman of the Space Committee.

Mr. MILLER. Brother, you asked for it. You ain't seen nothing yet.

RELATIONS WITH NASA

Quite naturally, NASA officials all preferred the Miller chairmanship to the Brooks chairmanship, even though it meant far more work on details than Brooks had required. Under Brooks, NASA never knew when they would receive a quick summons, after a space spectacular by either Russia or the United States, to appear before the committee in a public hearing with little time to prepare. It was Miller's habit to call Webb frequently on the phone, sit down with Webb in his office, have lunch with him, and even ask Ducander to call Webb personally to straighten out any problem.

Webb recalls:

We worked very closely together. As a matter of fact, Congressman Miller had asked me when he became chairman of the committee how we could best work together. After we had discussed this for some time, he decided to set up subcommittees patterned very much after the NASA structure. So there was a Committee on Manned Space Flight dealing with Dr. Mueller, and Mr. Brainerd Holmes before him, in the field of Manned Space Flight.

There was a Committee on Science which was dealing with the scientific side of NASA—our structure, our organizational structure, fitted in very neatly with the committee structure of subcommittees. And this meant that the people in the subcommittees that the committee would look to for final judgments were in close personal contact with people in NASA who were working in the same field. And they developed an intimate working relationship.

With relation to informal contacts and committee trips to installations, Webb also recalls:

We brought the subcommittees of the House Committee on Science and Astronautics which is now the Committee on Science and Technology together with our important leaders for a face-to-face contact at a small, intimate hearing. And it was after this basic laying of groundwork and understanding that Congressman Teague and many others went out then to the centers to see what was going on, to the contractors' plants that were doing the work.

MEMBERSHIP AND SUBCOMMITTEE ORGANIZATION

When the 2d session of the 87th Congress convened in January 1962, the following Members constituted the Science and Astronautics Committee:

Democrats	*Republicans*
George P. Miller, California, *Chairman*	Joseph W. Martin, Jr., Massachusetts
Olin E. Teague, Texas	James G. Fulton, Pennsylvania
Victor L. Anfuso, New York	J. Edgar Chenoweth, Colorado
Joseph E. Karth, Minnesota	William K. Van Pelt, Wisconsin
Ken Hechler, West Virginia	Perkins Bass, New Hampshire
Emilio Q. Daddario, Connecticut	R. Walter Riehlman, New York
Walter H. Moeller, Ohio	Jessica McC. Weis, New York
David S. King, Utah	Charles A. Mosher, Ohio
J. Edward Roush, Indiana	Richard L. Roudebush, Indiana
Thomas G. Morris, New Mexico	Alphonzo Bell, California
Bob Casey, Texas	Thomas M. Pelly, Washington
William J. Randall, Missouri	
John W. Davis, Georgia	
William F. Ryan, New York	
James C. Corman, California	
Thomas N. Downing, Virginia	
Joe D. Waggonner, Jr.,Louisiana	

Subcommittee on Manned Space Flight, Olin E. Teague, *Chairman*
Subcommittee on Advanced Research and Development, Victor L. Anfuso, *Chairman*
Subcommittee on Space Science, Joseph E. Karth, *Chairman*
Subcommittee on Applications and Tracking and Data Acquisition, Ken Hechler,
   Chairman
Subcommittee on Patents and Scientific Inventions, Emilio Q. Daddario, *Chairman*
Special Subcommittee on Women as Astronauts, Victor L. Anfuso, *Chairman*
Special Subcommittee on Solid Propellants, David S. King, *Chairman*

JURISDICTION

"We have some very grave responsibilities in the field of our work in connection with the Science Foundation, and very serious, responsible work to do with respect to the Bureau of Standards. * * * There are many matters pertaining to education and supply of scientists in this country that we have a grave and direct responsibility for," Chairman Miller told his committee in executive session on April 16, 1962. Miller was determined to broaden the work of his committee into scientific areas beyond space, and also to stress the development of unmanned activities without excluding the more spectacular Apollo program.

On May 2, 1962, at an executive session of the committee, Representative Karth asked what course of action should be taken on the communications satellite bill coming up before the Rules Committee. Chairman Miller responded:

"As far as I'm concerned, this is a matter that is fully within the jurisdiction of the Committee on Interstate and Foreign Commerce." Representative Chenoweth contended that he felt there was a "twilight zone" and the committee should not abdicate jurisdiction. Perhaps recalling the stiff challenge which Interstate and Foreign Commerce Committee Chairman Oren Harris had made in 1961 in a floor discussion of Science Committee hearings on communications satellite research, Chairman Miller backed off. Miller stressed that although research and development were clearly within the Science Committee jurisdiction, commercial use was not.

With the creation of the Science, Research and Development Subcommittee in 1963 under the chairmanship of Representative Daddario, Miller very positively asserted the jurisdiction of the committee in all areas of science. The circumstances of the establishment of the new subcommittee are covered in the next chapter.

In general, there was a marked difference between the policies of Chairmen Brooks and Miller in their approach to jurisdiction. With Brooks, it was a case of damn the torpedoes full speed ahead, push the jurisdiction upward and outward as fast and as far as time allowed. Brooks always reacted with bland and suave surprise when another committee hinted or came right out to say he was treading on forbidden jurisdictional grounds. Miller ran the committee in a far more orderly fashion, politely declining to get involved in jurisdictional squabbles, and shunning a stance that grabbing for power was the mark of a successful committee chairman. He was a team player.

At the first organizational meeting of the committee in January 1962, Chairman Miller was asked whether he would press for authorizing power for the committee in the case of the Weather Bureau and the Bureau of Standards. Chairman Miller answered in the negative, explaining:

> It isn't something you can go out to do with a bludgeon. It takes a lot of persuasion and we have to remove a lot of resistance that may come. * * *

It was Chairman Miller's view that the committee had plenty to do without reaching out for vastly expanded jurisdiction. He preferred the orderliness of good management to the frenetic, frantic efforts of his predecessor to flail out in all directions.

STAFF OPERATION

Morale under Chairman Miller rose immediately and sharply. Staff Director Charles Ducander had a bad case of the flu when Chairman Brooks died, and could not even attend the funeral. But Miller called and asked if he could come over to Ducander's house. There he asked Ducander to stay on in his capacity as staff director. On the rest of the staff, he asked: "What do you think? Should we keep everybody?" Ducander replied affirmatively.

Ducander described his relationship with Miller in the following way:

> I was in Mr. Miller's office no less than three or four times a day, every single day. * * * I was down in his office starting about 9 in the morning, and I always had a list of things: "George, tomorrow, this, this, this, this * * *." This came from my interrogation of staff members: "What did you all do yesterday? You had some hearings. What did you talk about? What did you do? Who said what? Did anything happen of importance?" * * * Well, now all these things were never written down and Phil Yeager would come in and tell me what he and Daddario were doing and this sort of thing, just in a conversational way, and I would take a few notes. And the next morning in my briefing with the chairman, it was like somebody comes in and briefs the President every morning. I thought this was my duty to keep him informed, and he liked this.
>
> That's the way it operated. No written memos. That's why you can't find any of this in the files, because it seems to me it's ridiculous that if Mr. Miller wanted to hear it, he didn't want to read it in a memo.

Chairman Miller's philosophy on staff was that a staff which was lean and hard-working was more efficient. He wanted good people who were paid good salaries and then expected to get the job done without a lot of independent pools of power, or lack of coordination. Both Brooks and Miller inherited their attitudes from the staff of the Armed Services Committee, on which they had both served, and where there was a very small professional staff which furnished support to both majority and minority Members.

Although Representative Martin was the ranking Republican, Representative Fulton assumed most of the prerogatives of the senior Republican and spearheaded an effort to obtain special staff for the minority. Chairman Miller's customary response was that no committee in the Congress was less partisan in its attitudes than the Science and Astronautics Committee. It is certainly true that the spirit of bipartisanship dominated the Science Committee. On the other hand, the minority, if properly staffed, would have been better equipped to fight against political decisions such as the location of the Electronics Research Center. Eventually, the Science Committee was provided with minority staff, largely because of powerful forces outside the committee itself. Once the Republican Members of the House presented a united front and made the issue of a minority staff their Holy Grail, the Science Committee bowed to the pressure and allocated separate staff for the minority. But it was a long and agonizing fight, stubbornly and narrow-mindedly resisted by Chairman Miller every step of the way.

In addition to Staff Director Ducander, the following staff professionals were on board during the early Miller years:

Spencer M. Beresford, a lawyer and veteran of the select committee staff, who left the Science Committee on June 30, 1962, and later joined NASA.

W. H. Boone, a technically trained electrical engineer, with experience in military applied research in the Department of Defense, who joined the staff August 6, 1962, and remained for ten years.

John A. Carstarphen, Jr., a Louisiana lawyer recruited by Chairman Brooks, who initially assisted on the Anfuso Subcommittee on Advanced Research and Technology, and later became chief clerk of the committee, remaining throughout the Miller years.

Frank R. Hammill, Jr., a lawyer with Pentagon and FBI experience who joined the committee February 29, 1960, worked primarily for the Karth Subcommittee on Space Science and Applications, and served until 1979 as counsel on the Science and Technology Committee.

Richard P. Hines, a writer and veteran of the select committee, who remained with the Science Committee until March 31, 1973, working mainly on tracking and data acquisition, and advanced research.

Raymond Wilcove, another veteran of the select committee, a journalist, who assisted in staffing the Advanced Research Subcommittee, and remained with the Science Committee until March 10, 1963.

Philip B. Yeager, counsel, also a veteran of the select committee staff, lawyer and journalist, who initially staffed the Manned Space Flight Subcommittee and the

Mitchell and Daddario Patents Subcommittees. Later served as staff director of the Subcommittee on Science, Research and Technology. Over the years, Yeager has been identified with perhaps more successful hearings and reports than any other staff member, and in 1979 was appointed General Counsel.

Chairman Miller continued the practice initiated by Chairman Brooks of asking for the detail of a series of Army, Navy, and Air Force officers who served tours of duty an average of one year apiece, and assisted the professional staff of the committee in its work. Among the abler and more effective military officers assigned to the committee were Col. Earl G. Peacock and Lt. Col. (later Col.) Harold A. Gould. (Gould became deputy director in 1975 and executive director in 1979.) Chairman Miller discontinued the practice of assigning military officers in 1964, and it has not been revived since.

At the close of 1961, when Representative Miller assumed the chairmanship, there were 11 professional and 6 clerical members of the staff. At the close of 1962, the staff had dropped to 10 professional and 6 clerical. The size of the staff increased very slowly in the ensuing years, and the number of the staff members under Chairman Miller reached a high point of 17 in 1971. One of the notable additions in 1963 was James E. Wilson, who was appointed staff director of the Manned Space Flight Subcommittee when Philip B. Yeager moved over to become staff director for the Subcommittee on Science, Research and Development. Wilson had been Director of Research and Development for the Naval Propellant Plant in Indianhead, Md.

DELEGATION TO SUBCOMMITTEES

Chairman Miller's wise decision to delegate responsibility to the subcommittees was universally applauded by all the committee members and staff. To be sure, some officials in NASA grumbled that the authorization hearings were too long and too detailed. Other critics tried to argue unsuccessfully that the Senate Aeronautical and Space Sciences Committee members in House-Senate conferences had a broader picture of NASA operations, but these critics quickly conceded that there were few areas of NASA operations that some House committee members didn't know best. But the predictable effect of subcommittee specialization was that more committee staff was desperately needed.

Nearly all the committee members indicated they needed more complete briefings in preparation for the hearings on authorization bills, investigations, and for general understanding of the issues involved in the policy decisions confronting them. Time after time, committee members stated that they had to "accept so much of the agency presentations on faith," instead of having the staff personnel to probe into the full justification of millions of dollars. In any area under the committee's jurisdiction, whether it was NASA, the National Science Foundation, the Bureau of Standards, or any other agency, the committee members felt that they were supporting a scientific and engineering program which Congress wholeheartedly wanted to be successful. After all, the committee had been prodding NASA for several years to spend more, go faster, get the job done with greater urgency. At the same time, cost-conscious members wanted to have the tools to differentiate between essential expenditures and waste or "padding."

The situation became critical in the early 1960's as NASA's budget ballooned upward. Committee members faced a billion-dollar budget at the beginning of 1961. By July, following the decision to go to the Moon, the administration was asking Congress for $1.7 billion. By the beginning of 1962, the new budget was over twice that big, and in the next few years the budget continued to soar until it leveled off at between $5 billion and $6 billion in the midsixties. To understand and grapple with these massively expanding programs required extensive staff assistance to do the job right.

"I'm all in favor of accelerating the space program, but I don't want to remain so ignorant about the program that I overlook these critical areas where investigations should be made," said Representative Karth early in 1962 in citing the need for more committee staff. Teague echoed the need for more staff, adding: "I studied animal husbandry. And nobody else on the committee is a scientist either. I just had to work overtime—reading all kinds of stuff and got help wherever I could—NASA, the Air Force, industry—every place I thought I could learn something."

Representative Anfuso joined his fellow subcommittee chairmen in pointing out that the Science Committee had a much tougher job than the Armed Services Committee, because the latter had a backlog of experience and knowledge to draw on, as well as guidelines to measure performance.

Representative R. Walter Riehlman (Republican of New York) stated in a committee executive session that "our committee has been lacking in a staff that is qualified to follow thoroughly these programs and be of assistance to members in evaluating them." Riehlman particularly noted the absence of scientists or engineers on the staff. Representative Mosher was somewhat blunter in his assessment. When his subcommittee chairman, Representative Anfuso, made the grand gesture of telling the subcommittee that "each member has become almost an expert," Mosher responded:

I would say that if I am an expert, then Lord help the Nation. I think it is a frustrating experience for all of us in this new world we are dealing with, that we have to accept so much on faith. I would like to echo what has been said earlier in the morning, that there is, I think, a desperate need for more technically trained staff assistants for us.

Chairman Miller adroitly sidestepped the many requests for additional staff, and adhered to his determined plan to keep the staff reasonably small and controllable. When the pressure got heavy Miller pleaded that there was insufficient space to house additional staff. But even when the committee moved to more spacious quarters in the Rayburn Building, when the space was available, the staff remained small.

The major functions of the staff did not differ materially from what they had been during the Brooks years. But the organization, tone, and logic showed marked improvement. Under Chairman Miller, there was more advanced planning of hearings, scheduling of witnesses, and less of a tendency to summon high-ranking officials to coincide with spectacularly newsworthy events. There was a marked increase in the number of inquiries from Members to staff. These were generally routed through Ducander, except of course for the subcommittee matters being handled by the regularly assigned staff members. Chairman Miller frequently expressed to those pressing for expanded staff that he was against using the staff to handle congressional constituent business (even though this was not contemplated).

Chairman Miller stressed the value of field trips and foreign trips, and he encouraged all committee members to visit not only NASA installations but also the Bureau of Standards and other scientific centers. Some committee members, notably Hechler and Fulton, made some attempts to expand the staff through a series of field investigators who could monitor the work at the various installations and then report back to the committee. As it turns out, field intelligence was derived almost entirely from trips to installations. And the staff continued to accomplish an almost superhuman amount of work in staffing hearings, conducting investigations, and producing a wide variety of useful reports.

ADDITIONAL ASSISTANCE

The committee made good use of the General Accounting Office, yet concluded that the accent of its investigations was on administrative performance and accounting, monetary, or budgetary matters. The committee members clearly felt the need for more technical, scientific, and engineering assistance beyond what was being supplied by the Legislative Reference Service of the Library of Congress. In later years, largely through the stimulus of the Committee on Science and Astronautics, LRS created a Science Policy Research Division in 1964. But in the early 1960's, when there were complex policy decisions to be made in rapidly expanding areas of space and science, committee members struggled with the problem of how to get objective information and advice.

As noted in chapter II, Chairman Miller strengthened the use of the Panel on Science and Technology which had been started by his predecessor in 1960. Under Miller's leadership, the meetings of the panel were held more frequently, the focus of the meetings was sharper and members had opportunities for airing some of the current committee policy decisions while panels were in session. One member suggested that the panel should be called on more frequently to backstop the staff, and to answer questions by members between the times the panel was actually meeting in Washington. It was felt that this would be too much of an extra burden on the hard-working panel members.

A similar suggestion was made by Representative Mosher: "Personally I like the idea, recently suggested by several people, that the standing committees should be able to employ science specialists for brief periods of time, and for rather intensive work during those periods—preferably scientists who have some knowledge of Government's relations to scientific activities and some understanding of the congressional process as such." Later, the Daddario Subcommittee on Science, Research and Development established a Research Management Advisory Panel, which proved very successful, yet its primary assistance was provided to the Daddario subcommittee. In the period when tough decisions were being made on the space program every day, a majority of the members groped and grasped for the tools to do the job.

Some members rationalized their lack of staff help by contending that Congressmen were supposed to react like their taxpaying constituents in measuring the value of complex programs. It was further argued that Congressmen should be generalists and not be armed with the specialized knowledge which might bias their decisions on behalf of one particular phase of a program. But most of these arguments seemed to apply to the qualifications of the Congressmen themselves, rather than go to the heart of the issues.

By its field trips to NASA installations, by its insistence that NASA witnesses express themselves in "everyday English," and by working long hours, the Manned Space Flight Subcommittee was probably as well informed as any subcommittee. Yet in the early 1960's, it was this subcommittee which really led the fight for more and better staff assistance. On April 11, 1962, the subcommittee had this discussion in executive session:

Mr. RIEHLMAN. I want to make the statement * * * that the chairman be advised that as far as I am concerned—and I think it should be unanimous that we should be provided with a staff sufficient to follow this program through and to see that we have continuous information as to the manner in which these funds are being spent, and that we be conversant from time to time with the progress being made in every one of these facilities and programs that are carried on.

Mr. TEAGUE. Yes.

Mr. MORRIS. Not only in facilities, but in research——

Mr. TEAGUE. The chairman of the committee has told me that as soon as we get through with this authorization that he wanted to go into the reorganization of the committee and staff. Now, might we write a letter from this subcommittee, signed by all members of the subcommittee, to the chairman, setting forth what you have said? Would you draw up a proposed letter, Duke, and give each member a copy? Let them make suggested changes, and let each member of the subcommittee sign the letter to the chairman.

Mr. DUCANDER. Yes; if I were asked, I would agree, that at one time, we didn't even have a man in the office to answer questions that Members of Congress call us on all the time. The girls had to just take messages down there and ask us.

Mr. RIEHLMAN. We have to impress on his mind that this subcommittee has had a heck of a lot of responsibility here, and we want to be sure from now on we will have help.

Mr. FULTON. I might say, at one point in these subcommittee hearings, the chairman and I have each pointed out that we need a staff of an investigatory nature, and need them to be competent, so they are able to evaluate. We need, really, a scientific contractor approach to it, that the General Accounting Office couldn't give us, and frankly said they couldn't, in these hearings. How many slots are open? Five?

Mr. DUCANDER. Mr. Fulton, it is not a matter of vacancies, it is a matter of money, and we have plenty of money. We are going to turn back, if we go as we are going now, about $95,000.

Mr. TEAGUE. Unless we do this, the Government Operations Committee will be over on this committee taking over our job.

Before the full committee five days later, Representative Fulton said: "I think Tiger Teague and I, as well as the other members of the committee, such as Tom Morris (Democrat of New Mexico) have recommended that we immediately get technically trained people with a background in this field, that we can follow these programs and follow them carefully."

In response, Chairman Miller stated: "May I say this, that one of the reasons you haven't got some technical assistance now is you haven't any place to put technical assistants."

On August 7, Chairman Miller introduced W. H. Boone, a graduate of Mississippi State College with an electrical engineering degree, as the "first purely technical member we have on the staff." This prompted Representative John W. Davis (Democrat of Georgia) to quip:

A great deal has been written about the fact that the Republican Party does not have an adequate representation on the staff. I would simply like to welcome Mr. Boone as a representative of the Confederacy.

Boone proved to be a good staff member on the scientific and technical side, but others showed that the greatest talent provided by the committee staff was in the area of management. The best staff members were those who sensed the right policy and management questions to ask, could challenge bureaucratic practices and see through efforts to gloss over problems, and could write clearly and simply—traits not always possessed by the technical "experts."

At the final executive session of the committee for 1962, held on September 26, Chairman Miller expressed the "hope that in the not-too-distant future we can get physical facilities that will allow us to expand the staff. We could have expanded the staff, but we would have no place for them to work. We examined this quite thoroughly. They would be sitting in one another's laps."

In the final analysis, Chairman Miller simply resisted the efforts of his committee members to gain more staff assistance, and he was chairman in the days when revolutions were generally unsuccessful.

THE COMMITTEE AND THE MERCURY PROGRAM

The Russian cosmonauts were the best thing the American space program had going for it. Gagarin jolted America toward speedier action leading to the May 25, 1961, decision to go to the Moon. After the successful suborbital flights of Alan Shepard and Gus Grissom, NASA hoped to go for a three-orbit flight to beat Gagarin's one-orbit effort. Then along came Cosmonaut Gherman S. Titov with a day-long, 17-orbit flight on August 7, 1961. Instead of a drop in public support for the Mercury program, the Titov flight seemed to rally public opinion behind John Glenn as he prepared for three orbits of the Earth.

Delays plagued the Glenn blastoff during January and early February. Some members of the committee were irked at some more headline-grabbing by the ranking Republican, Jim Fulton, who remarked after an unsuccessful launch attempt on January 27, 1962, that the Mercury capsule and Atlas booster were "a Rube Goldberg device on top of a plumber's nightmare." Trouble had developed in one of the bulkheads of the Atlas booster. Dr. Robert R. Gilruth, Director of the Manned Spacecraft Center, had no recollection of Fulton's statement, which made the front page of the Washington Post. But Gilruth did remember one Congressman who made no public statements or press releases at the time:

Mr. Teague was with us during those real key times—like just before we orbited John Glenn. We had so much trouble with the Atlas rocket with the bulkhead. * * * He was always right there, and he was always supportive. It was good to have somebody who could understand and help you like he did.

On February 20, 1962, Glenn completed his flight, and returned to the cheers of millions throughout the country. Just one week later, Glenn, Grissom, and Shepard made a dramatic appearance on behalf of the $3.7 billion NASA authorization bill as hearings were kicked off before the Science and Astronautics Committee.

Hoarsely, Administrator Webb rasped:

> I regret that my voice is not very good today. In common with many of our fellow citizens, I think I have almost worn it out cheering for Colonel Glenn and the tremendous achievements which the Mercury team has performed.

When Chairman Miller commended Webb for his "push and drive," Webb responded: "The atmosphere here is a little bit different than the day I appeared before you after the first Russian flight."

THE COMMITTEE AND NOVA

Weightlifting was the name of the game in the early days of the space program. It was a simple proposition understood by every schoolboy that Russia had a big lead because she had bigger boosters. To get to the Moon obviously required far bigger launch vehicles than we possessed. For the two-man Gemini missions, the Air Force Titan was used, while NASA was developing the Saturn for the three-man Apollo mission. At the same time, the gigantic Nova was designed for direct ascent to the Moon and return.

By early 1962, a clear-cut decision had not yet been reached on whether the manned lunar trip would be by direct ascent, by Earth orbit rendezvous or lunar orbit rendezvous. Nova was the alternative if direct ascent were the way to go. Nova was also projected as the big truck which would carry flights to the planets and perform deep space probes.

The Nova program was so massive as to defy the imagination. The idea was to cluster eight engines in the first stage with a thrust of 1.5 million pounds apiece for a 12 million pound thrust. Other versions increased the thrust up to 20 million pounds. The launch and test facilities required construction costs which ran into hundreds of millions of dollars.

The committee was appalled at the size and fuzziness of the justifications for the huge Nova expenditures, which in the fiscal year 1963 totaled $163 million for research and development, something over $60 million downpayment on launch facilities at Cape Canaveral, and over $12 million for test facilities in Mississippi.

Teague had his doubts about Nova from the start. On February 28, 1962, he asked Dr. Seamans: "How much would you lose if you cut down the Nova program to just surveying and engineer studies, but go ahead with your engine?" Seamans answered: "We will carry out the development of the Nova rapidly enough so that if the rendezvous does not prove to be a satisfactory method, we can still get to the Moon by the end of this decade by direct ascent. However, we think we can get there a year sooner using advanced Saturn in rendezvous than by direct ascent."

On March 26, 1962, D. Brainerd Holmes, Director of Manned Space Flight, testified before the committee that "It appears to be logical to carry on the parallel although somewhat later Nova approach due both to the uncertainties as to the difficulties which may be associated with rendezvous, and due to the fact that we will undoubtedly need these more powerful launch vehicles for explorations deeper into space."

Representative Richard L. Roudebush (Republican of Indiana) raised the question:

We find ourselves spending billions on a rocket, Nova, that would be old fashioned, if I could use that term, by the time the engineering was completed.

On April 4, Teague asked Milton W. Rosen, Director of Launch Vehicles and Propulsion:

Mr. Rosen, is there any wild guess what this Nova total will be?

Rosen responded:

I would say for vehicle development alone, assuming a 10-vehicle program, we should expect a program of about $2 billion.

Chairman Miller used this simile to describe transporting Nova:

This is almost like rolling the Empire State Building back and forth a couple of blocks.

During the debate on the NASA authorization bill, on May 23, 1962, Miller prophetically suggested that:

In the next year or two, if we meet with success in the orbital rendezvous techniques we will want to take another hard look at the Nova program to see if this vehicle is really needed for manned space flight explorations.

The Teague subcommittee moved ahead to prod NASA on a major policy decision. It was not a case of the committee substituting its judgment for that of NASA. Rather, it was a subtle type of pressure on NASA to make a decision on what kind of a mission they really had in mind for Nova and to relate that decision to a more precise definition of how they planned to get to the Moon. In the spring of 1962, NASA officials still had their options open among the various possibilities—direct ascent, earth orbit rendezvous or lunar orbit rendezvous.

In the decisionmaking process, Teague took his subcommittee to the Cape, to Huntsville, to Houston, to visit with contractors, to talk informally with the astronauts, and to listen, learn, question, argue, and challenge. Something had to give. Meeting in executive session on April 11, 1962, the Teague subcommittee took its first step to kill Nova. The subcommittee voted to cut out of the authorization bill $60,630,000 of construction funds for the gigantic Nova launch complex at Cape Canaveral. The subcommittee then slashed $12 million from the Nova test facility in Mississippi. Nova was not terminated, because funds were left in the bill to carry forward the research and development, but the committee sent a clear signal to NASA that Congress wanted a better justification for such a gigantic project with a loosely defined mission.

NASA responded to the committee prodding. On July 12, 1962, D. Brainerd Holmes, Director of the Office of Manned Space Flight, made a special appearance before the committee to present NASA's clear-cut decision to proceed with the lunar-orbital rendezvous method of landing on the Moon.

On both sides of the aisle, committee members praised NASA for the manner in which the decision was presented to the committee. Daddario stated:

I am sure that because of this candid approach that you will get better, and stronger support from the Congress.

Riehlman added:

I feel confident that it is this type of presentation that will assure you of closer cooperation with the committee and more favorable consideration in the future.

Hechler made these remarks to the committee, in commenting on the choice presented:

I think this will go down as a classic in decisionmaking. History will tell whether it is right.

Chairman Miller concluded the hearing by noting:

I think it evidences the good relationship existing between NASA and this committee.

The decision itself resulted from very thorough studies and excellent arguments by John Houbolt and others of Langley Research Center. It served to sharpen the committee's determination to push NASA on the issue of what was planned for Nova.

Nova did not die easily. NASA continued to request funds for advanced research on a post-Saturn vehicle. In 1963, von Braun clearly pointed out that Nova was "on the back burner." He did not give up on its future use, but confessed that "when we shall have enough money to go into high gear with Nova we would not want to base it on a concept that we developed in 1962."

The President's Science Adviser, Jerome Wiesner, as well as some members of the President's Science Advisory Committee, continued to press their opposition to lunar orbit rendezvous, even after NASA had announced its decision in July 1962. The opponents of lunar orbit rendezvous brought their case to Teague. This prompted Teague, who knew that von Braun had been one of the strongest early advocates of Earth orbital rendezvous, to put the question to von Braun during the March 18, 1963, committee hearings:

> Mr. Teague. Is there disagreement within NASA as far as the method of going to the Moon is concerned?
>
> Dr. von Braun. None whatsoever.
>
> Mr. Teague. In Houston, we were told that the astronauts were unanimous in their belief that this was quicker, cheaper, and safer.
>
> Dr. von Braun. We believe so too. I am aware that there have been some statements to the effect that it was a bit surprising that Marshall, after having advocated Earth orbit rendezvous, came around and recommended lunar rendezvous.
>
> Fact is that at first we put a great deal of work into the lunar orbit rendezvous mode also, and now we are convinced that this is the fastest and safest way to go.

Henceforth, the committee reacted negatively whenever the word Nova came up. In response to critical questions as to why funds for Nova advanced vehicle studies were included in the fiscal 1964 budget, a red-faced Brainerd Holmes confessed:

> Nova as used here is a little misleading. * * * I think the terminology in the (budget) book is a little unfortunate.

The committee wanted to be absolutely sure that money allocated for advanced research was not actually being used to revive the program which the committee had helped to kill. On May 8, 1963, Brainerd Holmes and his deputy, Dr. Joseph Shea, had to clarify this point:

> Mr. Daddario. Nova, as used here, does not mean what we understand Nova to mean a year ago?
>
> Mr. Holmes. That is correct——
>
> Dr. Shea. There is a wonderful definition that comes from science about Nova. I have forgotten exactly how it goes, but I think it says: "Nova is often the brightest object in the sky for a short period of time, but then it wanes."
>
> Mr. Fulton. This committee helped put that particular Nova program into history. It went into history around this table.

But even then, the Nova program dribbled along for another year. Finally, in 1964, Edward Z. Gray, Director of Advanced Manned Missions for NASA, walked into a lion's den by presenting the Manned Space Flight Subcommittee a chart which mentioned another advanced mission study of Nova. Representative Fulton was infuriated:

> From this chart it appears you have resurrected the Nova concept. I thought that concept had met a fast death before this committee some time ago. How did it

get back into operation again? * * * How much money are you spending on the Nova concept?

Mr. GRAY. We are probably spending about $75,000 trying to identify its characteristics as related to launch facilities and launch operations.

Maxime A. Faget, Assistant Director for Engineering and Development at the Manned Spacecraft Center, also worked on one version of Nova which would cluster huge solid rockets as a first stage. "We called the individual solid rocket 'the Tiger,' " explained Faget. "We figured it would be a noisy animal and would roar like a Tiger." Of course, Faget was not at the hearing to explain what he had in mind. But the real "Tiger" left no doubt where he stood on the whole question of Nova. The following exchange took place between Teague and Gray:

Mr. TEAGUE. If I were you, I would never use the word "Nova" again.
Mr. GRAY. We never will, so help me.

PROTECTING LAUNCH OPERATIONS

Ordinarily, Tiger Teague did not respond to anonymous phone calls. But this one had a strange ring of truth to it. "What happened was that I got an anonymous call from either Jacksonville or Atlanta," Teague explained. It was a tip that the Air Force was plotting to take over title to the expanded NASA launch facilities at Cape Canaveral.

Throughout the late summer of 1961, negotiations between the Air Force and NASA went on to define the details of NASA's proposal to buy over 80,000 acres of land for $60 million, to become the Nation's major space launch base. When agreement was reached, NASA authorized the Corps of Engineers to proceed with the purchase, using reprogramed funds left over from having abandoned the Nova program.

Kurt Debus, Director of the Kennedy Space Center, got together with Maj. Gen. Leighton Davis, the Air Force commander of the Atlantic Missile Range, to work out an agreement which was subsequently ratified by Webb and Deputy Defense Secretary Roswell Gilpatric. Once the land was purchased, General Davis surprised Debus by informing him that title to the land should be transferred to the Air Force because they owned all the previous Atlantic Missile Range land.

Teague called Webb and Gilpatric into his office to find out what had motivated this power play, but at this level both officials insisted that everything was sweetness and light. "So I just picked up the phone and called the Corps of Engineers," Teague said, and the corps confirmed the Air Force efforts to obtain title to the new land.

Meanwhile, down at the Cape, Teague and his subcommittee began to poke around some more. They discovered that space agency officials were very concerned about the safety factor of Titan III fly-overs. So Teague ordered his subcommittee to probe the whole issue of range management at the projected Gemini and Apollo launch facilities at Cape Canaveral.

On March 29, 1962, Teague summoned Assistant Secretary of Defense John H. Rubel and Dr. Seamans in an executive session before the Manned Space Flight Subcommittee. Teague told Rubel:

> The main thing that troubles the committee is, we go to the Cape, for example, we talked with some of your responsible people there, we talked with some of Dr. Seamans' responsible people and we came away confused, frustrated, disturbed, and they don't agree on this overflight matter, and they don't agree to a Titan sitting next to a Saturn * * *. We have some questions we are going to submit to you, Mr. Rubel and Dr. Seamans, which we want answered for the record.

Gen. B. A. "Benny" Schriever, head of the Air Force Systems Command, came up to lobby Teague in his office, but Teague pronounced: "I want NASA to administer the land and its launch center."

It took another NASA–DOD agreement, signed by Webb and Secretary of Defense McNamara, to establish that NASA was more than a tenant but could freely plan its own operations on the 87,000-acre Merritt Island launch area.

The average congressional committee would have received testimony from the responsible top officials, and tried to resolve any disputes at the very top. The Science and Astronautics Committee and its subcommittees, given free rein by Chairman Miller, went out to the contractors, the NASA centers throughout the country, sought the advice of independent experts, talked to the workers in the plants and their foremen, and had a real understanding of what was going on in every program.

Teague describes the efforts of the Air Force to prevent NASA from establishing a machine shop to repair minor parts needed at the Cape. "They would assign a bunch of bright young Air Force colonels to lobby the committee," Teague recalls. "They did everything on Earth to try * * * to get control of the space program." At one of the subcommittee parties at the Cape, an Air Force officer at Patrick Air Force Base was describing to one Science Committee member that his machine shop wasn't very busy, and that it would be a waste of money to set up a separate NASA shop. As Teague describes it:

> The next morning, I got up early, and went by cab to the machine shop. There was a major in charge. I told him who I was, and that I just wanted to go through.
> I asked him how busy they were. They were so busy they couldn't even begin to keep up. They would be running 24 hours a day. They could not do any more work.

Teague relayed the word to the Air Force officer at the base in no uncertain terms that "you say one more [deleted] word to my committee, I will put you under oath and make you testify. You just stay out of this."

Chairman Miller encouraged Teague and the other subcommittee chairmen to poke around whenever necessary, to travel extensively, and to expand their oversight functions on a personal basis. When Teague gave his subcommittee report in 1962, Miller commented:

Of course, I don't want to say too much about Teague, because I used to sit right below him on the Committee on the District of Columbia. I was in the lower tier. He was in the upper tier. Anytime I didn't vote right, he didn't do as I do with many, and try to reason with him. He used to just reach out and conk me on the head. I can show you the bumps.

Glancing toward two bald-headed members of Teague's subcommittee, Representatives Daddario and Tom Morris of New Mexico, Miller remarked: "You see, you haven't any padding like I have." On behalf of the bald-headed members, Morris had this to say about Teague:

He is a fine chairman. However, I am not too impressed with that habit he had of throwing the gavel at junior members of the committee at times.

Daddario added: "His aim is usually very bad, Mr. Chairman."

The inquiries by each subcommittee were searching, grueling, exhaustive for both members and witnesses, but not without their lighter moments. Maj. Rocco A. Petrone, Chief of the Heavy Space Vehicles Systems Office at Cape Canaveral, had been testifying at length one day early in 1962, when he was asked about his pay status. Then he was asked whether he would be eligible for hazardous duty pay. Major Petrone responded: "Probably only for appearing before the committee, sir."

KARTH AND SPACE SCIENCE

Hanging in Joe Karth's office was a letter to attest that he had once outdriven Arnold Palmer in a golf match. Aside from consistently winning the annual congressional golf tournament, Karth, a square-jawed, brawny, no-nonsense legislator with a firm handshake and clear-eyed gaze, was the workhorse of the Science and Astronautics Committee. A charter member of the committee, by early 1961 he was chairing a subcommittee in his sophomore term, participating in conference committees, and quickly making his mark as a tough, hard-nosed inquisitor.

In 1958, when Minnesota's Eugene McCarthy went to the U.S. Senate, Democrat Joe Karth captured McCarthy's St. Paul seat in the House of Representatives. A union organizer, he had studied engineering two years at the University of Nebraska. His backers wanted him to be on the Education and Labor Committee, but there simply weren't enough vacancies in 1959. He pitched into his Science Committee work with a vengeance, won the respect of his colleagues on the committee and in Congress as a man who did his homework, and exercised great influence over the unmanned side of the space program. Karth served as chairman of the Subcommittee on Space Science and Applications until he left the committee to join the Ways and Means Committee on October 6, 1971.

As Teague put it, "Manned space flight got all the credit and all the publicity, and the hardest working subcommittee was Joe Karth's and he got no publicity." Karth and his subcommittee were sort of like the battered and begrimed gridiron linemen who rarely were recognized, while Teague and his glamorous crew in the backfield got all the headlines and the glory of manned space flight.

The Karth subcommittee's investigation of the Centaur program in 1962 marked the first critical, independent analysis of NASA's management problems with private contractors. It was a healthy demonstration of legislative oversight, and revealed the committee at its best in probing and recommending how NASA could improve its administrative performance.

Centaur was a second-stage rocket, fueled by liquid hydrogen and liquid oxygen, and mounted on top of an Atlas missile. In the mid-1960's, Centaur was important as an intermediate link between the smaller Atlas-Agena class and the Saturn vehicles still being developed. The Centaur was needed for NASA's Ranger and Surveyor hard and soft instrumented landings on the Moon, as well as for the Defense Department's communications satellite program. Furthermore, its success was essential to provide the needed experience in handling and storing liquid hydrogen fuels for many other vehicles. The Karth subcommittee investigation revealed that inadequate supervision and quality control by NASA had been factors in the many delays. The committee report concluded that "Putting out fires is no substitute for effective program management. The subcommittee is forced to conclude that management of the Centaur development program has been weak and ineffective both at NASA headquarters and in the field, and that the program has suffered from a diffusion of authority and responsibility."

In its final report, the Karth subcommittee recommended that "NASA should exercise close, continuing, and centralized supervision and direction over the Centaur development program. This should result in a coordinated program in which contractors and subcontractors are required to exercise high-level quality control."

One of the results of the Karth subcommittee's investigation of Centaur was to transfer the responsibility for Centaur out of the

Marshall Space Flight Center. Down at Huntsville, the von Braun team had been concentrating its single-minded effort toward Project Apollo and the development of Saturn. It was evident that insufficient technical and managerial talent was being devoted to Centaur, as proven by the Karth subcommittee investigation. Accordingly, once the Karth subcommittee had made its recommendations, the management and supervision of the Centaur program were transferred from Huntsville and placed under the Lewis Research Center.

The Karth subcommittee hearings and report on Centaur were tough in their criticisms. Yet the effect was good. As Karth reported to his colleagues in executive session: "I had people from NASA come up to me after the hearings had been concluded and say 'We think this is one of the best things that has ever happened.'"

In addition to the Centaur investigation, the Karth subcommittee also held hearings on and issued a useful report on Project Anna, a geodetic satellite operated by the Department of Defense. Dr. Fred L. Whipple had recommended to the committee's Panel on Science and Technology that it did not make sense to classify Project Anna. Acting with remarkable and more than coincidental speed, the Department of Defense suddenly declassified Project Anna after the hearings were announced but just before they got under way. This was perhaps the greatest example in legislative history of anticipatory oversight.

The Karth subcommittee also exerted its influence over policy in the annual authorization hearings. In 1962, the subcommittee authorized funds for the instrumented lunar programs Ranger and Surveyor, but cut out $10,400,000 requested for an advanced, unmanned lunar exploration program termed Prospector. The committee reasoned that Ranger and Surveyor could gain all the necessary knowledge required for manned lunar landings and by the time Prospector was scheduled men would already be on the Moon and there was little that Prospector could do that men couldn't do themselves. The committee judgment was confirmed by subsequent events.

Karth's training as a labor negotiator, his exposure to engineering at the University of Nebraska, and his determination to get the facts rather than accept the fluff all made him an excellent subcommittee chairman. He had an imagery of expression which frequently spiced up a protracted hearing as, for example, he characterized the supervision of the Centaur as "Just like that prehistoric animal named brontosaurus who grew faster than his brain, and somebody nibbled off his tail and his defenses were down and despite his size he was gentle as a mouse."

Karth also had a manner of cajoling certain witnesses, and carefully measuring their reactions so he could psychoanalyze where his committee could justifiably recommend reductions, as noted in the following colloquy:

Mr. KARTH. You tell us where the soft spots are.

Dr. FELLOWS. In my estimation, there are no soft spots. You are proceeding on the assumption that a budget has soft spots.

Mr. KARTH. There is quite a feeling among the Members of Congress that there are some very inflated areas within the budget that could be removed from the budget this year without hurting anybody.

Dr. FELLOWS. My answer to that is simply the removal of funds must of necessity slow down some element.

Mr. KARTH. You know you are going to get cut by the Appropriations Committee. We know this, you know it, and better than that, the Appropriations Committee knows it. * * * We have got to either agree with you totally, or take an arbitrary amount. * * * We would like to look a little more responsibly * * *. If we were somewhat responsible and had some idea of what action we should take which would better indicate our responsibility, it might in fact save you people money. * * * Speaking as a friend of NASA, I just think you guys are heading for a little trouble now and in the future years ahead unless you help this committee which is your helper in the Congress.

As time went on, Karth became tougher in his questions, more challenging in his attitude, and even portrayed open skepticism to elicit reactions from witnesses. "Lengthy meetings such as this committee hearing could tend to make us brainwashed," Karth sternly advised Dr. Homer E. Newell on March 8, 1963, adding: "On occasion, the witness may even feel that we are abusive." A few days later, he warned Dr. Newell "to be completely candid and scrupulously honest," and threatened that "once I think the committee should lose confidence in the judgment or in the veracity of the statements or testimony that is being given, I think, then, just because of the nature of the beast that we do face—I think probably it would be extremely disastrous."

NASA witnesses reasoned that Karth's long background as a labor negotiator made him suspicious of the "initial pitch" which so often reveals labor and management presenting extreme positions. Karth frequently pointed out, as he did one day to Dr. Newell, that "There are few, if any, witnesses who ever appear before this committee in opposition to any one of Dr. Newell's programs * * * . If on occasions we appear to be critical in our pursuit of getting information, or in the manner in which we ask questions, it is only because I feel more often than not we feel frustrated for lack of having the information with which to properly compete with those who are sitting on the other side of the table."

CONSTRUCTION OF FACILITIES

In July 1961, Col. Earl G. Peacock of the U.S. Army Corps of Engineers was detailed to the committee staff. Working primarily with the Manned Space Flight Subcommittee, he was able to advise the committee during the early hectic days when many decisions had to be made on millions of dollars of construction for the space program. Colonel Peacock assisted in the preparation of an amendment the committee proposed in 1962 to require NASA to utilize the facilities design criteria and construction standards established by the General Services Administration, the U.S. Navy Bureau of Yards and Docks and the Army Corps of Engineers—until such time as NASA established its own standards.

When it was pointed out to Teague that NASA opposed the amendment, Teague remarked to the committee at a May 10, 1962 executive session: "I can see nothing wrong, and it seems to me it is protection for our committee until we know that they have standards. I am not going to permit them to build a palace some place to embarrass all of us."

The committee voted the amendment, NASA appealed to the Senate, and the conference report made the following notation:

> The Senate amendment contained language similar to the House provision, but inserted the clause, "to the fullest extent practicable." The managers on the part of the House agreed to this language change, based on information from NASA that a substantial portion of design work had been completed and the more restrictive language of the House bill would require a detailed review of the completed design work, thereby delaying the national space program, unless the flexibility provided by the Senate amendment was accepted.

The loophole driven by the Senate served to give NASA the leeway not to move very fast to develop their own construction standards. Colonel Peacock's tour with the committee ended at the close of 1962, and he was replaced by Lt. Col. Harold A. Gould, already well-known to Chairman Miller and other Members through his able testimony on Army construction projects before the Armed Services and other committees. "He was their talk man," recalled Ducander "so he knew a good number of the committee members casually and some of our Armed Services Committee staff because he was over there a good number of times."

"When I got here, they handed me two large justification books," Gould says. "I didn't know what the space program was all about, and I didn't know what the words were—all those acronyms—and finally after struggling in the office for three weeks with these two books involving $800 million worth of construction (authorizations), I went to George Miller and I said: 'Mr. Chairman, I don't even know where these places are, and I don't understand these words.' " He felt that only a personal inspection would produce the information the committee needed.

Col. Harold A. Gould (right) with Representative Richard L. Roudebush (Republican of Indiana) and an unidentified person. Colonel Gould was named deputy director of the committee in 1975, and executive director in 1979.

So Colonel Gould went out to visit every NASA center in the United States, and the Guaymas Mercury tracking station in Mexico. He reflects:

When I came back, I knew more about these facilities, about what was in the program, than the witnesses who later testified on the programs at (NASA) headquarters.

Having had long experience with orderly planning methods in force throughout the Department of the Army, being trained to observe and ask the right questions about construction programs, and having the advantage of his field investigations, Gould quickly spotted that NASA construction and design procedures were in quite a mess. There were 16 different facilities offices at the NASA headquarters, each going its own way and each in charge of its own construction. Meanwhile, out at the NASA centers, huge discrepancies were cropping up in the budgeted costs for construction. For example, there were some similar style buildings for which the cost figures were glaringly different. Gould discovered that no installation had a master plan; they had "as built" plans which showed where the buildings were, but lacked future projections. As Gould reported,

It was not unusual for NASA's construction projects to miss the mark by 100 percent in the amount authorized versus the actual cost.

Gould made a number of recommendations which were incorporated in both the statute and the committee report. NASA was instructed in the statute to develop its own "uniform design criteria and construction standards" and in 1963 the House succeeded in beating back the NASA-Senate effort to insert weasel-worded loopholes like "to the fullest extent practicable"—as had been done in the 1962 statute. The committee did not stop there, but continued to ride herd on NASA management until NASA in 1965 finally published its very own "Design Criteria and Construction Standards." Associate Administrator Seamans in his introduction to the 1965 volume gave due credit to the House Committee on Science and Astronautics for having initiated the fight for these home-grown standards. Gould also pointed out to the committee that NASA had two different "pots" hidden in the authorization bill out of which they drew facilities design money: (1) there was a separate overall lump sum for that purpose, and (2) each construction project included some design funds. On his field trips, Gould discovered that NASA's practice was that "someone would make a thumbnail estimate of what a facility would cost and put it in the budget, add a given percentage to design it and a given percentage for contingencies and come to Congress to ask for the money." To correct this haphazard practice, the committee clamped down by knocking out the multiplicity of design authorizations for all the construction projects, lumping all design money at one point in the authorization, and then forcing NASA through strict oversight to justify and utilize these funds for the real purpose they were intended.

At the end of 1963, Colonel Gould went off for a year's tour of duty at the Army War College. He returned to the committee in 1964 as a uniformed officer, and without shifting gears at all he moved easily into a civilian slot as one of the top staff members of the committee in 1965. As noted above, Gould was later named deputy staff director of the committee, and in 1979 executive director.

In contrast to his early criticisms of NASA, Gould looked back in 1978 on the committee actions to beef up NASA's internal management in the facilities area, and concluded:

> I think our actions in the oversight area helped to shape NASA's management. NASA has, for example, now one of the best facilities management organizations in the entire Government, in my opinion.

PATENTS AND INVENTIONS

One of the thorny issues which occupied the committee during its early years was how to treat patents and property rights in inventions. As noted in chapter I, the National Aeronautics and Space Act of 1958 required NASA to obtain ownership of inventions, developed through NASA contracts, unless the Administrator could show that the public interest would best be served by waiving title. In the space program, many companies felt their investments were not being protected if the Government contracted for certain types of work and then took title to company-sponsored inventions.

There was a rumbling of discontent with the 1958 provisions, not only because there had been little public or agency input into their formulation, but also industry, the legal profession, NASA, and other interested parties felt the patent provisions might slow down the space program. Under the chairmanship of Representative Erwin Mitchell (Democrat of Georgia), hearings were held in 1959, with most witnesses urging changes in the 1958 law. The bill produced by the Mitchell subcommittee did not clear all the legislative hurdles, passing the House but not being acted on by the Senate.

When Representative Mitchell left the committee in 1961, Representative Daddario was made chairman of the subcommittee. Chairman Miller rounded up most of the lawyers to serve on Daddario's subcommittee, but since there weren't enough Republican lawyers, Representative Alphonzo Bell (Republican of California), a special species—a nonlawyer—was drafted to join the group. The Daddario subcommittee recommended that NASA should have more flexibility in determining whether to obtain a royalty-free license on each invention developed in the course of NASA research contracts, or to try to obtain full title. In separate views differing from the majority, Representative William Fitts Ryan (Democrat of New York) advocated continuation of the policy set forth in the 1958 act as the best protector of the public interest.

In practice, NASA interpreted the 1958 act liberally, which quieted many of the fears of aerospace contractors.

Although strongly favoring the Daddario committee recommendations, NASA did confess it was not really having any "major difficulties" with the 1958 Act. So it was not the end of the world for the space program when Congress, despite the favorable committee recommendation, failed to take action in either the House or Senate to implement the report. In fact, the Space Act itself was never changed to reflect the Daddario committee's recommendations.

What makes the Daddario committee recommendations significant, however, is the fact that President Kennedy on October 10, 1963, issued a memorandum on patent policy which paralleled and restated many of the Daddario committee recommendations. The memorandum stipulated that all Federal agencies with work performed under Government contract would normally be guided by the terms of the memorandum. Therefore, even though the extensive labors of the Mitchell and Daddario subcommittees were not actually frozen into statute, they did serve as sounding boards for patent policy and also surfaced in statements of Presidential directive which stood, with only minor subsequent revisions by President Nixon.

In the happy-go-lucky early days of the space program, NASA sent up to Capitol Hill a bill which they insisted was necessary to provide the "flexibility" for emergency construction during periods when the slow congressional funding process might really put a crimp in speedy progress. Thus originated the authority to reprogram R. & D. into construction not to exceed 3 percent. The committee soon discovered that reprograming was going on at a merry pace, so in the authorization bill passed in 1963, the reprograming limit was cut down to 2 percent—and subsequently squeezed down to ½ of 1 percent by degrees over the years. Also, the committee had moved to require NASA to give House and Senate committees 30 days' notice before reprograming would become effective, unless approval were given immediately prior to the expiration of the 30 days. Those actions tightened the reins of oversight available to the committee, enabled the committee to review and assess reprograming recommendations, and served notice that oversight meant something more meaningful than overflights.

Still another committee-induced reform rendered committee oversight over NASA more effective. In the early years, NASA preferred to present its funding requests with "Research, development, and operations" lumped together. The committee insisted that to make an intelligent review of funding, NASA should separate out administrative operations from research and development. The committee also insisted on "1-year" authorization instead of allowing a blanket authorization for administrative operations stretching over several years. NASA complied with the committee request in the authorization requests from fiscal year 1965 onward, and this enabled more meaningful congressional action in the interests of economy and efficiency.

Among other oversight actions taken by the committee, started in the early Miller years, was to rescind all unused or excess authorizations after three years' time. So as early as 1963, the Committee on Science and Astronautics was taking some of the first steps in the direction of what in the next decade became popularly known as sunset legislation.

OTHER AREAS OF ACTIVITY

Chairman Brooks was never happy unless he could be doing something, and he was always unhappy if he did not observe his staff doing something. Chairman Miller was much more philosophical, yet he encouraged both the committee and staff to get involved in a wide ranging series of different activities. He was also far more interested than Chairman Brooks in the international aspects of scientific work, and kept closely in touch through travel, speeches, and conferences, with work being done in other countries and the assistance which the United States could give to encourage better exchange of scientific information.

The committee held significant hearings on and issued meaningful reports on the NASA development of weather and communications satellites. However, the committee stopped short of getting involved in areas beyond its jurisdiction over research and development. The committee tippy-toed around the fringes of the raging controversy over the public interest in the new Communications Satellite Corporation.

Following up hearings held by Chairman Brooks, Chairman Miller also held investigative hearings on what research and development was being done in new modes of air, land, and sea transportation—including hydrofoils, monorails, aircars, and vertical takeoff planes. Special hearings and reports were issued on radio and radar astronomy, international scientific activities, United Nations discussions and negotiations on the peaceful uses of outer space, military astronautics, the Bureau of Standards, the National Science Foundation, qualifications for astronauts, the Soviet space program, and the development of solids for propulsion.

THE END OF THE HONEYMOON

On May 23, 1962, the day after Maj. Robert A. Rushworth flew the X–15 at top speed of 3,477 miles per hour, and the day before Astronaut M. Scott Carpenter orbited the Earth in his Mercury capsule, the House of Representatives passed a $3.742 billion authorization for NASA by a rollcall vote of 343 to 0. It was the last time that the committee was able to achieve such unanimity on the House floor and among its own members.

Yet harbingers of things to come showed in the acid remarks of that inveterate watchdog of the Federal Treasury, Representative H. R. Gross (Republican of Iowa). After the debate had proceeded for several hours, Gross interrupted to inject the first note of discord into the proceedings:

Mr. Chairman, I hesitate to barge into this mutual admiration meeting that has been going on all afternoon, but there are a couple of questions I would like to ask concerning the bill.

Gross wanted to know why so much money was being spent in the southern states and in California, and whether anybody was watching the high salaries which space contractors were paying their executives. Although Gross wound up voting for the bill—the last time he voted for a NASA authorization—he grumpily suggested:

It would be my hope that if and when we do get to the Moon, we will find a gold mine up there, because we will certainly need it.

Chairman Miller confers with Vice President Lyndon B. Johnson at NASA's Ames Research Center in California.

Representatives Olin E. Teague (Democrat of Texas), left, and Charles A. Mosher (Republican of Ohio) inspect Vehicle Assembly Building at Kennedy Space Center, Fla.

Representative Emilio Q. Daddario (Democrat of Connecticut), left, confers with Dr. Harvey Brooks of Harvard University, who frequently testified before and consulted with the Daddario subcommittee.

Representative Alphonzo Bell (Republican of California), left, who served for several years as ranking minority member of the Daddario subcommittee. At right is Dr. Antonie T. Knoppers, senior vice president of Merck & Co., Inc., who appeared before a committee panel.

CHAPTER V

Science, Research and Development, 1963–69

The year 1963 brought many changes to the Committee on Science and Astronautics. The NASA budget request soared again from $3.7 billion up to $5.7 billion, but then for the first time serious opposition was mounted in the Congress against the Moon program as well as the balance of the space budget. Several new Members joined the committee in 1963, including Representative Don Fuqua (Democrat of Florida) and Representative John Wydler (Republican of New York) who 16 years later had risen to become chairman and ranking minority members of the committee. Also in that year Majority Leader, and later Speaker, Carl Albert began the first of four years of service on the committee, modestly accepting a seat on the bottom rung of the seniority ladder. Albert joined the committee at the request of both President Kennedy and Speaker of the House John W. McCormack. "President Kennedy wanted me to go on the committee because he was very interested in his 10-year Moon project," Albert relates. And McCormack, who had been on the committee as Majority Leader, was eager to have Majority Leader Albert continue the tradition. When he became Speaker, Albert also took a strong and active interest in the work of the Science Committee.

The death of President Kennedy and the accession of Vice President Johnson did not have a material effect on the committee's mode of operation or influence, even though it was President Kennedy who dramatically focused national and world attention on the manned lunar landing program. The Apollo program still constituted the biggest chunk of the annual authorization bill with which the committee had to wrestle. President Johnson, as one of the architects of the 1958 NASA legislation, had a paternal interest in both the space program and scientific development in general. In addition to his many visits to space installations and personal encouragement given to the astronauts, President Johnson steadfastly supported the efforts of the Science Committee—at least until the budget squeeze caused by the war in Vietnam.

One of the most significant developments within the committee in 1963 was the establishment of the Subcommittee on Science, Research and Development, chaired by Representative Emilio Q. Daddario (Democrat of Connecticut).

The Subcommittee on Science, Research and Development in 1963, from left: Representative J. Edward Roush (Democrat of Indiana), Representative Emilio Q. Daddario (Democrat of Connecticut), chairman; standing, Philip B. Yeager, subcommittee staff director; and Representative R. Walter Riehlman (Republican of New York).

Among committee members active in the mid-1960's were, from left: Representatives Weston E. Vivian (Democrat of Michigan), Gale Schisler (Democrat of Illinois), William R. Anderson (Democrat of Tennessee), Brock Adams (Democrat of Washington), and Lester L. Wolff (Democrat of New York).

At the start of 1963, the following constituted the membership of the full committee:

Democrats	*Republicans*
George P. Miller, California, *Chairman*	Joseph W. Martin, Jr., Massachusetts
Olin E. Teague, Texas	James G. Fulton, Pennsylvania
Joseph E. Karth, Minnesota	J. Edgar Chenoweth, Colorado
Ken Hechler, West Virginia	William K. Van Pelt, Wisconsin
Emilio Q. Daddario, Connecticut	R. Walter Riehlman, New York
J. Edward Roush, Indiana	Charles A. Mosher, Ohio
Thomas G. Morris, New Mexico	Richard L. Roudebush, Indiana
Bob Casey, Texas	Alphonzo Bell, California
William J. Randall, Missouri	Thomas M. Pelly, Washington
John W. Davis, Georgia	Donald Rumsfeld, Illinois
William F. Ryan, New York	James D. Weaver, Pennsylvania
Thomas N. Downing, Virginia	Edward J. Gurney, Florida
Joe D. Waggonner, Jr., Louisiana	John W. Wydler, New York
Edward J. Patten, New Jersey	
Richard H. Fulton, Tennessee	
Don Fuqua, Florida	
Neil Staebler, Michigan	
Carl Albert, Oklahoma	

Subcommittee on Manned Space Flight, Olin E. Teague, *Chairman*.
Subcommittee on Space Science and Applications, Joseph E. Karth, *Chairman*.
Subcommittee on Advanced Research and Technology, Ken Hechler, *Chairman*.
Subcommittee on NASA Oversight, Olin E. Teague, *Chairman*.

At the organizational meeting of the committee on February 5, 1963, after the customary introduction of the newly elected members, Chairman Miller stressed one of his favorite themes:

We must expand our operations to give more attention to the National Science Foundation, the National Bureau of Standards, and to other facets of science which are our responsibility, than we have in the past.

In practice, the committee tried to do this in between the most time-consuming job of all—exercising oversight over NASA and holding careful and lengthy authorization hearings every spring.

In 1959, the committee sponsored a resolution empowering the National Science Foundation to produce a status report on scientific manpower and education. Subsequently, the committee actively engaged in efforts to expand the training of more scientists and engineers. The committee also conducted many hearings and issued useful reports in related scientific areas such as weather modification, progress toward the metric system in the United States, dissemination of scientific information, and agency reviews of the National Science Foundation and the National Bureau of Standards.

As noted in chapter II, the Panel on Science and Technology, meeting on the average twice a year, began to assume greater significance outside the committee as it served to forge closer ties with the scientific community. The panel began to concentrate on central themes, was well attended by a wide number of invited scientific guests, and received wide attention by the news media.

WHY THE DADDARIO SUBCOMMITTEE WAS FORMED

In 1963, the convergence of several events, and the emergence of just the right congressional personality, sparked the creation of a new and significant subcommittee—the Daddario Subcommittee on Science, Research and Development.

The naming of the subcommittee provided an interesting twist. It was originally planned to call it the "Subcommittee on Scientific Research and Development." Mrs. Eilene Galloway, of the Library of Congress Legislative Reference Service, argued strenuously that a broader charter would result if a comma were placed after the word "Science" in the title. She won her point.

Congress and the Nation were becoming uneasily aware that Federal spending for research and development was rocketing upward. From $74 million in 1940, the Federal price tag had mounted to $2 billion in 1953. Many Congressmen, despite the research which went into the development of the atomic bomb in the Manhattan project, were uncomfortable with the doling out of such huge amounts for research.

The chairman of the House Committee on Rules, that crafty manager of the abattoir of liberal legislation, Representative Howard W. Smith (Democrat of Virginia), decided to set up a select committee to investigate where these research dollars were going. Because of the huge increase of Federal research spending from $2 billion in 1953 to $12.2 billion in 1963, the Smith resolution gained widespread support editorially, in letters from home, and within the Congress itself. At the same time, the groundswell of popular support caused many committee chairmen to shake in their jurisdictional boots. Chairman Oren Harris of the Interstate and Foreign Commerce Committee, with the huge expenditures for the National Institutes of Health research under his jurisdiction, voiced concern behind the scenes, as did Chairman Miller whose Science and Astronautics Committee had over 25 percent of Federal research spending under its jurisdiction. Chairman Carl Vinson of the House Armed Services Committee took another defensive tack: he set up a special subcommittee for military research.

During early August, there were many huddles among members of the Science Committee. Daddario, Miller, and other senior members of the Committee seriously considered openly opposing the resolution.

Speaker McCormack tapped Representative Carl Elliott (Democrat of Alabama), a moderate southerner and highly respected in the House, as chairman of the new select committee designed by Representative Smith.

FORMATION OF THE ELLIOTT COMMITTEE

Daddario soon became the major spokesman in favor of creating a new subcommittee of the Science and Astronautics Committee, not only because it was needed, but in the hope that it might, in conjunction with other research subcommittees, head off the rush to set up the Elliott committee. Chairman Miller decided to move, and on August 23, he announced the formation of a nine-member subcommittee to be chaired by Daddario. On September 11, the Elliott resolution came up for a rollcall vote. Miller expressed his faint praise mixed with damns in these words on the House floor:

> Investigation into research and development has to begin someplace, and perhaps this is as good a place as any. * * * I am certain the Committee on Science and Astronautics will cooperate with the new committee, but it will protect its own interests and will fight against any duplication of effort in those areas in which the House of Representatives has given it statutory jurisdiction.

The resolution was passed, 336 to 0, because this was a kind of motherhood issue that was difficult to vote against. But it was significant that three powerful subcommittee chairmen of the Science Committee—Teague, Karth, and Daddario—did not vote, and neither did one of the high-ranking Republicans on the new Daddario subcommittee—Representative Charles A. Mosher of Ohio.

Out of this somewhat foggy atmosphere, Daddario emerged as a real leader. Fifth-ranked member of the full committee, Daddario was a charter member of the science and astronautics group. He had already played an active role as an expert in the life sciences, and had won a good reputation as a fair and thorough pilot of the Subcommittee on Patents and Inventions.

"I talked to George Miller and to Tiger," Daddario recalls. "I talked to both of them about the importance of broadening out the committee jurisdiction. They were both very willing to listen (and) as we talked about it, the ordinary course of events built up some requirements about what ought to be done." Daddario was convinced that the time for talk was over, and the time for action was at hand.

The relationship between Miller and Daddario was always very warm and close. They shared mutual interests, socially as well as intellectually. One day, after he became chairman, at an executive session of the committee in room 214–B of the Longworth Building, Miller, as was his custom, was spinning an account of some of his early background and reminiscences. He remarked that his father was Irish and his grandmother was Italian, and then related some of the early history of the times when Italy was composed of independent city states which eventually evolved into one nation. He wound up his little history lesson with the remark: "And that's how the Italians were born."

Daddario immediately asked for recognition, and proclaimed:

Mr. Chairman, I do not think you should give the wrong impression. Italians are born just like everyone else.

DADDARIO BACKGROUND

The new chairman was just short of 45 years old when he took the reins of the subcommittee. A Massachusetts native, he gained fame at Wesleyan University in Middletown, Conn., where he captained both the football and baseball teams. His senior year he was selected as quarterback on the Little All American Team, and Sports Illustrated honored Daddario by naming him to their All-Time Little All American Team. He also played professional football for Providence while earning his law degree at the University of Connecticut.

Assigned to the Office of Strategic Services as a military officer in World War II, Daddario drew the delicate and dangerous assignment of negotiating the surrender of German troops in northern Italy prior to the arrival of the American forces. He also served as a major in Korea during combat. Mayor of Middletown, Conn., municipal judge, practicing attorney in Hartford, he had a richly varied background when elected to the House of Representatives in 1958.

The Committee on Science and Astronautics was an easy choice for Daddario for these reasons:

I had a great interest in matters affecting technology and science because of my own involvement in intelligence activities in World War II and the Korean conflict where the leadership of science and the applications of technology were so important. I thought that this committee offered me a place to participate, a brand new committee, a wide area of jurisdiction, and involvement in development of applications important to science.

Bald-headed and bushy-browed, Daddario always walked with the springy step of an erstwhile athlete. Fluent in Italian and French, he had a catholicity of interests which covered not only his own art collection, but also music, the theater, education, economics and international relations. Philip B. Yeager, who served as Daddario's chief of staff for over 10 years on both the Patents and Science Subcommittees, characterizes him in the following terms:

Mr. Daddario, to use current parlance, is unflappable. * * * He tends to accept people at face value until circumstances dictate otherwise, but his instincts for distinguishing within a very short time between the genuine and the phony are pronounced. * * * Mr. Daddario never lost a bill on the floor of the House for which he was responsible as manager. No mean trick. * * * Mr. Daddario tends to be a political liberal. He has been a lifelong Democrat. By and large, his leanings have been on the liberal side, yet always leavened with common sense. During his term in the House he worked equally well with conservatives. * * * During my service with Mr. Daddario I found him to have a remarkable memory and almost uncanny ability to keep many balls in the air at the same time without dropping any of them. He is a thoughtful, philosophical man, tough-minded but always willing to listen. There is never any doubt as to who is boss in a situation where Mr. Daddario has been placed in charge.

PHILIP B. YEAGER

Yeager himself had been a very productive staff director for the Manned Space Flight Subcommittee since its organization when Miller became committee chairman, and Teague subcommittee chairman. By the time the new Subcommittee on Science, Research and Development was formed, James E. Wilson had come aboard and was assigned immediately to the Manned Space Flight Subcommittee. This made it easier for Daddario to persuade both Miller and Teague that for his new committee to succeed, it would need Yeager as chief of staff.

It was a fortunate choice, not only because Daddario and Yeager shared a mutual respect, but also because of Yeager's indefatigable energy, facile writing ability, and talent for organization. A graduate of the University of Arizona and George Washington University Law School, Yeager had served as Capitol Hill correspondent for a number of newspapers. One of his freelance articles was noted by the then Representative Kenneth B. Keating, who was serving on the House Select Committee on Astronautics and Space Exploration. Keating asked Yeager to draft a speech for him. Then Yeager also approached House Republican Whip Les Arends of Illinois, who was also a member of the select committee. Yeager was soon hired by the select committee in 1958, and helped draft NASA's organic act which Congress passed that year. Initially, Yeager was by mutual consent the Republican staff member who handled inquiries from Republican Congressmen, as well as serving the entire committee.

Yeager proved a tower of strength, not only to Daddario but also to his successors who chaired the subcommittee and many other members of the full committee where Yeager still serves in 1979, as "dean of the staff," and General Counsel.

Daddario let no grass grow under his feet when he received the new assignment. He called a meeting for his new subcommittee on August 27 and outlined his plans to the charter members:

Democrats	*Republicans*
Emilio Q. Daddario, Connecticut, *Chairman*	R. Walter Riehlman, New York
	Charles A. Mosher, Ohio
J. Edward Roush, Indiana	Alphonzo Bell, California
Thomas G. Morris, New Mexico	James D. Weaver, Pennsylvania
John W. Davis, Georgia	
Joe D. Waggonner, Jr., Louisiana	
Edward J. Patten, New Jersey	

Chairman Miller dropped in on the organizational meeting, to underline his full and personal support of the new venture.

FIRST HEARINGS OF DADDARIO SUBCOMMITTEE

Between October 15 and November 20, 1963, the subcommittee held a series of nine basic hearings whose objectives were defined by Daddario as follows:

First, to review the nature of the country's overall scientific effort, and second, to locate and identify the major problem areas which exist or may soon exist within the science relationship of the Federal Government to industry, the universities, foundations, professional societies, and among Federal agencies.

The leadoff witness was Dr. Frederick Seitz, President of the National Academy of Sciences. Many of the Nation's most prominent scientists testified.

Daddario also asked members of the Panel on Science and Technology to evaluate scientific research and development throughout the country, how to strengthen congressional sources of information, and how to more effectively utilize the Nation's scientific and engineering resources.

The responses were provocative, and helped form the basis for additional hearings and reports by the committee. One response from Dr. G. B. Kistiakowsky, former science adviser to President Eisenhower, struck a responsive chord with the committee:

While I cannot speak officially for the National Academy of Sciences, I feel confident that it would be eager to discharge its obligations under its congressional charter and render assistance to your subcommittee request.

Some of the replies warned against too much emphasis on cost-consciousness, and the need to give more priority to basic research. As Dr. H. Guyford Stever, professor of aeronautical engineering, Massachusetts Institute of Technology, Cambridge, put it:

Basic scientists need time and freedom to think and work if they are to produce.

Dr. Lee A. DuBridge, president, California Institute of Technology, Pasadena, warned the committee:

Government funds made available for research purposes in colleges and universities are not at all analogous to purchase orders which the Government may issue for supplies and equipment. These contracts and grants are not for the purchase of services or commodities, but for the stimulation of intellectual endeavor in a chosen scientific field.

Roger Revelle, director, Scripps Institution of Oceanography, University of California, bubbled over with enthusiasm as he reported:

During the past 15 years, Federal policy in support of basic research has been to assist all first-rate scientists to do the research they wanted to do, particularly when this research also involved the teaching of graduate students. This emphasis on excellence and on freedom has produced remarkable results. It is not an exaggeration to say that the flowering of American science since the war is as spectacular an outburst of human creativity, though on a far larger scale, as the outpouring of art and literature in Florence during the days of Lorenzo the Magnificent.

Perhaps the most interesting advice came in this form from Dr. Harold C. Urey, professor of chemistry, University of California:

Outsiders should not try to plan the work, or say what is important. Do you really think that any outside group, congressional committee or otherwise, in 1931, could have told a rather unknown scientist by the name of Harold C. Urey that it was important to work on the discovery of heavy hydrogen? * * * I think it is entirely probable that if outsiders had attempted to direct my research at that time that they would have advised that the work be discontinued as unnecessary.

Many other thoughtful and stimulating replies were received, which were summarized in the committee's first report, "Government and Science—A Statement of Purpose."

Dr. Frederick Seitz, President of the National Academy of Sciences, testified on October 15, 1963 that "if your committee were to ask us to make a study, we would regard the report which emerges as your property to be used as you desire." Accordingly, a contractual agreement was made with the Academy which resulted in several excellent studies, the first of which was "Basic Research and National Goals."

RESEARCH MANAGEMENT ADVISORY PANEL

During its first 6 weeks, the subcommittee also assembled a Research Management Advisory Panel. The central purpose of the panel was to point the way to improve research management and policy control of some of the large and costly applied science research programs. The panel was initially composed of the following members:

James B. Fisk, president of Bell Telephone Laboratories, Inc.
James M. Gavin, chairman of the board, Arthur D. Little, Inc.
Samuel Lenher, vice president, E.I. Du Pont de Nemours & Co.
Wilfred J. McNeil, president, Tax Foundation, Inc.

Don K. Price, dean, John F. Kennedy School of Government, Harvard University.
C. Guy Suits, vice president and director of research, General Electric Co.
Jerome B. Wiesner, president, Massachusetts Institute of Technology.

The Research Management Advisory Panel met with the subcommittee members three or four times a year to discuss issues and procedures involving the relationship between government and science. The panel also aided the subcommittee in identifying and isolating problems requiring priority attention by the Congress. Michael Michaelis of Arthur D. Little, Inc., a research management consultant firm, was retained by the subcommittee as executive director of the Research Management Advisory Panel, commencing in 1964.

By the end of 1963, the Legislative Reference Service of the Library of Congress had published two studies which outlined the aids and tools available for Congress in the area of science and technology, with suggestions on how to strengthen them. The committee itself also published a study on "Scientific-Technical Advice for Congress: Needs and Sources." The committee underlined the rising importance of the subject on which it was focusing by noting that in 1964 total Federal expenditures on research and development had risen to $14.9 billion— as contrasted to $12.2 billion when establishment of the committee was first under consideration. This staggering total dwarfed the $74 million being spent in 1940.

SCIENCE POLICY RESEARCH DIVISION

Largely through the influence of the Daddario subcommittee, the Library of Congress established a Science Policy Research Division in 1964. This new division had one of its closest relationships with the entire Committee on Science and Astronautics, and it also helped strengthen the scientific and technical assistance needed by all Members of Congress.

On November 5, 1963, the Daddario subcommittee assembled most of the regular members of the Panel on Science and Technology for a two-day Government and Science Seminar. Special guests were Dr. Wernher von Braun, Dr. Alan M. Thorndike of the Brookhaven National Laboratory, and Dr. S. Fred Singer, Director of the National Weather Satellite Service. At the opening of the meeting, Daddario mentioned that Representative Riehlman, the ranking Republican on the subcommittee, and the staff "have been meeting with a number of scientists in Government and out in order to chart our course." He added that the hearings were pointing toward trying "to locate and identify the major problem areas which exist or may soon exist within the science relationship of the Federal Government to industry, the universities, foundations, professional societies, and, indeed, amongst and between its own agencies." Daddario noted:

I am also pleased to say that the chairman of the full committee, Mr. Miller, has given the subcommittee outstanding support. He has not only given us the kind of staff that we need, the type of office help that the staff must depend upon to do its work, but he has participated in many of the meetings and has given us the highest moral support.

In his response, Chairman Miller revealed:

I can assure you that the subcommittee was carefully selected. Each member has demonstrated his interest and sincerity and in many cases has some background for the work. Likewise, the staff of the committee is a good one, and they are the ones on whom we depend greatly.

"GOVERNMENT AND SCIENCE" SEMINAR

The "Government and Science" seminar surfaced enough ideas, suggestions and advice to fill several volumes. After all, what do you expect when some of the most intelligent people in the Nation get together and brain-storm with each other? Certainly few people could claim to have as exacting a managerial job as Wernher von Braun. Yet, with timetables staring him in the face, well over 4,000 employees to supervise, superbusy with putting out fires and appealing for more firefighting equipment in his effort to meet deadlines, he still had time to philosophize:

I would make a request on behalf of our working scientists. They are not magicians, they have no crystal ball. Therefore, they should not be expected to precisely predict the practical benefits of their research. It is no more possible for them to make such a prediction than it is for a historian, a social scientist, or I presume a Member of Congress to predict history.

The seminar erupted into a raging debate over the proper geographical distribution of Federal research funds, which gave Roger Revelle the chance to indicate perhaps one of the reasons for maldistribution:

The difficulty with many sections of the country is that there are Representatives in Congress representing the citizens and the leaders of those communities who have in fact not emphasized the right things.

If we look at some sections of the country, the amount of Federal money spent in those sections is very large, but it is spent for military bases, and it is spent for a great many activities other than education, other than real research or support of science and the support of higher education.

NATIONAL SCIENCE FOUNDATION

A few weeks before Christmas in 1963, Leland Haworth, the Director of the National Science Foundation, received a memorandum which intrigued him. He was thoroughly acquainted with the Daddario subcommittee, and had testified before it and taken part in the "Government and Science" seminar. Yet like many scientists, despite his position of responsibility, Dr. Haworth related:

> I could not bring myself to go and see a Congressman, unless I had something to see him about. I would not take his time just to butter him up, so to speak.

But the memorandum put the shoe on the other foot. "Representative Daddario, Chairman of the Subcommittee on Science, Research and Development, would like to meet with you next Monday or Tuesday for 'up to an hour probably—morning, afternoon, or evening'—to talk over with you some of the subcommittee's plans for the future," the memorandum started off. Most contacts with Capitol Hill, particularly with a committee or subcommittee chairman, were confined to the frosty atmosphere of the formal hearing where you brought a thick, prepared statement which you hoped would not leave too much time for questions. The memorandum went on to say that "with Mr. Daddario would be Representative Riehlman, the ranking minority member, and Philip Yeager, the committee's counsel. They would welcome your bringing anyone you wish, and no preparations or materials would be necessary." Daddario said he wanted to discuss the possibility that the National Science Foundation might prepare a report for the subcommittee on scientific education at the secondary school level, the extent to which the NSF should be involved in the support of research in the social sciences, and the question of future authorization review of the NSF by the committee.

Daddario's initial, informal contact with the National Science Foundation as with the National Academy of Sciences, developed into a pattern which proved very successful. He recognized that stronger and more personal bridges had to be built with the scientific community. The agencies involved were more executive-oriented, and only dealt with Congress at arm's length when the annual appropriation time came up. "I recall that they would have preferred not to deal with Congress," said Daddario with reference to a minority feeling among some scientists and administrators of science policy. Concerning another experience of a meeting at the National Academy of Sciences, Daddario related that there were "a couple of people there who had never met a Member of Congress before."

The subcommittee also operated in a strictly nonpartisan fashion. As with Riehlman, and later with Representatives Bell and Mosher, Daddario went out of his way to make sure they were involved in all the important decisions of the subcommittee. Whenever a new program was being worked out, Daddario exercised great care to be sure it was talked out with the Republican members before any public hearings were held or statements made. As a result, there was a better than average cooperative spirit within the subcommittee.

The informal discussion between the committee and the NSF personnel was long overdue. Chairman Miller felt that "it appeared best not to begin a general review of the Foundation until the incoming group had a chance to get its bearings." Haworth, replacing Alan T. Waterman as Director, was entitled to a honeymoon. So the committee broke him in rather easily by asking the NSF to conduct a series of studies on science education, to show (1) what had happened to science education in the 20th century; (2) where the country stands; and, (3) what should be done in the future to overcome the difficulties. The reports were delivered as follows:

"Science Education in the Schools of the United States," March 1965.
"Higher Education in the Sciences in the United States," August 1965.
"The Junior College and Education in Sciences," June 1967.

GEOGRAPHIC DISTRIBUTION

During 1964, the committee itself produced a series of useful reports on the allocation of Federal research funds and the geographic distribution of Federal R. & D. funds. Geographic distribution was a subject which every Congressman understood; the Congressmen from California and Cambridge, Huntsville and Houston, clamored that their disproportionate shares were only due to the fact that they had created "centers of excellence," and that those suggesting there should be more equitable geographic distribution were only attempting to tear down the centers of excellence and replace them with mediocrity. At the same time, there was a hue and cry from the Middle West, from the Appalachian area, and other underfed sectors whose Congressmen articulated the fact that enhanced employment always followed Federal research dollars.

The discussion of geographic distribution got so hot that the Daddario subcommittee asked the National Science Foundation to produce yet another report on the subject. The NSF reported factually on the geographic trends in Federal research dollars, 1961–64, in a report released to the committee in 1964. The report and preliminary groundwork laid the basis for the committee hearings from May 5 to June 4 on both geographic distribution and the issue of allowable indirect costs in Federal grants.

The committee brought out some of the glaring inequities in the geographic distribution of Federal contracts. One of the recommendations made in its October 1964 report was that "the country should work to raise the level of all our colleges and universities without lessening the support of those strong schools which are recognized as being centers of academic excellence." The committee later was pleased to note that President Johnson on September 14, 1965, set forth a new policy in a memorandum to the heads of all Federal departments and agencies, stating in part:

> Research supported to further agency missions should be administered not only with a view to producing specific results, but also with a view to strengthening academic institutions and increasing the number of institutions capable of performing research of high quality.

The Daddario subcommittee also had a salutary influence in revising the standard procedure for allocating indirect (administrative) costs in awarding Federal research grants and contracts. In its hearings the subcommittee brought out that too many appropriation bills set statutory limits on the ratio of indirect to direct costs of federally sponsored research, and also that the older Bureau of the Budget policy regulations were so inflexible as to hurt both the grantee and the Government's interest. The hearings resulted in the issuance of more flexible regulations on indirect costs by the Bureau of the Budget, as well as having an educative effect in the subsequent appropriation bills.

"BASIC RESEARCH AND NATIONAL GOALS"

In March 1965 the National Academy of Sciences furnished the committee with its comprehensive study of "Basic Research and National Goals." The study was made through a committee-Academy contract—the first such contract with Congress in the 102-year history of the Academy. The committee asked the Academy to furnish a report on two questions:

> (1) What level of Federal support is needed to maintain for the United States a position of leadership through basic research in the advancement of science and technology and their economic, cultural, and military applications?

> (2) What judgment can be reached on the balance of support now being given by the Federal Government to various fields of scientific endeavor, and on adjustments that should be considered, either within existing levels of overall support or under conditions of increased and decreased overall support?

As reported in the magazine Science on April 30, 1965, "the vagueness of the questions and their essential unanswerability inspired a fair degree of despair behind the Academy's marble facade. But there were the questions—reasonable ones from the point of view of legislators who must appropriate money—and the Academy accordingly turned to the task of answering them."

The assignment was given to the Academy's Committee on Public Policy, which was headed at that time by Dr. George B. Kistiakowsky (who had been Science Adviser to President Eisenhower). The Committee on Public Policy then set up an ad hoc committee of 15 members, which in turn produced a set of 15 individual papers rather than attempting to produce a consensus. Yet there was one dominant theme which cropped up time and again in the 315-page report on "Basic Research and National Goals": That the future of basic research in the United States was closely tied to the fortunes of the National Science Foundation, and that increased support for NSF was essential for the future strength of the Nation. The report also emphasized the need for a more stable level of funding for basic research on a more dependable, incremented basis.

The committee experience with the Academy was a healthy one, even if it did not produce the kinds of quick, pat answers which some impatient observers seem to demand to write newspaper stories or answer TV quiz questions. Science magazine, frequently a critic of the committee, editorially observed following the Academy report that one thing favoring a closer relationship between the Academy and the Congress "is the scientists' respect for Representative Emilio Q. Daddario.* * * It is generally agreed that Daddario has been running his subcommittee in a responsible and intelligent fashion, and that the subcommittee is developing into an important channel of communication between the scientific community and the Congress."

THE MOVE TO THE RAYBURN BUILDING

After six years cramped in the small quarters of the inadequate committee room in 214–B of the Longworth Building—where frequently the committee members and agency witnesses outnumbered the spectators by space necessities—a red letter day for the committee occurred on January 26, 1965. It was the first hearing held by the committee in its spacious new basketball-court sized hearing room in 2318 Rayburn Building. It also marked the very first hearing that any committee had held in the newly opened Rayburn Building.

Before the committee moved into its new quarters, a few adjustments had to be made. The Architect of the Capitol laid the plans for the new area on Executive Director Ducander's desk one day. They were laid out to preserve very tight security in the manner of the Joint Committee on Atomic Energy. When these plans were scrapped, it was discovered that the plans had been radically altered in the opposite direction—with four huge rooms allowing no privacy for any individual staff member. So there was much hammering and partitioning before the committee was ready to move into its new location.

To mark the occasion, Chairman Miller and Representative Daddario arranged for a discussion by the committee's Panel on Science and Technology, which marked the sixth meeting of the panel. In opening the two-day meeting on January 26, Chairman Miller stated:

This is the first formal meeting in the Rayburn Building, and I think it is only fitting that Speaker McCormack, who has done so much for this building and this committee, be the first man to preside over a meeting here.

With a cheery smile, Speaker McCormack cracked an out-sized gavel, and brought laughter when he proclaimed in an exaggeratedly authoritative fashion: "The committee will come to order." Speaker McCormack celebrated the occasion very briefly with these words:

It seems to me to be singularly appropriate that the Committee on Science and Astronautics should be the first committee to meet in this magnificent building of public service, named for the late beloved Speaker of the House, the Honorable Sam Rayburn of Texas.

Speaker McCormack also praised Majority Leader Carl Albert, noting that "it was with his support that this committee was established, the first standing committee without precedent to be created by the House since the turn of the century."

Chairman Miller also used the occasion of the first meeting of the new Congress to introduce the new membership of the Committee on Science and Astronautics, which in 1965 included the following:

Democrats	*Republicans*
George P. Miller, California, *Chairman*	Joseph W. Martin, Jr., Massachusetts
Olin E. Teague, Texas	James G. Fulton, Pennsylvania
Joseph E. Karth, Minnesota	Charles A. Mosher, Ohio
Ken Hechler, West Virginia	Richard L. Roudebush, Indiana
Emilio Q. Daddario, Connecticut	Alphonzo Bell, California
J. Edward Roush, Indiana	Thomas M. Pelly, Washington
Bob Casey, Texas	Donald Rumsfeld, Illinois
John W. Davis, Georgia	Edward J. Gurney, Florida
William F. Ryan, New York	John W. Wydler, New York
Thomas N. Downing, Virginia	Barber B. Conable, Jr., New York
Joe D. Waggonner, Jr., Louisiana	
Don Fuqua, Florida	
Carl Albert, Oklahoma	
Roy A. Taylor, North Carolina	
George E. Brown, Jr., California	
Walter H. Moeller, Ohio	
William R. Anderson, Tennessee	
Brock Adams, Washington	
Lester L. Wolff, New York	
Weston E. Vivian, Michigan	
Gale Schisler, Illinois	

Subcommittee on Manned Space Flight, Olin E. Teague, *Chairman*.
Subcommittee on Space Science and Applications, Joseph E. Karth, *Chairman*.
Subcommittee on Advanced Research and Technology, Ken Hechler, *Chairman*.
Subcommittee on Science, Research and Development, Emilio Q. Daddario, *Chairman*.
Subcommittee on NASA Oversight, Olin E. Teague, *Chairman*.

At the beginning of 1965, the Subcommittee on Science, Research and Development included the following members:

Democrats	*Republicans*
Emilio Q. Daddario, Connecticut, *Chairman*	Charles A. Mosher, Ohio
J. Edward Roush, Indiana	Alphonzo Bell, California
John W. Davis, Georgia	Barber B. Conable, Jr., New York
Joe D. Waggonner, Jr., Louisiana	
George E. Brown, Jr., California	
Weston E. Vivian, Michigan	

THREE-YEAR REVIEW OF THE NATIONAL SCIENCE FOUNDATION

Congress is forever facing deadlines dictated by the calendar, such as the necessity of completing authorization legislation in time, hopefully, for the appropriation bills to reach the floor and be enacted before the opening of the fiscal year. The scientists who contributed to the National Academy of Sciences study "Basic Research and National Goals" were somewhat appalled that the Science Committee pressed them for an early deadline. In its report, the Academy states:

The National Science Foundation is viewed as playing a decisive role. The National Science Foundation is the sole agency of Government whose purpose is support of science across the board and without regard for immediate practical gains. If there is good basic science ready to be done but which does not as yet command support from some mission-oriented agency, then the National Science Foundation must be equipped to step in, if it chooses, to pick up the tab.

The summary also noted the possibility that contemplated making—

the National Science Foundation a much larger agency than it now is—so large that it can eventually become the "balance wheel", or even the main "umbrella", for the support of basic research—especially in the physical sciences—that is too remote to merit support from the mission-oriented agencies. Such a specific policy with respect to the future growth of the National Science Foundation involves a major political decision by Congress and by the executive branch, as formidable and as far-reaching as its decision has been with respect to expansion of the National Institutes of Health.

The Academy study brought into sharper focus the need for writing a new charter for the National Science Foundation—which the Daddario subcommittee proceeded to do in the next three years. If the Daddario subcommittee had followed the practice of some congressional committees of rushing in, thrusting a sweeping solution at the Congress and the NSF within a self-imposed deadline, and insisting on early action, it is entirely possible that the whole exercise might have been futile.

Theodore Wirths of the National Science Foundation, in a June 5, 1978, letter to Chairman Teague, described the approach of the Daddario subcommittee both on this issue and in general:

Its operating style is in many respects a model for producing good legislation. * * * Entirely in keeping with its interests, the Committee's approach has certain essential elements of good science and good scholarship. Its approach to problems involves research, analysis, recommendations, examination of those materials, presentation of them for general discussion and assessment and then a repetition of this process at least once and sometimes many times. Eventually, the Committee brings forth a recommended legislative package that has been studied with great care, has an intelligent and credible record and is trusted by those involved or interested as a responsible approach.

At the request of the Daddario subcommittee, the Science Policy Research Division of the Library of Congress made a thorough study of the NSF's legislative authority, organization, funding and programs which the committee published in May 1965, as "The National Science Foundation—a General Review of Its First Fifteen Years." After this report was published, Daddario's subcommittee held hearings from June through August of 1965, designed to provide a critical evaluation of the Foundation, its functions and operations as they then existed and in the future.

In an extensive article published by the International Science and Technology magazine, senior editor David Allison sized up the work of the Daddario subcommittee in its hearings and gave it high marks. Concerning the investigation itself, Allison noted that, first, the Congressmen had prepared for it through the National Academy of Sciences and Library of Congress studies. Allison also concluded:

The second distinguishing feature was the subcommittee itself: Daddario and his colleagues, whose questions were often superior to the answers they evoked, and whose responses to those answers often showed a deeper understanding of the place of science in our society than did the responses of some of the committee's witnesses.

The NSF by 1965 was spending $420 million annually, as contrasted with only $3.5 million in 1951, its first full year of operation. Director Haworth was a little troubled by the fact that he felt some of the NSF grants might be considered as being applied research. Haworth related in a June 30, 1978 letter to Chairman Teague that he had several discussions with Mim Daddario in Mim's garden after work. "Indeed it was on one such occasion that I first suggested to him and Phil Yeager the idea of including authority for applied research in any bill to modify the National Science Foundation Act." In Haworth's words, it was a suggestion to Daddario that Congress "make the Foundation honest." This was indeed one of the recommendations in the committee report which was subsequently incorporated into the new legislative charter for the NSF.

On December 30, 1965, the Daddario subcommittee issued its landmark report on "The National Science Foundation—Its Present and Future," which formed the basis for the legislation developing a new charter for NSF.

RESEARCH IN SOCIAL SCIENCES

One of the difficult issues in revising the NSF charter was how to broaden NSF's authority to support research and education in the social sciences. When the initial bills to establish the NSF were first debated in the House in 1947, Representative Clarence J. Brown (Republican of Ohio) had declared that support of the social sciences would result in "a lot of short-haired women and long-haired men messing into everybody's personal affairs." Much the same type of opposition was voiced on the House floor in the 1960's. Although Representative Fulton voted for the bill, he spoke against broadening the authority to cover social sciences. Representative Durward Hall (Republican of Missouri) cited several titles of grant studies such as "Food Gathering in a Primitive Society." Representative Mosher in the debate countered:

> I very strongly urge that all of us resist the temptation, which so often overcomes reporters, editors—and yes, politicians—to evaluate a piece of scientific research merely on the superficial fact that it might have a curious or silly-sounding title.

Daddario, in presenting the bill to the House on April 12, 1967, pointed out that his committee's hearings and analysis found that the NSF had followed a role which was too passive, which had not kept pace with the demands of society, and which should be dealing more actively with emerging problems faced by industry and society as well as the academic community. He indicated that "the National Science Foundation is presently required to collect and collate data on national scientific technical resources. The bill would have the Foundation analyze and interpret those data as well." Daddario also pointed out that the new charter authorized the NSF to "undertake the support of scientific activities relating to international cooperation on foreign policy." This concept had first been suggested by Dr. Roger Revelle at one of the committee's panel meetings.

The legislation which furnished a new charter for the National Science Foundation clearly shows the initiative of the Science and Astronautics Committee in the making of public policy. In opening the debate in the House on April 12, 1967, Daddario observed:

I do believe it is important we recognize here in the House that the changes which are being offered are changes which emanate and initiate here in the House. The impetus comes from us, and the recommendations are not those which are summarily sent to us by the executive branch. * * * Mr. Chairman, this legislation was born and raised in the House of Representatives. It is my opinion that it represents the ability of the House of Representatives to meet the challenge of adaptation to the present needs of our society.

As a matter of strategy, Daddario and Mosher were concerned with the fact that Representative Ford, who was then minority leader, habitually opposed the extension of annual authorizing power to House committees, as he had when the Science Committee first received authorizing power over NASA. They knew that Ford, as a long-time former member of the Appropriations Committee, felt that this not only diluted the power of the Appropriations Committee but also delayed the passage of appropriations bills. Yet they felt that from the standpoint of effective oversight, it was vital to obtain annual authorizing power over the National Science Foundation.

Several confabs among Chairman Miller, Daddario, Mosher, and Yeager finally concluded that Senator Edward M. Kennedy (Democrat of Massachusetts), who chaired the subcommittee handling the issue on the other side of the Capitol, might be interested in inserting this provision into the Senate bill. Since this was Senator Kennedy's only subcommittee, it was reasoned that Senator Kennedy might welcome the opportunity to have such authorizing authority for his subcommittee. Accordingly, the House bill which passed on April 12, 1967, by a 391–22 vote was silent on the issue of annual authorizations for NSF. As planned, the Senate included the annual authorization amendment, and the House accepted the Senate version without the necessity for a conference committee.

Miller talked with Ford on the House floor about the Senate changes, and persuaded him not to oppose the Senate version. Although Ford did not really like the concept, his public reaction was revealed in the following colloquy:

Mr. GERALD R. FORD. Mr. Speaker, would the chairman of the committee at this point in brief terms explain what the Senate amendments do?

Mr. MILLER. The Senate amendments are procedural. The only amendment that is important is, this will set up authorization for the National Science Foundation and will give to the Committee on Science and Astronautics the right to review annual requests for authorization legislation, something which the committee long felt should be done——

Mr. GERALD R. FORD. May I ask one other question. The annual authorization requirement does, I think give to the legislative committee new responsibility, but with that new responsibility comes the need and necessity for prompt action on the annual program of the agency. Can we have the assurance of the committee that the annual authorization legislation will be handled promptly at the beginning of each session?

Mr. MILLER. I can give that assurance. It will be handled with the greatest of facility we can give it.

After the President signed the bill on July 18, 1968, Daddario, in an address on the House floor, assessed the significance of the achievement. Once again he underlined the fact that it was "a bill conceived in the Congress, shaped by a cooperative and concerned effort of both legislative and executive branches, and approved overwhelmingly by the House and Senate on a bipartisan basis." He noted that the bill had been enacted and signed with little fanfare:

It carried none of the emotional or political fervor to which we have become accustomed while dealing with such trying matters as crime, urban redevelopment, welfare, foreign aid, pollution, gun control, and the like.

Yet Daddario expressed the thought that "without the kind of research and frontier thinking for which the new law provides, it seems unlikely that we will solve the sobering dilemmas—physical or social—which now face us."

He added:

I feel it is important to emphasize that new and fundamental knowledge must be obtained in all fields of science if we are to make any real progress toward a better life for our citizens. In fact, we will require better knowledge and understanding merely to keep our present standard of living from crumbling.

VICE PRESIDENT HUMPHREY'S VISIT

In the course of its history, the Science Committee has dealt closely with Presidents, Vice Presidents, Chief Justices, Governors, and even U.S. Senators on occasion.

Partially in response to the interest generated by the National Academy's report on "Basic Research and National Goals," the seventh meeting of the committee's Panel on Science and Technology dealt with general science policy. The keynote speaker for the first of the two-day sessions on January 25, 1966, was Vice President Hubert H. Humphrey.

The Vice President noted in his introductory remarks the presence of a special guest, Lord Snow, Joint Parliamentary Secretary of the British Ministry of Technology:

I want you to know, Mr. Chairman, how proud all of us are, and in particular how proud is President Johnson, of the work which your committee has performed in the past and now performs today and will in the future. The committee has provided a model of congressional oversight. The word "oversight" is one which is used frequently, Lord Snow, in the parlance of American congressional government, and it is a way of indicating not that you just glance over something, but the way in which you take deep perception into what the Government is doing.

Vice President Hubert H. Humphrey addresses the committee's Panel on Science and Technology, as Congressman Daddario looks on.

The challenge posed by the Vice President was contained in his peroration:

> We can either rebuild and make a new world, or destroy the old one, and I suggest that we build on the foundations that we have, but build anew and direct our great knowledge, our great fund of knowledge in science and technology with a spiritual dedication that all of it has but one purpose: the emancipation of mankind from his fear; from his hunger; from his despair; and to imbue him with faith, confidence, optimism, love, and hope. I believe that is what we mean when we put together public policy and science.

STANDARD REFERENCE DATA LEGISLATION

> The careful textbooks measure
> Let all who build beware
> The load, the shock, the pressure
> Materials can bear
> So, when the buckled girder
> Lets down the grinding span
> The blame for loss, or murder
> Is laid upon the man.
> —RUDYARD KIPLING, "Hymn of Breaking Strain."

On July 13, 1966, Chairman Miller introduced H.R. 15638 to provide a comprehensive standard reference data system within the Department of Commerce to be administered by the National Bureau of Standards. The Miller bill was referred to the Daddario subcommittee, which held hearings June 28–30, 1966.

The bill in essence sought to make data of known reliability conveniently available for use by scientists and engineers. The aim was to reduce the time-consuming necessity of searching the available literature and attempting to evaluate data where the searcher might not be expert; for example, measurements describing the properties and ingredients of different types of materials, and the rates of chemical reactions. The bill provided for an integrated, comprehensive system to replace the work being done in a piecemeal, uncoordinated, and less efficient manner by individual members of the scientific and technical community.

The Daddario subcommittee and staff did its usually thorough job of hearing the affected Government agencies and private industry, plus soliciting the opinions of the General Accounting Office, Copyright Office, and a number of individuals and business and professional groups. By the time the legislation was ready for the House, the skids were very well greased, and the bill went through the House of Representatives easily on August 15, 1966. Since the Senate failed to act in

1966, Ch airman Miller reintroduced the legislation in 1967 and after the Senate finally acted, the President signed the legislation July 11, 1968.

FIRE RESEARCH AND SAFETY

On March 6, 1967, Chairman Miller introduced H.R. 6637, a bill to authorize the National Bureau of Standards $10 million for a fire research and safety program. The Daddario subcommittee held hearings on the bill during May and June of 1967. The subcommittee also recommended a National Commission on Fire Prevention and Control which had been suggested by the professional firefighting organizations. The bill cleared the legislative process and was signed by the President on March 1, 1968.

Action during 1972 on several innovative fire research proposals will be dealt with in a subsequent chapter.

APPLIED SCIENCE AND TECHNOLOGICAL PROGRESS

The speed of the work of the Daddario subcommittee and the volume of excellent reports turned out by the committee and its close allies like the National Academy of Sciences sometimes got them far ahead of the scientific community. Witness the letter which came in to Chairman Miller on September 23, 1969, from Dr. Harold Brown, later Secretary of Defense, and at that time president of the California Institute of Technology.

Dr. Brown related that he had read the committee-sponsored report of March 1965, on "Basic Research and National Goals."

I read it at the time and considered it a most useful and provocative study. Recently I had occasion to see it again, and learned that it is no longer in print. I am writing now in the hope that you can look into the possibility of another edition.

Dr. Brown, in his 1969 letter, noted that the "questions raised and answers given in this report have become even more relevant and useful since 1965." He added:

Concerns about the intellectual contributions of basic science and about its medium and long-term utility to the material advantage of the United States and of mankind have grown since then. Indeed, we are experiencing an adverse tide of popular and congressional opinion, for a number of reasons. It would help us to have easily available in complete form once again the arguments advanced in the 1965 report; we could use them to educate more people about the specific reasons for governmental support of science.

Dr. Brown went on to describe the number of colleges and universities using the 1965 report as textbooks in science and policy seminars.

Early in 1966, Chairman Daddario negotiated a new contract with the National Academy of Sciences for a follow-on report to deal with applied science and its relation to the national well-being. Dr. Harvey Brooks, of Harvard University, headed an ad hoc panel of 20 Academy members who undertook the study and it was published in June 1967, under the title of "Applied Science and Technological Progress." The report dealt with the special problems of effective application of the resources of science to advances in technology. The various views dealt with the nature and strategy of applied research, the environment and institutions in which applied research is carried out, and the individual scientist and the role of the Federal Government.

Needless to say, in response to Dr. Brown's letter, the earlier 1965 report was reprinted along with sufficient copies of the 1967 report.

On June 7, 1967, the Science Policy Research Division of the Library of Congress produced, at the Daddario subcommittee's request, a report summarizing science activities during the 89th Congress. The report was so startling in its comprehensiveness that Chairman Daddario states in his letter of transmittal to Chairman Miller:

> Science and technology is being latticed into the structure of government and the patterns of everyday life. From the report comes explicit evidence that science, in its broadest terms, is now one of the largest, most powerful, and most important forces with which Congress must deal.

INSTITUTIONAL GRANTS FOR SCIENCE EDUCATION

In 1966 and subsequent years, Chairman Miller introduced legislation to promote the advancement of science and the education of scientists through a national program of institutional grants to colleges and universities. The Education and Public Welfare Division as well as the Science Policy Research Division of the Library of Congress prepared studies on this issue. The Daddario subcommittee held hearings on this legislation in 1968 and 1969, and the full committee favorably reported a Miller-Daddario bill in 1969. But the Committee on Rules declined to clear the bill for debate by the full House of Representatives.

Nevertheless, the discussion, studies, and reports provided good sounding boards for the critical needs in higher education. NSF Director Leland Haworth, in a letter to Chairman Teague on June 30, 1978, commented:

> I have always been sorry that this bill did not succeed in passage, both because its provisions were generally good ones and because it would have established permanently a national policy of institutional support for our institutions of higher education. Unfortunately, it came just at a time when support for academic institutions generally and academic science in particular was waning in popularity in the face of other pressing problems. Indeed, it might well be thought of as a casualty of the Vietnam war. I still hope that sometime ideas of this general sort may again be brought forward.

At the start of 1967, the members of the full Committee on Science and Astronautics included the following:

Democrats	*Republicans*
George P. Miller, California, *Chairman*	James G. Fulton, Pennsylvania
Olin E. Teague, Texas	Charles A. Mosher, Ohio
Joseph E. Karth, Minnesota	Richard L. Roudebush, Indiana
Ken Hechler, West Virginia	Alphonzo Bell, California
Emilio Q. Daddario, Connecticut	Thomas M. Pelly, Washington
J. Edward Roush, Indiana	Donald Rumsfeld, Illinois
John W. Davis, Georgia	Edward J. Gurney, Florida
William F. Ryan, New York	John W. Wydler, New York
Thomas N. Downing, Virginia	Guy Vander Jagt, Michigan
Joe D. Waggonner, Jr., Louisiana	Larry Winn, Jr., Kansas
Don Fuqua, Florida	Jerry L. Pettis, California
George E. Brown, Jr., California	D. E. (Buz) Lukens, Ohio
Lester L. Wolff, New York	John E. Hunt, New Jersey
William J. Green, Pennsylvania	
Earle Cabell, Texas	
Jack Brinkley, Georgia	
Bob Eckhardt, Texas	
Robert O. Tiernan, Rhode Island	

The members of the Subcommittee on Science, Research and Development at the start of 1967 included:

Democrats	*Republicans*
Emilio Q. Daddario, Connecticut, *Chairman*	Alphonzo Bell, California
John W. Davis, Georgia	Charles A. Mosher, Ohio
Joe D. Waggonner, Jr., Louisiana	Donald Rumsfeld, Illinois
George E. Brown, Jr., California	D. E. (Buz) Lukens, Ohio
William F. Ryan, New York	

At the start of 1969, the members of the full Committee on Science and Astronautics included:

Democrats	Republicans
George P. Miller, California, *Chairman*	James G. Fulton, Pennsylvania
Olin E. Teague, Texas	Charles A. Mosher, Ohio
Joseph E. Karth, Minnesota	Richard L. Roudebush, Indiana
Ken Hechler, West Virginia	Alphonzo Bell, California
Emilio Q. Daddario, Connecticut	Thomas M. Pelly, Washington
John W. Davis, Georgia	John W. Wydler, New York
Thomas N. Downing, Virginia	Guy Vander Jagt, Michigan
Joe D. Waggonner, Jr., Louisiana	Larry Winn, Jr., Kansas
Don Fuqua, Florida	Jerry L. Pettis, California
George E. Brown, Jr., California	D. E. (Buz) Lukens, Ohio
Earle Cabell, Texas	Robert Price, Texas
Bertram L. Podell, New York	Lowell P. Weicker, Jr., Connecticut
Wayne N. Aspinall, Colorado	Louis Frey, Jr., Florida
Roy A. Taylor, North Carolina	Barry M. Goldwater, Jr., California
Henry Helstoski, New Jersey	
Mario Biaggi, New York	
James W. Symington, Missouri	
Edward I. Koch, New York	

The members of the Science, Research and Development Subcommittee at the start of 1969 included:

Democrats	Republicans
Emilio Q. Daddario, Connecticut, *Chairman*	Alphonzo Bell, California
John W. Davis, Georgia	Charles A. Mosher, Ohio
Joe D. Waggonner, Jr., Louisiana	D. E. (Buz) Lukens, Ohio
George E. Brown, Jr., California	Larry Winn, Jr., Kansas
Earle Cabell, Texas	Jerry L. Pettis, California
Bertram L. Podell, New York	
James W. Symington, Missouri	

CHAPTER VI

Gemini and Apollo

The night before Neil Armstrong, Buzz Aldrin, and Mike Collins blasted off on their trip to the Moon, NASA arranged a large dinner party at the Cocoa Beach, Fla., Country Club. As a prelude to the highly successful Moon flight on July 16, 1969, the dinner was a memorable affair because it brought together once again many of those who had worked for years toward the goal about to be realized. House Majority Leader—later Speaker—Carl Albert recalls the occasion vividly.

NASA Administrator Thomas O. Paine was at the microphone introducing the leading Members of the House and Senate who had played a big part in the program, as well as those NASA officials who could spare some time away from the blockhouse or control room. Warm applause greeted some of the congressional leaders and their wives. Then something very unusual happened, according to Carl Albert. When Tiger Teague came in, Tom Paine raised his voice to proclaim, "And now, Mr. Manned Space himself, the guy who really put this show on the road * * * Tiger Teague!" The crowd went wild with sustained applause. It was basically a tribute by those who worked on Project Apollo, who were expressing their appreciation not only for the unflagging support and leadership given to the program, but for the man himself who had done so much to put Apollo across in Congress and the Nation.

As chairman of the Manned Space Flight Subcommittee, Teague was in a position of leadership where he and his subcommittee could make or break the manned lunar landing program, not only so far as NASA was concerned, but even more important in the rest of the Congress and the Nation.

Dr. Wernher von Braun, who always measured his words carefully when assessing a contemporary, put it this way: "Without 'Tiger' Teague's unwavering support our Apollo astronauts would never have landed on the Moon."

One of Teague's most important achievements as chairman of the Manned Space Flight Subcommittee was his ability to educate other Members of Congress so they would understand and vote for funding the space program. His leadership, strategy, and tactics closely resembled the pattern he had developed as a battalion commander in combat, when he went out personally to reconnoiter enemy positions and went up and down the frontlines talking with his men prior to issuing combat operations orders.

After finishing Texas A. & M. College in 1932 and being commissioned a reserve second lieutenant in the infantry, Teague assumed a full-time job at the post office in College Station, Tex. During the depression years, he rose to become superintendent of that office until his enlistment as a first lieutenant in the Army on October 5, 1940. With the famous "Cross of Lorraine" 79th Division, Teague went into combat almost immediately after the Normandy invasion in 1944. As a battalion commander, his outfit was engaged in intensive combat in the Normandy hedgerows after landing on Utah Beach. For 120 successive days of bloody fighting, Teague's battalion had no rest, and in the battles across France toward the German border one-third of the battalion was killed and one-third injured. Teague himself was wounded six times, the most serious occurring on December 18, 1944, as he was reconnoitering alone near the Siegfried line. Shrapnel tore his left ankle, and another shell's fragments entered his lower back. Fashioning his own tourniquet from a lace from his right shoe, he crawled back on his own power, but that was the end of combat for Tiger. He then had many operations and two years in Army hospitals.

His war wounds eventually led to the loss of his left leg, in 1977, but before then a special rocker-type shoe enabled Teague to master his disability and also become one of the undisputed paddle-ball champions of the House of Representatives.

While still in McCloskey Army Hospital in Temple, Tex., in 1946, Teague made his decision to run for Congress. Congressman Luther Johnson from the Sixth Congressional District in Texas was named a Federal judge, and a special election was called. "Some of us in the hospital had done a lot of talking about the war and the Government, and I just thought I would try it," Teague said casually, starting toward a legislative career which would span 32 years of service. As one of the most-decorated combat veterans in Congress, Teague was a natural to rise to become Chairman of the Veterans' Affairs Committee. Along the road to the top, he opposed and helped narrowly defeat a giveaway pension bill which would have milked the Treasury for an eventual $125 billion. From the time he became Veterans' Affairs Committee chairman in 1955 at the age of only 44, he shepherded through Congress over 200 bills which he sponsored, almost all of which went through by crushing majorities.

Stocky and barrel-chested, Teague's wide popularity stems from a variety of sources. He has the knack of going right to the heart of issues without a lot of the palaver which is the bill of fare of some politicians.

Whether fishing, playing gin rummy, or presiding over a committee meeting, Teague is frank and direct. But he can be blunt and sharp, too.

Teague's sense of humor is robust and unforgettable. Early in 1959, Maj. Gen. Bernard A. Schriever, who headed the Air Force Ballistic Missile Command, was briefing the committee in a classified session. Suddenly, Teague lowered his voice and asked in a confidential tone: "May I say one thing in complete secrecy?" A sudden hush fell over the committee room, as several Members leaned forward expectantly. In a dramatically sepulchral voice, Teague announced:

> The general went to the same school I went to and graduated one year ahead of me. It is one of the best schools in the country.

Teague has successfully preserved nonpartisanship and bipartisanship on the Science Committee. Thus, when Teague chose Astronaut Jack Swigert as the committee staff director in 1973, Swigert warned him: "Before you hire me you should know that I'm a registered Republican." Teague's immediate response was: "Jack, I don't give a damn but if you ever mention Republican or Democrat in that committee, I'll fire you. You should know that."

Chairman Teague (left) and Neil A. Armstrong, the first man to set foot on the Moon.

Teague's rapport with Republicans as well as Democrats is best evidenced by President Ford's remarks on February 27, 1975, at the unveiling of a portrait of Chairman Teague in the Science Committee hearing room. The President said:

> I think you are all familiar with the slogan, "Put a tiger in your tank. * * *" I think America can be mighty grateful that 29 years ago some Texans put a tiger in the House * * *. From what we know of those who dealt with him, words of trust and honor—they were the sort of thing that Tiger believed in and acted on and respected * * *. In my younger days, there was a popular song with the words "Hold that tiger." Ladies and gentlemen, here is one Tiger you will never hold.

The nickname "Tiger" had come naturally to Teague, a 125-pound quarterback on his high school football team at Mena, Ark. Born April 6, 1910, on a wheat farm near Woodward, Okla., he and his family moved fairly early to Arkansas where his father ran lumber camps in the Ozarks. Tiger spent his summer vacations while in high school loading log wagons, driving mule teams, or firing the boilers which powered the saws. He worked his way through Texas A. & M. College, where he studied animal husbandry and for 25 cents an hour fed the college's show calves, shoveled out the stalls, did other odd jobs at the local post office and sold tickets for the Missouri Pacific Railroad.

One observer, commenting on the difference between Chairman Miller and Chairman Teague, mentioned that with George Miller "He liked to discuss so many different things. If you went in and you wanted to get an answer from him, you would often spend 30 or 40 minutes in his pleasantries and his discussions about history or to see his latest gadget or model and listen to him explain all that, and then in two or three minutes you would explain your problem and then you would get your business done. You always got your business done even though it took a long time. Mr. Teague says: 'Come on in, say what you have to say and get out.' And he does it in such a way that you don't mind it at all."

In 1963, NASA's Associate Administrator for Manned Space Flight, Brainerd Holmes, had an internal disagreement with Administrator James E. Webb. Holmes, who was very popular with both the committee and Tiger Teague, was replaced by Dr. George E. Mueller who explains that:

> I was aware that Holmes had considerable support in Congress. Anticipating major difficulties working with the committee, I flew to Washington and had my first meeting with Tiger Teague. He didn't pull any punches.
> "I don't like what happened to Brainerd Holmes," he said without any preamble, "but I believe in supporting the job, not the man. I don't have any personal opinion about you, but as long as you do the job, I'll support the office. You should know one thing, though. If you double-cross me once, it's your fault. If you double-cross me twice, it's my fault—and I never have that problem."

Those few words certainly cleared the air, and from that moment forward, Tiger Teague never wavered in his support for me or for the program. Those early days couldn't have been easy for him, though, because I was proposing some major changes in the way the manned space program was to be run.

Dr. Mueller in 1963 proposed what he termed a "politically explosive" reorganization which would take away the autonomy of NASA's three operating centers at Huntsville, Cape Canaveral, and Houston, being run by three very strong-minded individuals— Wernher von Braun, Kurt Debus, and Robert Gilruth. Mueller suggested to the committee that he proposed to centralize authority and direct the program from above. He relates:

I certainly couldn't predict how Wernher von Braun, Kurt Debus, and Robert Gilruth would react to my reorganization plans, but it would have been naive not to expect strong—and loud—opposition.

When I told the committee what I planned to do, there was a long silence as each Member considered the nasty situation which might develop. Finally, Tiger broke the silence: "If that's what you think you have to do, go to it."

BRIEFING CONGRESS ON THE SPACE PROGRAM

High on Teague's personal priority list was his very strong emphasis on providing other Members of Congress with information on the value and importance of the space program. Not only were Congressmen made aware of the contracts and dollars which were pouring into their districts, but Teague also made sure that Members were briefed on all the up-to-date details on the new plans and projects which affected their areas.

As Dr. Mueller indicated:

The committee perceived that one of its primary functions was to provide Congress with a window into the manned space program. This was no easy task because the program was incredibly complex and involved the cutting edge of technology * * *.

Each year, just before congressional hearings, Teague and his subcommittee would go on a fact finding trip. They would visit the operating centers and major contractors throughout the country. It was a grueling trip, but it enabled Tiger to find out where things stood and what was needed. The effort paid off, too; the committee had outstanding success in influencing Congress to vote for the needed appropriations. The appropriations proved to be reasonable, too. It must be remembered the entire $26 billion manned space program was performed within the budget originally set in 1961.

Another facet of the education process which Chairman Teague emphasized was to persuade Congressmen who were critical of the space program to visit NASA installations, especially for the exciting manned space flights. One of the sharpest critics of the NASA program was Representative Ben Jensen (Republican of Iowa), a member of the Thomas subcommittee which handled NASA appropriations. Tall, blond-haired, inclined to be sarcastically cynical about almost every new or expanding Federal program, Jensen was also bitterly anti-Kennedy and therefore even more anti-Apollo. Not long after the John Glenn flight in 1962, Teague persuaded Jensen to come down with him to Cape Canaveral one Sunday night in April 1962. According to Maj. Rocco Petrone, Jensen came in "just absolutely going to tear us apart."

Teague briefed Petrone on how to handle Jensen. "Hey, look, this guy can be rough, can be gruff, can be mean—take it," Teague advised. According to Petrone: "He was going to make sure we didn't say anything mean back to him. He was giving us fatherly advice." Jensen observed a test firing of the first stage of Saturn generating 1.3 million pounds of thrust. He was not only impressed, but also agreed to pose with Teague in front of the gantry, smilingly demonstrating his approval. More important, when the NASA authorization bill was debated on the floor on May 23, 1962, the following colloquy occurred:

Mr. JENSEN. Mr. Chairman, will the gentleman yield?

Mr. TEAGUE of Texas. I am glad to yield to the gentleman.

Mr. JENSEN. I want to commend the gentleman from Texas (Mr. Teague) for the great interest he has taken in this space program. It was my pleasure to be in the gentleman's company at Cape Canaveral a couple of weeks ago when the Saturn was launched.

There, for the first time, I had the pleasure of meeting and visiting with Dr. von Braun and Dr. Debus, two German scientists who are perhaps the greatest authori-

Representative Olin E. Teague (Democrat of Texas) talks with Dr. Kurt H. Debus (right) on one of Chairman Teague's many visits to the John F. Kennedy Space Center, where Dr. Debus served as Director.

ties on missiles and space exploration in the world. Whenever the gentleman from Texas (Mr. Teague) has a day to spare, he is there visiting and getting more information about this great space program. I compliment the gentleman most highly for the great interest he has taken and for the fine presentation he has just made which I am sure is of the greatest importance to the future of our country.

Mr. TEAGUE of Texas. I thank the gentleman.

PERSONNEL OF MANNED SPACE FLIGHT SUBCOMMITTEE

When the Manned Space Flight Subcommittee was first established in early 1962, the following members were assigned:

Democrats	*Republicans*
Olin E. Teague, Texas, *Chairman*	James G. Fulton, Pennsylvania
Emilio Q. Daddario, Connecticut	R. Walter Riehlman, New York
Thomas G. Morris, New Mexico	Richard L. Roudebush, Indiana
William F. Ryan, New York	

After the 1962 elections, the subcommittee was enlarged with the addition of new members, and when the committee was organized early in 1963, the following were assigned to the Manned Space Flight Subcommittee:

Democrats	*Republicans*
Olin E. Teague, Texas, *Chairman*	James G. Fulton, Pennsylvania
Emilio Q. Daddario, Connecticut	R. Walter Riehlman, New York
Bob Casey, Texas	Richard L. Roudebush, Indiana
Joe D. Waggonner, Jr., Louisiana	Alphonzo Bell, California
Edward J. Patten, New Jersey	Edward J. Gurney, Florida
Don Fuqua, Florida	

As the most glamorous, most senior, and most active subcommittee with the biggest budget and the greatest focus for publicity, it was natural that all members of the full committee muscled a little with each other to try and gain assignment to the Manned Space Flight Subcommittee or get transferred from one of the other subcommittees.

1963: THE FIRST YEAR OF STRONG OPPOSITION

Less sophisticated observers, as well as some officials in NASA itself, viewed the role of the Science Committee essentially in terms of a group of laymen who were educated through briefings on technical details; who then voted certain changes—almost always downward—in the budget figures presented to them; and who occasionally expressed opinions on certain priorities. If anything were done too slowly, or in a fashion to cause adverse publicity, or if there were excessive cost overruns, or public washing of any dirty linen, congressional investigations were warranted.

Almost all observers and critics of the committee's work, including most of the personnel in NASA itself, failed to recognize one of the most important roles of the Science Committee which Teague always stressed: the education of Congress and the country on the value of the space program. This in particular meant the persuasion of a majority in the House of Representatives that the program merited continued support. For several reasons, the role of the Manned Space Flight Subcommittee was crucial. In the first place, every Congressman understood that the objective of a manned lunar landing by the end of the decade, first enunciated by President Kennedy on May 25, 1961, was the top priority of the space program. Second, the Manned Space Flight Subcommittee was assigned the major hunk of the NASA budget—about $3 billion, or more than half of the entire NASA expenditures. Third, unmanned space science and advanced research, while not directly related to the manned lunar landing, were certainly assisted and spurred along by whatever popular support could be generated by the lunar program.

Aside from the launching visits and direct contacts with the astronauts, which he constantly encouraged for all Congressmen, Teague also began on an informal basis to talk with as many noncommittee members as possible to help forge the majority necessary to win the authorization and appropriations battles. He also deputized his subcommittee members, and other members of the full committee, to undertake as much missionary work as they had time to do.

Selling the space program to Congress was no easy task, and Teague and his subcommittee shouldered the heaviest share of the burden. Up to 1962, this was comparatively easy; the shock of Sputnik and Gagarin's flight had not yet worn off, and John Glenn and the other Mercury astronauts had made the program easy to sell. But in 1963, the first real opposition surfaced in Congress.

Congress in 1963 was reflecting incipient dissent from many groups and areas throughout the country, and this dissent expressed itself in several different ways. A large group of scientists began vocal criticism of the Moon program, advocating reallocation of NASA's resources to the unmanned aspects of space, including more emphasis on instrumented landings on the Moon. Writing in Science magazine on April 19, 1963, Philip H. Abelson editorialized:

> If a scientist is not among the crewmen, the alternative of exploration by electronic gear becomes exceedingly attractive. The cost of unmanned lunar vehicles is on the order of 1 percent of the cost of the manned variety; unmanned vehicles can be smaller and need not be returned.

The mood in the country was gradually changing also. The Bay of Pigs disaster and the Gagarin flight in 1961 shocked the Nation to demand positive action to overtake the Soviets. Somehow, the success of making Russia back down during the Cuban missile crisis in 1962, plus American successes in the Mercury program, had a slightly lulling effect on our gung-ho enthusiasm for a crash program in space. The successful Russian flights by 1963 were viewed with more mature and objective reactions.

House Republican Leader Charles A. Halleck (Republican of Indiana) released a letter of protest from former President Eisenhower, which was printed in the April 2, 1963, Congressional Record and contained this warning:

> The space program, in my opinion, is downright spongy. This is an area where we particularly need to demonstrate some common sense. Specifically, I have never believed that a spectacular dash to the Moon, vastly deepening our debt, is worth the added tax burden it will eventually impose upon our citizens. This result should be achieved as a natural outgrowth of demonstrably valuable space operations. But having made this into a crash program, we are unavoidably wasting enormous sums. I suggest that our enthusiasm here be tempered in the interest of fiscal soundness.

The New York Times of June 13, 1963, reported that former President Eisenhower, at a breakfast meeting with Republican Congressmen, had bluntly characterized the projected Moon flight as "nuts."

The very size and steep increases in the NASA budget alarmed many Congressmen. To leap from $1.7 billion to $3.8 billion and then to $5.7 billion over the calendar years from 1961 to 1963 terrified those accustomed to pruning budgets and cutting out waste.

Despite the fact that Teague's subcommittee slashed NASA's manned space flight requests by some $300 million—close to a 10-percent cut—opposition began to form in preparation for the floor fight over the authorization. Six Republican committee members filed "Additional Views" on the bill, even though they voted for the bill both in committee and on the floor. The six were as follows:

Richard L. Roudebush, Indiana	James D. Weaver, Pennsylvania
Thomas M. Pelly, Washington	Edward J. Gurney, Florida
Donald Rumsfeld, Illinois	John W. Wydler, New York

In their additional views, the six committee members attacked the emphasis on outer space to the exclusion of more stress on the military advantages of "inner space"—between 100 and 500 miles from the Earth. They also opposed NASA facility and training grants, and succeeded in cutting NASA grants to educational institutions from $55 million down to $30 million in an amendment on the floor. Further opposition surfaced to the proposed Electronics Research Center.

The fight over these two issues is detailed in the next chapter. Among other items of opposition, the six members also attacked both lack of committee staff and lack of a minority staff. Their additional views pointed out that NASA had the fourth largest budget of any Government agency, "yet the Science and Astronautics Committee, with the task of overseeing the operations of NASA, has but 10 professional staff members, the smallest staff in Congress. * * * This situation constitutes a weakness in the system of checks and balances. Here is an instance where the legislative branch, because of inadequate staff, is unable to keep watch on a huge executive agency."

In pleading for specific staff assigned to minority, the six members noted that "It is absolutely vital that all staff members are reasonably available to all the minority members of the committee. The present staff is overburdened with the result that it is difficult for them to be of assistance to minority Members." Representative James G. Fulton (Republican of Pennsylvania) joined in the plea for a special staff assigned to the minority, a reform which was resisted for many years by the committee majority.

As the leadoff speaker to open the critical debate on the NASA authorization bill, on August 1, 1963, Teague brought models of the Saturn boosters and spacecraft onto the House floor. He refuted the arguments that we were just going to the Moon to collect some rocks and lunar soil. He pointed out:

I do not favor the program because it is a glamorous technological exercise, or simply because it would flatter our vanity to beat the Russians at the space game. There would be no excuse whatsoever for such a frivolous expenditure of the taxpayers' money.

No, Mr. Speaker, I am heartily and completely in favor of this program because it is an essential part—but only one part—of our entire space program.

Because the idea of putting human beings on the Moon is so glamorous, too many people think of it as an entire program in itself. That is wrong. Our goal is to be first in every area of space research, development, and exploration. Our goal is to be the leader in space, just as we always have on land, in the air, and on and under the sea. * * *

There is a further reason why Moon exploration is so important to us. In making the prodigious effort to put a man on the Moon, we are going to have to move forward dramatically in many important fields: science, engineering, industrial development, design, mathematics, biology—the whole spectrum of scientific and technological accomplishment.

Teague also stressed the military aspects of space, and the danger of yielding the mastery of space to the Soviet Union. He then reviewed the practical achievements and benefits which already constituted a spinoff from space spending in the areas of medicine, new fabrics, new metals and alloys, and the whole field of miniaturization as well as the development of computer technology. Commencing in 1960, the Science

Committee published studies of "The Practical Values of Space Exploration" which were frequently updated to reveal productive new spinoffs from the space program.

Chairman Miller next took up the cudgels for the lunar landing program. He compared the pessimism of opponents to the opponents of exploring the land beyond the Mississippi River early in the 19th century. With obvious relish, Miller quoted Daniel Webster:

> What do we want with this vast, worthless area, this region of savages and wild beasts, of shifting sands and whirlpools of dust, of cactus and prairie dogs? * * *
>
> I will never vote one cent from the Public Treasury to place the Pacific coast one inch nearer to Boston than it now is.

In supporting the majority of the Science Committee in its 1963 bill, former Speaker Martin again underlined the fact that "it is not a partisan committee. They give equal treatment to all, no matter what party may be involved. The decisions are fair and impartial. The gentleman from California (Mr. Miller) has always been fair and generous, and he is a good leader."

One of Teague's major accomplishments during the 1963 and subsequent congressional debates was to convince his colleagues that as a consistent supporter of the antispending bloc in Congress, he was not about to vote for wasteful expenditures. He also could demonstrate, through the cuts voted by his own subcommittee and the other subcommittees, that the Science Committee was rigorously investigating the NASA budget request and was taking the initiative to make the necessary reductions. Teague's credentials as a conservative on spending were known and respected. The thousands of pages of hearings, visits to NASA and industrial installations, conferences with contractors, and investigative reports bolstered his case. He also won support by taking a middle-of-the-road position between those who felt the committee was embarked on a wild spending spree through a crash program, and those in NASA who professed that the committee was cutting the space program too deeply. Teague won many friends and supporters by this line of argument:

> I would like to take a moment to try to dispel several extreme notions that a lot of people have about our man-to-the-Moon program.
>
> One of these notions is founded on the allegation that we are proceeding on a crash basis, that we are thereby spending a lot more money than we otherwise would need to and are greedily consuming the bulk of the Nation's scientific talent in the process.
>
> The other notion is based on the allegation, which we have recently heard from NASA's Administrator, that the amount of money requested for the manned lunar landing is a sacrosanct bare minimum which must be left totally intact if we are not to slip badly in our lunar landing schedules and lose money in the bargain.
>
> In my opinion, neither of these allegations will win any awards for accuracy.

Representative Thomas Pelly (Republican of Washington) took the position, even though he finally voted for the bill, that "a great many thinking people are questioning whether the projected date of landing a man on the Moon could not be delayed to better advantage." Pelly was one of the opposition Congressmen whom Teague persuaded to visit Cape Canaveral, where Rocco Petrone took him in hand. Petrone quickly discovered that Pelly was upset because so much of the space budget was being concentrated in Florida rather than in the northwest or the Seattle area which he represented. So Petrone made sure that Pelly rode up the elevator to the top of the Vehicle Assembly Building, where he could see the huge crane marked "Colby Crane Corp.," and Pelly knew immediately that there was a hometown flavor to it. "Those cranes were built by the Colby Crane Corp., and I happen to know they were built in Seattle," Petrone explained.

Miller and Teague effectively lined up their supporters for the final vote, assisted by some strong statements by freshmen Congressmen. For example, Representative Don Fuqua (Democrat of Florida) painted the challenge of the future in these terms:

> Space is the challenge of our time. We stand on the threshhold of advancements such as the world has never seen. As Columbus charted new worlds, as the Wright brothers ushered in a new era, so the American people today, united in a gigantic effort, are charting new worlds of scientific advancement.

When the roll was called, the members of the Science Committee all returned to the reservation and voted for the bill. But some powerful opposition reared its head for the first time. The chairman of the Committee on Rules, Representative Howard Smith (Democrat of Virginia) voted no, as did the Republican whip and former member of the select committee which established NASA, Representative Leslie Arends (Republican of Illinois). Nevertheless, the Miller-Teague forces carried the day on August 1, 1963, by a majority of 335–57.

JOINT U.S.-U.S.S.R. EXPEDITION TO THE MOON?

Before the end of 1963, NASA got into some more funding trouble on Capitol Hill as a result of President Kennedy's recommendation, in a September 20, 1963 address to the U.N. General Assembly, that there should be United States-U.S.S.R. cooperation in space. President Kennedy was more specific, advocating the possibility of a "joint expedition to the Moon." The President asked: "Why should the United States and the Soviet Union, in preparing such expeditions, become involved in immense duplications of research, construction and expenditures?"

President Kennedy's statement hurt NASA's support among some of the strongest friends of the space program. In the midst of considering NASA's appropriation, Chairman Albert Thomas (Democrat of Texas) of the Independent Offices Appropriations Subcommittee (which was responsible for the NASA budget) protested to the President on September 21, and Teague followed up on September 23 with a stinging letter to the President. Quoting the President's May 25, 1961, establishment of the lunar landing goal, Teague asked:

> In view of your statement to the United Nations supporting the possibility of a joint venture with the Russians to reach the Moon, I am very anxious to know whether or not this national goal is being abandoned or changed.
>
> I was disappointed in the suggestion. I have been a very strong supporter of the space program, believing we can be the first nation to put a man on the Moon and knowing that we must achieve this goal if we are to help establish the fact that space will be used for peaceful purposes. Also, I believe that our national security and the security of the rest of the free world is very dependent upon the success of our space program.

Representative Bob Casey (Democrat of Texas), Representative Thomas Pelly (Republican of Washington), Representative Richard Roudebush (Republican of Indiana), and numerous other members of the Science Committee joined in the clamor of opposition to the President's suggestion. When the President answered Representative Thomas' letter with the thought that cooperation would not slow down the space program, a copy was sent to Teague, who again reacted sharply:

> That letter says nothing as far as I'm concerned. * * * I'd just as soon attempt to cooperate with any rattlesnake in Texas.

Teague then fired off a letter to Larry O'Brien, at the White House, who had forwarded to Teague a copy of the President's reply to Thomas:

> Larry, I am the chairman of an 11-man-subcommittee which has the responsibility of the authorization for the manned space flight program. In my opinion, ten of the eleven members of that subcommittee support our manned space program almost completely on the basis of national defense and national security. It is my opinion that this year except for the national security aspect, the subcommittee would have cut this budget in half.
>
> I do not believe the President's letter to Albert Thomas is responsive to the last paragraph of my letter. For that reason, I would appreciate a letter which I may distribute to my subcommittee and which may be placed in the subcommittee authorization hearings which will begin again in January.

President Kennedy knew enough not to start or continue a personal feud with the powerful chairman of the Manned Space Flight Subcommittee. To Teague's September 27, 1963, letter to O'Brien, McGeorge Bundy, the President's adviser on foreign policy, responded on October 4:

Larry O'Brien and I have talked with the President about your letter of September 27th, and the President asked me to send you an interim answer to the important question which you raise.

The relation between national security and the space program is very clear and important in the President's judgment, and he is currently engaged in a major review of the relative roles of different agencies, precisely with the programs for next year in mind. I think, therefore, that we can assure you that there will be new expressions of the administration's point of view in good time for your subcommittee authorization hearings in January.

In the midst of all this furor, NASA's appropriation bill came before the House. The atmosphere was ripe for a $250 million cut below the figure authorized, and when the appropriation process was completed NASA wound up with $5.1 billion; $500 million short of its budget request. Also added was a provision that no funds could be used for "expenses of participating in a manned lunar landing to be carried out by the United States and any other country without consent of Congress."

President Kennedy never did get around to answering Teague's letter directly. But he was obviously stung by the charge that he had abandoned the lunar landing goal. Perhaps this is why, in San Antonio, Tex., on November 21, 1963, the day before he was assassinated in Dallas, he reaffirmed his commitment in these words:

Frank O'Connor, the Irish writer, tells in one of his books how, as a boy, he and his friends would make their way across the countryside; and when they came to an orchard wall that seemed too high to climb, too doubtful to try, too difficult to permit their journey to continue, they took off their caps and tossed them over the wall and then they had no choice but to follow them.

My friends, this Nation has tossed its cap over the wall of space and we have no choice but to follow it. Whatever the difficulties, they must be overcome.

VISITOR CENTER AT THE CAPE

When Cape Canaveral began functioning as a launch center in the 1950's the Atlantic Missile Range was controlled by the Air Force. For reasons of security and safety, the Cape was usually under wraps and even the working press had difficulty in covering activities and launches, many of which were classified. In addition to bringing many Members of Congress to the Cape to educate them through the excitement of seeing actual launches and a chance to get a firsthand feel of the complexities of the program, Teague brought many other visitors on numerous occasions. With some maneuvering, it was usually possible to invite a very limited number of guests to view launches at the cramped facilities available. But for millions of Americans whose tax dollars were supporting the space program, it was either a case of sleeping on the nearby beaches or watching television—which could never quite convey the precise size of the monster boosters which progressed in size from Redstone, Atlas, Titan to Saturn.

Teague's conviction was that the more people who could see what was going on, the more they had an opportunity to learn through asking questions, the better understanding of and support for the program would result. In the spirit of the 1958 Space Act, he constantly lobbied for more openness in the space program, and more liberal policies toward admitting the general public to the Cape. On December 16, 1963, the Defense Department bowed to pressure and began to allow motorists to drive through portions of its 17,000-acre reservation. They were only allowed to drive through during a 3-hour period from 9 a.m. until noon on Sundays, and then only along a marked route a mile or so from the launch pads. Nobody was allowed to stop. Photographs? Yes, if you took them from your moving car without stopping. Even with these restrictions, the public response was enthusiastic.

Teague brought up with Webb the whole issue of public access in 1963. Webb countered that this was a Defense Department responsibility, but that NASA would consider the question of public visits when construction was completed at the Merritt Island spaceport. On New Year's Day 1964, Teague had one of his small-scale persuasion tours of Cape Kennedy for Representatives Joseph Karth (Democrat of Minnesota) and Thomas Pelly (Republican of Washington). Shortly after returning to Washington and before the 1964 hearings had gotten underway, Teague decided to formalize his campaign with a lengthy letter to Webb, dated January 10, 1964, which said in part:

I would like to bring up a matter which, it seems to me, is of increasing importance—and that is the problem of permitting visitors to make a tour of the general Cape area and the Space Center. There is no question that the pressures are growing for a more liberal policy in this respect, and to permit the average American to get a glimpse of what is going on at this major center of our space effort. Indeed, it seems to me that the Cape has already become an area of national interest and that if more people were permitted to visit it and see for themselves, our space program might receive much benefit in the way of public support.

I realize that there have been good reasons for the limitations imposed thus far and, also, that a plan to open the Cape to tourists would have to be carefully conceived so as not to disturb operations there or result in danger to visitors. Nevertheless, it seems to me that plans of this nature could be studied, produced, and put into operation. I would like to discuss this matter with officials at the Cape during the forthcoming visit of my subcommittee and I would appreciate your views on the subject and would be grateful for any comments you might have to make prior to that time.

In an extended reply, Webb concurred with Teague's suggestions. He promised that when the Merritt Island launch area became operational, a year hence, tours would be supplemented by written and oral explanations of the work in progress and of our programs.

To follow up the recommendations, the Teague subcommittee recommended in its March 11, 1964, report:

The public interest in the Kennedy Space Center is now of such proportions and of such a nature as to place the Cape almost in the category of a national monument or park. With this in mind, the subcommittee has added $1 million to be applied to the construction of facilities authorization for the Kennedy Center for fiscal 1965. This amount would provide for the construction of public facilities which NASA must have if it is to make the Cape available to the public in any real sense.

Early in 1965, Dr. George E. Mueller, NASA Associate Administrator for Manned Space Flight, asked the National Park Service to recommend an appropriate visitor program. There were some queasy feelings in NASA (as sometimes happens when an agency or individual does not think up a good idea first themselves), but after several years of shifting from one foot to the other the idea began to take form through designs and advance planning work. While all this was going on, and before the Visitors Center was constructed, an interim program of bus tours was started on July 22, 1966. It just so happened that Chairman Teague was at the Cape that day, and he was delighted to note that despite having to stand in the rain there were 1,500 people who took the tours. Within a few years, the proper facilities were constructed, and the Kennedy Space Center contracted with TWA to operate the Visitors Center.

Starting in 1969, the number of visitors topped 1 million, reached a peak of 1,736,302 in 1972, and has exceeded well over 1 million every year. Teague's determination to prod NASA also had its effect in the establishment of similar visitor facilities at other centers, where they were overwhelmingly successful, and also carried out the spirit of the program "for the benefit of all mankind."

Mueller had these conclusions on the public impact of the visitor program first pushed by Chairman Teague:

Teague and the committee believed in the manned space program and worked tirelessly on its behalf. Recognizing that public support was essential, the committee stressed the importance of the program and sought to make its complexities comprehensible to the public. Tiger would, from time to time, introduce groups of influential private citizens to key figures in the space program. He was convinced that the more people knew about the program and its goals, the more they would support it. This conviction, and the steps he took to put it into action, was a vital element in the total success of our program to place the first man on the Moon.

MANNED SPACE PROGRAM COMES OF AGE

When Chairman Teague opened the hearings of his subcommittee on February 18, 1964, he made a brief and pointed introduction:

Dr. Mueller, it is a pleasure to welcome you and the members of your staff to the hearings on manned space flight.

I believe that this year, above all other years, represents a historic turning point in the program. In your hands rests the task of bringing to final reality objectives which have now reached a sharp focus.

Plans of the past must become the crucial experiments and hardware of today.

I encourage you to use these hearings to maximum advantage—to get the program before the committee and the people.

The one-manned Earth orbiting Mercury program had been concluded in 1963. The two-manned Gemini program had slipped a year. But the three-manned Apollo lunar program was progressing nicely, with the first unmanned Saturn I launch on January 29, 1964. For the first time, the United States could claim the ability to orbit a heavier payload than the Russians. (Thinking in appropriations terms, Representative Albert Thomas, who did not customarily commit slips of the tongue, referred to the "payroll" orbited by the Saturn). By 1964, Mueller had taken a firm grasp on a reorganized NASA manned space effort. And President Johnson in his State of the Union and Budget Messages made it clear that he intended to support the goal of a manned lunar landing before the end of the decade.

After its usual round of extensive visits to contractors and field installations, the Manned Space Flight Subcommittee in 1964 recommended only minimal reductions of $41 million out of a grand total of $3.541 billion requested. This prompted Daddario to observe how lean the budget actually was, and that "as we have looked at it, there is no question but that further cuts will in fact be expensive. They will cost the country more * * *. There is no doubt in our minds but that cuts at this time will stretch out the program to the point where it will not only be more costly, but perhaps will prevent us from accomplishing our objectives before the end of this decade."

Mosher agreed, pointing out that for the first time witnesses were able to give more concrete answers, thus eliminating some of the guesswork. "In the beginning, we were doing some drastic cutting because every one of us were doubtful as to exactly what they really needed, and today they are giving us the information necessary for us to come to conclusions and I think it is a very healthy and good sign," Mosher reported to the full committee in executive session.

In 1964, the opposition votes on the floor showed an increase, as the House sustained the recommendations of the Science Committee by a vote of 283–73 on March 25. But the committee succeeded in beating back the only two amendments which were offered, further indication of the faith which the House had in the committee's thorough groundwork.

CRITICAL ISSUES COUNCIL

At Cape Kennedy on May 28, 1964, Chairman Teague and members of his subcommittee witnessed the successful firing of a Saturn I two-stage launch vehicle, which boosted into orbit an Apollo "boilerplate" spacecraft. The roar of the Saturn on blast-off failed to drown out the noise and excitement created by a statement on the same day by the Critical Issues Council, of the Republican Citizens Committee (a group presuming to speak for elements of the Republican Party), calling on the United States to abandon the 1970 goal of a manned lunar landing. One sentence in the declaration particularly infuriated Chairman Teague:

> The exploration of our universe is a goal too vast, too hazardous, too costly, and too important to all mankind to be financed and conducted by one country alone, and least of all in an atmosphere of unfriendly competition.

Having just emerged from the bitter fight over a President's suggestion that the United States and the U.S.S.R. undertake a joint lunar landing program, Teague was in no mood to have the issue revived. On the same day as the Saturn shot and the same day of the council's statement, Teague fired off a telegram to Dr. Milton Eisenhower, chairman of the council:

> This wire is to invite you and your committee or any member of your committee to appear before the Manned Space Flight Subcommittee of the House Committee on Science and Astronautics and present any evidence which you have which would indicate that our space exploration program is too vast, too hazardous, too costly, and too important to all mankind to be financed by one country alone.

Speaking from Cape Kennedy, Teague blasted the council as "doing a disservice to a very successful American space program." He pointed out that "it is not a crash program, but a very austere program and a program that is making excellent progress." He added that any slowdown in the program would eventually increase its costs to the country.

In his response to Teague's telegram on June 2, 1964, Dr. Eisenhower neatly sidestepped the issue and suggested that President Johnson should convene "leading experts" to reevaluate the program:

> When the Critical Issues Council decided to study the space program, two of its members agreed to form a study group with the understanding that they would be free to consult with leading scientists, engineers, and science administrators, regardless of their party affiliations. Most of those who were consulted wished not to be identified publicly. The statement issued by the Council represented a consensus of these specialists.
>
> The members of the Critical Issues Council believe profoundly that the space program should be kept out of partisan politics and a careful reading of its statement will show that it adheres to this view. I believe and am so recommending to him, that President Johnson should bring together leading experts in the field, with no reference to political affiliation of the advisers, and ask these specialists to reevaluate the program, a major portion of which is praised by the Council's statement.
>
> I realize that the responsibility which your committee has in this matter and my hope would be that the judgments of a Presidential advisory group would be presented to your committee in harmony with normal governmental procedures.

After his return from Cape Kennedy, Chairman Teague would not let the matter rest there. On June 10, 1964, he wrote the following letter to Adm. Arleigh A. Burke and other members of the Critical Issues Council:

> Recent newspaper articles quoted members of the Critical Issues Council as stating that "our space exploration is too vast, too hazardous, too costly and too important to all mankind to be financed by one country alone." On behalf of the Manned Space Flight Subcommittee, I wish to invite you to appear before that subcommittee and present any evidence which you have that would support this statement.
>
> The subcommittee would be pleased to convene at your convenience to discuss further your views.

In addition to Admiral Burke, Teague's June 10 letter was also sent to James H. Douglas in Chicago, Dr. T. Keith Glennan (the first Administrator of NASA) in Cleveland, Lewis L. Strauss in Washington, and Gen. Lauris Norstad (retired) in New York. There is no record that either Mr. Douglas or Dr. Glennan ever replied formally.

Admiral Burke and General Norstad both called attention to Dr. Eisenhower's June 2 response to Chairman Teague and indicated their agreement with that response. General Norstad added: "I, myself, can claim no particular competence in this field." Lewis Strauss, a former member of the Atomic Energy Commission, deplored the fact that more attention was not being paid to the military aspects of space, and stated: "I believe that the only basis upon which our current large expenditure of funds can be justified is national defense."

Since none of the combatants wanted to come out and fight, Teague had called their hand successfully. But the phrase "too vast, too hazardous, too costly, and too important to all mankind to be financed by one country alone" stuck in Teague's craw for many years after 1964. In fact, as long as Teague was associated with the space program, he never forgot it.

For example, during the floor debate on the NASA authorization bill on May 2, 1968, Teague made this observation:

> Mr. Chairman, if the pioneers who settled this country and made it great had been modern-day Republicans, they would never have crossed the Ohio River. The Pacific Ocean would be still an unconfirmed rumor and any attempt to reach the manifest destiny of America could have been a project too vast, too hazardous, too costly, and too important to undertake.

BIPARTISAN SUPPORT FOR MANNED SPACE FLIGHT

Despite the brief scuffle in 1964 with the Critical Issues Council, the work of the Manned Space Flight Subcommittee—and, in fact, the full committee also—remained essentially bipartisan. Even on issues where the Republicans on the committee took the lead, like opposition to the Electronics Research Center, there was also substantial Democratic cooperation.

Throughout his tenure as subcommittee chairman, Teague continued to go out of his way to enlist Republican interest and support, and the Republican members reciprocated by working vigorously toward the committee objectives. In presenting the recommended funding for manned space flight in both 1964 and 1965, Teague told his House colleagues:

This bill * * * is not a partisan bill. The space program was begun in a Republican administration and continued in a Democratic administration. Republicans and Democrats alike have supported it through the years. It is an American program, designed to place our country in its rightful position before the nations of the world.

Nevertheless, there were some issues which troubled Republicans on both the Manned Space Flight Subcommittee as well as the full committee. Throughout the 1960's, Republicans on the committee assumed leadership in agitating for a more adequate staff to arm all Members with the tools necessary to exercise meaningful oversight. Coupled with this request was the recommendation that there should be a staff for the minority members of the committee. The arguments advanced for both of these objectives overlapped somewhat, and were annually repeated in "Additional Views" printed every year in the back of the authorization reports. In the early years, there were many Democrats who both openly and behind the scenes advocated an increase in the size of the staff. As it became apparent that Chairman Miller strongly opposed staff expansion, most of the Democrats quietly abandoned the issue and left it to the Republicans annually to beat their chests in futile anger.

Chairman Miller favored the Daddario Subcommittee on Science, Research and Development not only with high quality permanent staff, but encouraged the use of contracts and outside scientific assistance which did the job needed for the Members. In the case of the Manned Space Flight Subcommittee, Chairman Teague made up for a great deal of the staff shortage through his very active, personalized custom of subcommittee visits to contractors, NASA installations and monthly conferences at the Office of Manned Space Flight in NASA Headquarters.

MINORITY STAFF

As has been noted, Fulton annually brought out in committee executive sessions, in "Additional Views" in committee reports, and during House debates on the authorization bill his opinion that there should be more committee staff, there should be a clearly defined minority staff, and that legislation should establish an "Inspector General" for NASA. Fulton proved to be a man ahead of his times; the committee staff eventually grew in the 1970's, a minority staff was authorized, and a 1978 statute provided for an "Inspector General" for NASA and other Federal agencies. Although former Speaker Martin remained the ranking Republican on the committee until his defeat in the Massachusetts primary of 1966 and his subsequent retirement, Martin yielded to Fulton on most questions of minority policy. Fulton's abrasive personality and tendency to shoot from the hip exasperated his fellow Republicans, not to mention his Democratic colleagues. Thus the full force of a unified minority bloc was rarely brought to bear on behalf of resolving the staff issue, except in the forum of a committee report.

During the 1960's, younger and more aggressive Republican members like Rumsfeld, Wydler, and Roudebush joined Bell and Mosher to raise a chorus of protests against a lack of minority staff. Fulton, as the senior Republican spokesman, repeatedly badgered Chairman Miller on the issue. When the subject was brought up in public, Miller usually tried to change the subject, displaying either angry irritation or amused tolerance in unpredictable mixtures. Miller and Ducander both had been trained under the tutelage of "Admiral" Carl Vinson, who would never deign to allow a minority staff and decreed that the staff should serve members of both parties equitably.

A combination of factors, including some developments totally outside the committee, finally helped achieve a breakthrough for the Republicans. The Madden-Monroney Joint Committee on the Organization of Congress reported its recommendations for congressional reform on July 18, 1966, including the stipulation that two professional and one clerical staff be assigned to the minority on each standing committee on request. Although the Senate passed the reform recommendations, they remained bottled up in the House Rules Committee until 1970 when the law was finally passed. But even then, with an almost solid Democratic vote, the House acted quickly to repeal the minority staff provisions before they could actually take effect in January 1971. Minority staff was a priority Republican goal. "Effective criticism from the loyal opposition is essential to good government," stated Representative James A. Cleveland (Republican of New Hampshire), in the book *We Propose: A Modern Congress.*

Journalists and political scientists interested in congressional reform began to turn out articles and stir discussions which generally favored the concept of minority staffs. Fulton and his allies became bolder and more frequent in challenging Chairman Miller. Finally, according to Ducander, Miller confided to him:

"I cannot stand that man coming to me and worrying me. Let's give him one goddam minority staff member. * * *" This went right against George Miller's grain. He could not stand to have Fulton on his back any more.

Ducander added:

When you get right down to the nitty gritty, (Fulton) just caused so much goddam trouble that George Miller said: "I cannot stand it." So I called Fulton and said: "The chairman has approved a staff member." He said: "Who do you recommend?" I said: "I do not know."

Fulton hired Richard E. Beeman on June 1, 1968. Beeman resigned on March 19, 1969, to be succeeded by James A. Rose, Jr., who came aboard June 2, 1969, and remained until August 15, 1970. Not until the arrival of Carl Swartz on February 23, 1971, and Joseph Del Riego in October 1971 did the minority have an organized unit which included more than one professional staff member.

During the 1960's, the following were some of the issues which were raised by Republican members of the committee:

Establishment of an Inspector General for NASA.
Opposition to the Electronics Research Center (discussed in the next chapter).
Increased emphasis on military space development.
Creation of Aerospace Safety Advisory Panel.
Greater emphasis on aeronautics, and the fight against aircraft noise.
Opposition to NASA-controlled tracking ships which would constitute a NASA "Navy."
End duplication of Apollo Applications and Manned Orbiting Laboratory.
Opposition to large nuclear rocket (NERVA).
Opposition to M–1 engine development.
Opposition to rapid increase of NASA training grants.
Strong effort to insure that NASA keep the committee better informed in advance of plans and actions.

VICTORIES IN 1965

At the start of 1965, the Manned Space Flight Subcommittee was reorganized to include the following:

Democrats	Republicans
Olin E. Teague, Texas, *Chairman*	James G. Fulton, Pennsylvania
Emilio Q. Daddario, Connecticut	Richard L. Roudebush, Indiana
Joe D. Waggonner, Jr., Louisiana	Alphonzo Bell, California
Don Fuqua, Florida	Edward J. Gurney, Florida
Gale Schisler, Illinois	Donald Rumsfeld, Illinois
William J. Green, Pennsylvania	
Earle Cabell, Texas	

The year 1965 saw one of many victories for the Science Committee members, as they savored the results of their earlier efforts bearing fruit and sustained progress was made toward the lunar landing goal. In contrast to 1964, which had been a year of some frustration, budget slashing, and the long span of inactivity between the last Mercury flight and the first two-manned Gemini flight, 1965 was a banner year.

On February 16, Chairman Miller told his cheering colleagues in the House of Representatives that "this morning the United States took another giant stride in the exploration of space. At 9:37 a.m., a Saturn rocket * * * with its 1,500,000 pounds of thrust, lifted off the launch pad at Cape Kennedy, Fla., on a mission to place in orbit around the Earth, the Pegasus satellite. This was the eighth launch of the Saturn rocket out of eight attempts, a truly outstanding scientific and engineering accomplishment of the men of the National Aeronautics and Space Administration and, of the many contractors who worked so long and hard to make this event a success." Pegasus was the meteoroid detection satellite, which stayed aloft until 1978.

On March 23, Gus Grissom and John Young completed their successful three-orbital flight of Gemini. This set the stage for the debate on the authorization bill on May 6, 1965, and a spirit of great optimism prevailed.

GEOGRAPHIC DISTRIBUTION OF RESEARCH CONTRACTS

The authorization bill which was passed on May 6, 1965, included an amendment which Representative J. Edward Roush (Democrat of Indiana) had inserted in the committee markup of the bill:

> It is the sense of Congress that it is in the national interest that consideration be given to geographical distribution of Federal research funds whenever feasible and that the National Aeronautics and Space Administration should explore ways and means of distributing its research and development funds on a geographic basis whenever feasible and use such other measures as may be practicable toward this end.

The location of NASA installations and the geographic distribution of research contracts were issues which were intensely debated within the committee from the start of the space program. Bobby Baker, in his book *Wheeling and Dealing* alleges that he worked through Senator Robert Kerr (Democrat of Oklahoma) and Vice President Lyndon Johnson to persuade NASA Administrator Webb to intervene on behalf of North American Aviation for the multibillion-dollar Apollo-Saturn contracts, thus enabling Baker to install his Serv-U automatic vending machines in North American plants. No proof of this allegation has ever been forthcoming. But the awarding of large contracts was frequently accompanied by intense argument over whether certain sections of the country were being favored.

Every member of the committee, with varying degrees of success, vigorously represented his own district and State when it came to the awarding of contracts or the funding of programs. Thus it was not unusual to see Miller and Bell active on behalf of some California projects, Mosher raising the flag for Lewis Research Center and Plum Brook, Downing standing up for Langley Research Center, Teague and the Texans plugging for Houston and Dallas, while Fuqua, Gurney, and Frey were interested in pushing everything which happened at Cape Canaveral-Kennedy.

Perhaps the most fascinating story about geography, Congress, and the space program occurred with respect to the location of the Manned Spacecraft Center in Houston. Keith Glennan, NASA's first

Administrator, reported that soon after taking office he had a telephone call from Representative Albert Thomas (Democrat of Texas), chairman of the Independent Offices Appropriations Subcommittee handling NASA's appropriations. Thomas said to Glennan:

> Doctor, I just want you to know how grateful I am that you're willing to come down here and take on the duties that you're taking on, and I want to be as helpful as I can. Now, Doctor, you're going to need some more research and development. I just want to tell you that down there in Houston, there's an institution known as Rice.

Glennan said: "Albert, I know Rice very well indeed."

At this point, Thomas mentioned that Rice had a sizable chunk of land, adding:

> I know Rice would give this land to the Government as a location for a research development laboratory.

Glennan replied:

> But Albert, we don't need any more laboratories. We have all that we need at the moment, and we're building one which was started before I came * * * out at Beltsville.

On subsequent visits to Capitol Hill, Glennan dropped in to see Thomas, and "each time he would bring up this same matter and I would turn it off."

Glennan then reported:

> Finally, I had a call from him one day, and I would have to guess this would be in the spring of 1959.
>
> "Doctor, about that research center matter down in Houston."
>
> I said, "Albert, you know, we've been over this several times and I have told you very frankly that I can see no reason for spending money for this and until there's a need—there may be at some point in time—I'm just not going to think about asking for money for a research center. We're going to finish the one that we've started."
>
> (Thomas responded): "Now, Doctor, let's stop all the horseshit. I've got your budget in front of me." I've forgotten the number but I think he said: "There's $14 million in there for Beltsville."
>
> I said, "Well, you know more about it than I do. I don't remember it in that detail."
>
> (Thomas): "Well, let me tell you, Bud, you won't get a nickel of that unless you put a research center at Houston."
>
> And I had sense enough to react by laughing and saying: "Now, Albert, I think it's about time you and I went out and had a drink."
>
> Well, that ended it, and I never did have any more arguments with him about that.
>
> Now, time passes and I go back to Case in Cleveland and Jim Webb takes over * * *.
>
> So the word was out that there was to be a manned space center someplace, and the Governor of Ohio called me one day and asked me if I would undertake to * * * put together a story nominating Ohio as the site for this center. And I just broke out laughing. I said: "You know, I suppose that there are 25 States doing just this at the present time, and I'll lay you a year's salary that that center is going to Houston."

When Webb succeeded Glennan as NASA Administrator, Thomas enlisted the aid of his old college roommate, George R. Brown, a heavy contributor to Lyndon Johnson's campaigns and a close personal confidant of Johnson. Brown, head of the big Houston engineering-construction firm of Brown & Root, had already been one of the most active consultants for Johnson in the Vice President's mission to prepare a space program prospectus for President Kennedy. On May 23, two days before the President's personal appearance before Congress to announce his Moon decision, Webb wrote Vice President Johnson:

> In other discussions with Congressman Thomas, he has made it very clear that he and George Brown were extremely interested in having Rice University make a real contribution to the effort, particularly in view of the fact that some research funds were now being spent at Rice, that the resources of Rice had increased substantially, and that some 3,800 acres of land had been set aside by Rice for an important research installation. On investigation, I find that we are going to have to establish some place where we can do the technology related to the Apollo program, and this should be on the water where the vehicles can ultimately be barged to the launching site. Therefore, we have looked carefully at the situation at Rice, and at the possible locations near the Houston Ship Canal or other accessible waterways in that general area.

Webb went on to say that California, Chicago, and the research triangle in North Carolina ("in which Charlie Jonas as the ranking minority member on Thomas' appropriations subcommittee would have an interest") were other candidates for space installations. Webb wrote Johnson that "I am convinced, and believe you should consider very carefully, that the merit of this program will attract the kind of strong support that will permit the President and you to move the programs on through the Congress with minimum political infighting."

Thus when many States and communities other than Ohio began to burn the midnight oil to put their proposals together, they had no Keith Glennan to break out laughing and bet: "I'll lay you a year's salary that the Center is going to Houston." Not until after President Kennedy's May 25, 1961, decision to go to the Moon was clearance received for budget approval of the funds. On September 19, 1961, Webb confirmed the fact that Houston had been selected in preference to 20 other cities submitting proposals. Glennan's scenario was accurate.

On April 5, 1962, Teague's Manned Space Flight Subcommittee held a public hearing which included a review of the Houston land deal. As Teague pointed out during the hearing, Humble Oil Co. conveyed most of the land to Rice University free of charge, on condition it would be reconveyed to NASA. When NASA found the need for additional acreage, NASA purchased additional land from Rice which had been conveyed through Humble Oil Co.

As will be noted in the next chapter, political considerations also certainly accompanied NASA's decision to build an Electronics Research Center adjoining MIT in Cambridge, Mass.

Other geographical prizes soon surfaced. Mississippi's Senator John Stennis was understandably pleased that a test facility had been established in his State, thus making more inviting his task of approving NASA's appropriations through the subcommittee that he chaired. In 1961, Science Committee Chairman Overton Brooks was upset that the Michoud launch vehicle assembly plant was located near New Orleans (Representative F. Edward Hébert's district) instead of Shreveport, but at least it was located in Louisiana.

The lion's share of contracts went to the coastal States, and as the have-nots began to grumble in the early 1960's, the haves polished up their rhetoric. "This is no WPA program," Chairman Miller frequently commented, pointing out that even if California were well endowed with space installations, his own congressional district was not being benefited. The coastal States pointed to the need for water transportation and argued on the lofty plane of taxpayer and national interest, demeaning those "grubby" Congressmen who would stoop to snatch at "pork."

The case for fairer distribution of NASA's billions did not come out in the open until 1964. It was sparked by hearings held in Daddario's Subcommittee on Science, Research and Development. Representative Roush started his long fight in 1964. At his own expense, he visited the northeastern office of NASA in Boston, and discovered that NASA personnel were being used to go out and assist contractors and universities in that region to formulate proposals leading to NASA contracts. Roush also pointed out that there was a western regional office which served the same purpose on the Pacific coast, and that Florida, Texas, Mississippi, Alabama, and Louisiana were well-represented by NASA installations in those States. He urged that more balanced attention be given to the Middle West.

In an executive session of the committee on March 17, 1964, Chairman Miller castigated Roush with this comment:

I will say, Mr. Roush, I hate to think of NASA and its activity being put on a parochial ground for any one section or sections of the country. * * * My own section of the country has less work in it than the State of Indiana.

In 1965, Roush did a lot of missionary work and lined up enough support in the committee to incorporate his amendment into the authorization bill. The bill with the Roush amendment passed the House of Representatives.

When the conference committee met, NASA officials approached the conferees and urged that the amendment either be deleted or watered down. As a result, the conference voted to make the amendment read:

> It is the sense of Congress that it is in the national interest that consideration be given to geographical distribution of Federal research funds whenever feasible, and that the National Aeronautics and Space Administration should explore ways and means of distributing its research and development funds whenever feasible.

The conference report, largely through Miller's and Teague's influence, slipped in a sentence indicating that the Senate had "modified the House language to avoid the implication that present governmental procurement philosophy, derived as a result of years of experience, will be materially altered by an overriding consideration being given to geographic distribution of Government funds."

In vain did Roush try to protest that his "very gentle, nudging amendment" was not in any way intended to make geographic distribution an "overriding consideration." But Chairman Miller assured Webb that he had little to worry about if he were concerned about the amendment.

Webb himself, in a letter to Science Committee staff member, Frank R. Hammill, Jr., outlined his philosophy on the amendment:

> To base the award of contracts on geographical considerations, rather than on competition for all companies regardless of location, would be inconsistent with the statutory procurement authority currently applicable to NASA. Moreover, limiting competition to geographical areas might mean that the company with the best capability for a project of importance would not be awarded a contract because of its location.

It was obvious that NASA quickly put its wagons in a circle, and called on assistance from Capitol Hill whenever the issue of fairer geographic distribution came up. In 1966, when NASA sent up the suggested text of a new authorization bill, the geographic section for some strange reason had been quietly dropped. Although the committee then restored the Roush amendment, in practical fact, given the attitude of NASA and Chairman Miller, it didn't amount to a hill of beans.

PASSAGE OF THE NASA AUTHORIZATION BILL IN 1965

The Manned Space Flight Subcommittee made one of its strongest and most convincing presentations on the House floor in 1965. Representative Alphonzo Bell (Republican of California) went into considerable detail to describe how the Committee over the years had pushed hard toward booster capability and rendezvous capability. Representative Don Fuqua (Democrat of Florida) described the process the committee followed in reaching its recommendations:

We were fully aware of our responsibility to the Nation to economize * * *. During the hearings, we probed the justification for each line item. We were guided by this question: "Is each budget item absolutely justified on its own merits to meet minimum program needs for fiscal year 1966?" We questioned each witness extensively in an attempt to uncover soft areas or unjustified expenses.

One measure of the effectiveness of the presentation was the fact that in 1965 the committee was rewarded by a 389 to 11 majority on the authorization bill.

FUTURE PLANNING

Over four years before Neil Armstrong first set foot on the Moon, Chairman Miller and Subcommittee Chairman Teague became concerned about planning for future programs beyond the lunar landing. Miller delegated the responsibility for planning studies to the Subcommittee on Oversight, headed by Teague. From March through June 1965, Teague dispatched letters to all the NASA centers asking them to set forth their goals for the future. Similar letters were sent to all major aerospace contractors, and the replies poured in throughout 1965 and early 1966. Replies were also solicited from Ed Welsh at the National Aeronautics and Space Council, from the Space Science Board of the National Academy of Sciences, and the Department of Defense.

With well over 400,000 people employed in 20,000 companies throughout the country working on the space program, billions of dollars invested in facilities and equipment, and very expensive and sophisticated hardware and flight systems available, there was a deep interest in what would happen after Apollo. Jim Wilson and the staff of the Oversight Subcommittee held a lengthy brainstorming session with Drs. Seamans, Mueller, and the top NASA brass on September 2, 1965.

When Webb appeared before the Science Committee for authorization hearings early in 1966, he confessed that funding for future planning and post-Apollo programs had been severely slashed by the Bureau of the Budget, reflecting the President's decision "to hold open for another year the major decisions on future programs—decisions on whether to make use of the space operational systems, space know-how, and facilities we have worked so hard to build up or to begin their liquidation." Teague made no secret of his displeasure with Webb's testimony on March 10, 1966:

I daily become more disturbed at the attitude of the executive branch of the Government as to whether they really want an aggressive program like we should be carrying out, and why we should wait another year to make major decisions, I don't understand. To me it is like telling a child that we are going to make you crawl another year before you can walk * * *.

In July 1966 the committee published the results of its future planning studies under the title of "Future National Space Objectives." The most important single recommendation made in the committee report was the opening gun which the committee fired in support of the Space Shuttle:

Immediate planning for a new generation of spacecraft capable of recovery at low cost and which are ground recoverable is a requisite to attaining lower total mission cost.

The report also made the recommendation—

that NASA report to the Congress not later than December 1, 1966, its recommendations on possible major national space objectives; the combination of missions included under such objectives; its expected total and annual cost; the benefits of such a program; and its composition in terms of the combined manned and unmanned building blocks required.

Had NASA taken this report requirement seriously, it is possible that the space program would have fared better budgetwise in the Congress. Instead, the idea was dismissed with a two-page letter dated December 1 with Webb's name typed at the bottom but signed by Seamans for Webb. NASA pleaded inability to formulate detailed future plans for the following reasons:

Because of the difficult budgetary situation resulting from the war in Vietnam and other factors, we are uncertain at this time as to what the President will approve for our fiscal year 1968 budget. Even in the absence of these uncertainties, of course, we would be precluded by the regular budgetary procedures from presenting specific statements on our future plans at this time.

To members of the committee who had been attempting to force NASA to put down its ideas about its future, it seemed almost as though NASA was refusing to admit it had much of a future.

PROGRESS IN 1966

Five highly successful Gemini flights during 1965, and the successful completion of space walks—extravehicular activity—and experience at rendezvous and docking of spacecraft set the stage for another successful year in which the Science Committee won a thumping majority for the NASA bill on the House floor.

Once again Chairman Teague put his subcommittee through a grueling schedule which included the usual round of visits to contractors and NASA installations. There was also a regular monthly visit to the Manned Space Flight Office in NASA Headquarters where George E. Mueller and his staff engaged in very frank, off-the-record sessions with as many committee members as could get away from Capitol Hill.

As an example of the type of interchange between the Manned Space Flight Subcommittee—plus other interested committee members—and George Mueller's staff at the NASA Manned Space Flight Office, the following memorandum excerpts were prepared at NASA:

Subject: Teague review, January 20, 1966.
Attendees: Congressmen Teague, Daddario, Casey, Rumsfeld, Schisler, and Adams.

Congressman Teague asked, Would we break the Apollo schedule (that is, slip the basic program beyond 1970) to have Apollo applications in order to absorb the cuts we are expecting. Answer, No, we owe it to the world and we have to keep these people working. Asked next question, How the program was cut by the Bureau of the Budget and the President. The answer was, A very selective cut rather than general but the Bureau of the Budget made general cuts and the agency had the opportunity to reconstruct its budget to conform to the cuts. He wanted to know what the impact of a severe cut would be, and asked for an analysis of cuts of 2 percent, 5 percent, and 10 percent of the budget [Loenig asked to do this]. [Teague said] Administrative Operations and Construction of Facilities will be covered in the full committee. The subcommittee will hear the research and development. Teague asked about the Russians attempting soft landing on the Moon. Vis-a-vis ours. The main thrust was, Did the Russian failures have any effect on our planning or thinking. Reply, No, it is a difficult problem and we are going as fast as we can and our pace is not influenced by the Russians. Mueller said we will keep the option open 1 more year if we lose Apollo Applications. Teague asked, Did the removal of the suits have any effect. The answer was it improved the general situation in the spacecraft. The astronauts were comfortable. Was the configuration of the suit different, the answer, Yes. Adams asked about the effect of the 5–7 PSI (pounds per square inch) atmosphere, The answer was no effect. Also, the effect of tumble—answer, if it's below 1 revolution per minute, no problem. In order to improve the situation, they covered the window with paper or something else. Daddario, Why did the Agena show up bad late. The reason is we had 185 good ones with plenty ground tests and checkouts. This was really an unexpected failure. Teague, How has your construction at the launch sites been delayed, was this caused by labor. The answer, On 37B we lost 37 days. The spacecraft was the pacing item on 201. Delays also caused by weather or other changes. There was no loss on pad 39 due to labor. Daddario asked if the cut caused a stretchout, will it cost more? The answer, It was no cut to basic Apollo, only the follow-on program. Daddario, Is the MOL (Manned Orbiting Laboratory) another Apollo. Answer, No, MOL is designed for DOD missions only. Crawler question was raised by Daddario, by asking has it moved anything, and the answer was, Yes, the LUT (Launch Umbilical Tower). Did we pay for the changes? Daddario asked, and then Teague became quite upset over the fact that we allegedly bail the big companies out and let the little companies go bankrupt.

The meetings with George Mueller and his staff were helpful, off-the-record opportunities for both groups to let their hair down, get to know each other better, and to get frank answers to questions and issues concerning the subcommittee.

As 1966 wore on, Teague began to worry about the pressures which Vietnam and the poverty program placed on the funds necessary to achieve the lunar landing goal. He remarked during the NASA authorization debate on the House floor on May 3, 1966:

The war in Vietnam has already forced a substantial reduction in the NASA budget for the coming fiscal year. Fortunately, however, thanks to our abundance of resources, it has not yet forced us to abandon our goals and our national requirements in space.

And Teague had reason to be concerned. The budget for NASA had already passed its peak, and it was touch and go whether the spending plateau could be maintained high enough to enable a successful Moon landing by the end of the decade. On August 17, 1966, Astronaut Lt. Col. Edward H. White wrote Teague from Houston to tell him that "We are coming along rather well in our preparations for the first manned Apollo flight and should be shipping our spacecraft to the Cape next week. If all goes well, which it usually doesn't on the first flight, we shall be ready for launch in about 100 working days after the spacecraft reaches the Cape. * * * I hope that your schedule will permit you to attend our launch as you did for the flight of Gemini IV. I would like you to feel that you have a personal invitation from the crew. Enclosed is a picture from the first Apollo crew."

Teague responded August 26, 1966:

DEAR ED: Thank you so very much for your letter of August 17 and the wonderful picture of you, Gus Grissom, and Roger Chaffee. I certainly expect to be at Cape Kennedy for your launch and appreciate your invitation.

At the moment, Ed, I am very depressed over our space program—more so than at any time since I have been working on it. There are so many things happening which indicate that the administration will make a serious cut in money this next year. To me it would be a great shame if we do not complete our space program because of money and not because of technology. It seems that billions have to go into the poverty program. It is my personal belief that the space program and the poverty program could be tied together very well. * * *

Only a few months later, tragedy struck on pad 34 at Cape Kennedy.

"FIRE IN THE COCKPIT!"

On February 18, 1960, when the Science Committee was in its infancy, Dr. Abe Silverstein, NASA's Director of Space Flight programs, was testifying before Chairman Teague's subcommittee on the subject of the first planned Mercury suborbital flight. After listening for awhile to the engineering complexities which were involved, Teague suddenly observed:

I am one who wants that first flight to be a successful flight, and I don't care how long we wait to do it.

From the start, the priority of the Manned Space Flight Subcommittee was always placed on human safety. Yet there was a feeling of high confidence within both the committee and NASA, as well as among the hundreds of thousands of contractor personnel, plus a "can-do" spirit which dominated the entire program outlook. The searching inquiries which the committee and the staff repeatedly made were all based on the assumptions that the program was a sound one and that someone was asking the right questions. The danger of fire was recognized, and studies were made on space rescue, but they were primarily directed at rescue in space. Few, if any, in NASA, on the committee or anywhere in the country ever asked the question which occurred to everyone by hindsight: Why was a lot of flammable material allowed in a pure oxygen pressurized atmosphere at a time when the secured hatch made it so difficult and time consuming to escape?

Apollo I was scheduled to fly with its premier crew of Virgil "Gus" Grissom, Edward White, and Roger Chaffee in February 1967. All three astronauts were strapped down in the spacecraft, simulating a launch, in their bulky space suits. A scheduled test of "emergency egress practice" was on the list but by 6:30 p.m. on January 27, 1967, they had not quite reached that point on the checkout. After the hatch on Grissom's Liberty Bell 7 Mercury capsule had prematurely blown off just after his splashdown, nearly drowning him, it was decided to design the Apollo hatches so they could not be blown off with explosive bolts. Hence it took a very strong man at least 90 seconds to turn the lever and lift the hatch from the inside.

In Fort Worth, Tex., on January 27, Teague addressed 2,000 high school seniors at a Career Conference at Texas Christian University. As examples of outstanding careers, he introduced Audie Murphy, the most decorated soldier in World War II; his son, Jack, an Air Force jet pilot; all-American football star Eddie LeBaron; and Astronaut Vance Brand. Teague and Brand were having dinner in Fort Worth that evening when they received the stunning news that a fire in the spacecraft had snuffed out the lives of three of their good personal friends. NASA Administrator Webb telephoned Teague with the grim news.

Immediately after the tragic fire, and during the investigations which followed, the attitudes and decisions of the Science Committee were very significant in their effect on the future of Apollo and the entire space program. In some quarters there was a feeling after the fire that the entire Moon flight program should be reappraised, stretched out and changed in emphasis. Had the accident occurred in space without sufficient means to investigate the circumstances, it is probable that the effect would have dealt an even more serious blow to the program. But the members of the Science Committee immediately rallied to the defense of the program after the fire.

This occurred in several ways. On the day following the fire, most members of the committee were interviewed by the news media. The consensus of their comments was that it was vital to get to the bottom of why the accident had happened, what changes should be made to insure that a similar type of accident would not happen again, and a determination to press on with the program.

THE COMMITTEE INVESTIGATION

Three days after the fire, Chairman Miller assigned the responsibility for a committee investigation to Teague's Oversight Subcommittee, stressing that he wanted a "comprehensive and impartial investigation." Although the Senate Committee on Aeronautical and Space Sciences began investigative hearings shortly after the fire, Teague decided that more meaningful hearings could be held once the full report of the NASA Review Board became available early in April.

In a colloquy on the House floor on January 30, numerous members of the Science Committee made it clear that they favored a vigorous continuation of the program, despite the tragedy. Chairman Miller stated on that occasion:

If the Almighty were to grant them the privilege of communicating with us, they could not help but say—carry on, you must not stop now, do not let our deaths be meaningless, and do not throw away what we have worked so hard to accomplish up to now.

Teague added:

If the meaning of their lives is to be sustained, we must take up the challenge of space they faced unafraid. Their quest for mastery of space must now be carried forward by their fellow astronauts. There can be no greater memorial to Grissom, White, and Chaffee than realization of the goals which they sought.

Majority Leader Albert told the House:

They have paved the way. Their brave companions, and men like them in the future, will carry on until the job they helped start is done.

A frequent critic of the Moon program, Representative Thomas Pelly (Republican of Washington), was affirmative in his advice to his colleagues:

I suggest that despite the accident the program will go forward and succeed.

And Representative J. Edward Roush (Democrat of Indiana), asked and answered a key question:

Where shall we go from here? We shall do just as these astronauts, whom we now honor, would want us to do. We shall continue to press forward with the determination that we shall attain this national goal for which they gave their lives.

Finally, Representative Don Fuqua (Democrat of Florida) related to the House that he had discussed the risks with the astronauts, all of whom understood very clearly the dangers involved. Fuqua added:

The task for which they gave their lives they knew was worth the sacrifice. This Nation could not honor their memory more than to continue its quest for knowledge.

In a letter to the members of the Oversight Subcommittee on March 22, 1967, Chairman Teague outlined the scope of the investigation and scheduling of witnesses. He bluntly stated:

It is my intention to conduct full and complete hearings on all matters relating to the accident. If additional testimony is needed to clarify any issue, such testimony will be taken. However, it is also my intention to complete the hearings as expeditiously as possible, including night sessions if necessary, in order that the public may have all the facts as soon as possible, and in order that the United States may get on with the program. * * *

In view of the recent press coverage concerning alleged statements of inadequacies in the Apollo program, I am inviting any member of the public, including employees of the Federal Government, to submit to the subcommittee for consideration any relevant statement or evidence concerning the subject under inquiry.

Teague stunned NASA by his initial reaction to the Review Board report. He said he was "outraged and hurt" at the carelessness and poor workmanship revealed. He said he was "surprised and disappointed at the number of mistakes" by both NASA and North American Aviation, Inc. He labeled the report as "shocking" and "unbelievable" and said it was a "broad indictment" of both NASA and its contractors. Teague followed through on his determination to keep the subcommittee in session mornings, afternoons, and in evening sessions until 10:30 p.m. Staff members like Jim Wilson slept in their offices in the Rayburn Building. It was clear from the chairman's attitude that the inquiry would be thorough and that everybody would get his say.

The reaction of most members of the Science Committee, and the manner in which they conducted the House investigative hearings, did a great deal to help stabilize the program and public reaction thereto. Some members challenged NASA's decision to set up a primarily internal investigative Board of Review, which Administrator Webb persuaded President Johnson was necessary in order to get the quickest evaluation of what must be done to get the Moon program back on schedule. For example, on the opening day of the hearings, this interchange took place between Representative Larry Winn, Jr. (Republican of Kansas) and Webb:

Mr. WINN. Mr. Webb, do you think it might have been wiser now, under the circumstances, and in the face of criticism, to have picked a completely outside investigating board?

Mr. WEBB. No, I do not. I do not think that the United States of America would have as complete information about this accident and all circumstances related to it or be in as good a position to move on with the next phase which is to get ready to fly the Apollo Saturn system.

A majority of the Science Committee defended Webb's decision, particularly in view of the thoroughness of the Board of Review report, which was completed in minute detail and combed over thoroughly by the committee both in public hearings and on-the-spot investigations at Cape Kennedy on April 21, 1967.

On the opening day of the committee hearing, Science Committee members clashed with Webb over his allegation that appropriation cuts had caused the deficiencies which the Board of Review pointed out. Daddario, in particular, documented the fact that there was no evidence to bear this out, and, furthermore, that the most severe reductions had occurred at the Bureau of the Budget level.

Members were angered at the suggestion by a North American Aviation witness that Grissom may have kicked a wire to cause the spark which ignited the fire. Representative William Fitts Ryan (Democrat of New York), Representative James G. Fulton (Republican of Pennsylvania), and others pointed out that Grissom would have had to be a contortionist to have kicked the wire.

The news coverage of both the fire and the investigation was as intense as the fire itself, and no doubt prompted some members of the committee to take very critical stances to attract publicity. Ryan, in particular, conducted a vendetta against NASA on virtually every point which the press seemed interested in headlining. On the other hand, NASA bungled its own public relations with reference to the so-called Phillips report. At first NASA denied it existed, then refused to release it, then tried to indicate it had acted fully on its recommendations.

THE PHILLIPS REPORT

Maj. Gen. Samuel C. Phillips of the Air Force, as NASA's trouble-shooting program director of the Apollo program in the Office of Manned Space Flight, presented to North American Aviation, Inc., a caustic review of management deficiencies as a result of his 1965 investigations. In a covering letter dated December 16, 1965, to J. L. Atwood, President of North American Aviation, General Phillips had made these comments concerning poor quality control and inferior workmanship on the Apollo spacecraft and Saturn:

I am definitely not satisfied with the progress and outlook of either program and am convinced that the right actions now can result in substantial improvement of position in both programs in the relatively near future.

Enclosed are ten copies of the notes which we compiled on the basis of our visits. They include details not discussed in our briefing and are provided for your consideration and use.

The conclusions expressed in our briefing and notes are critical. Even with due consideration of hopeful signs, I could not find a substantive basis for confidence in future performance.

Attached to General Phillips' letter, as he indicated, were his notes on the deficiencies uncovered at North American. At one point, this comment was made under the heading of "Summary Findings":

There is no evidence of current improvement in NAA's management of these programs of the magnitude required to give confidence that NAA performance will improve at the rate required to meet established program objectives.

As the hearings were getting underway, Webb and General Phillips called on Teague and persuaded him that it would not be in the best interests of NASA's frank and confidential relationships with their contractors to release publicly the notes which General Phillips had prepared. When Representative Ryan first raised the issue in the committee hearing with North American's president, J. L. Atwood, on April 11, the response was evasive:

Mr. ATWOOD. The Phillips report to whom?
Mr. RYAN. Has not that been discussed with you?
Mr. ATWOOD. I have heard it mentioned, but General Phillips has not given us a copy of any report.

Representatives Wydler and Rumsfeld joined Ryan in efforts to pry the Phillips report out of either North American or NASA. Wydler had this exchange with Dale D. Myers, vice president of North American:

Mr. WYDLER. Do I understand that no one in North American Aviation has ever seen General Phillips' report?
Mr. MYERS. We will have to identify the date or something that will give us an opportunity to check on it.
Mr. WYDLER. I have read about the report. You mean you never have heard of this report?
Mr. TEAGUE. Will the gentleman yield to the Chairman?
Mr. WYDLER. Yes.
Mr. TEAGUE. I have asked about the Phillips report. It is my understanding this is nothing more than a group of notes that General Phillips kept in the audit management of working with North American. There really is no Phillips report. You will certainly have a chance to ask General Phillips if he has a report.

In the Senate hearings, Senator Walter Mondale (Democrat of Minnesota), tipped off by Jules Bergman of ABC (who had seen a copy of the Phillips report at NASA Headquarters), unsuccessfully attempted to obtain a copy for the Senate committee. Although a month later NASA did supply a copy to that committee on April 12, Webb instructed his subordinates to stonewall requests for the report. Nevertheless, Teague asked NASA to sketch in the background of the report. On the evening of April 12, Teague at one stage of the hearings turned to the NASA witnesses and said:

Gentlemen, will you tell Mr. Wydler what the Phillips report is?

Answering, General Phillips read to the committee a carefully prepared statement explaining his review of North American's operations in 1965. The atmosphere became tense between Ryan and Teague with the following bitter exchange:

Mr. RYAN. General Phillips, did the notes which you handed to Mr. Atwood in December of 1965 relate to workmanship?.

General PHILLIPS. As I recall, in regard to their manufacturing——

Mr. TEAGUE. The Chair can advise General Phillips he can answer whatever he wants to. If I were in your position and asked that kind of question, I wouldn't answer. If you want to, you can. * * *

Mr. RYAN. I object to the instruction by the chairman to the witness.

Mr. TEAGUE. You can object all you want. The chairman will make his ruling; and he has made it. * * *

General PHILLIPS. May I check with counsel?

Mr. RYAN. Did the lawyer also write the statement, General?

General PHILLIPS. I didn't ask him.

Dr. MUELLER. Mr. Ryan——.

Mr. RYAN. My question, with all due respect, was addressed to the general——

Mr. TEAGUE. Would the gentleman submit his request in writing, and I will transmit it to the agency for what answers they think are appropriate?

NASA, however, continued to refuse to submit either the notes or the report to the House committee. In response to a question by Representative Rumsfeld, Mueller told the committee that there was no correlation between the findings of the Phillips report and the findings of the Apollo Review Board. Aside from some criticisms of slipshod quality control, Mueller's general conclusions are sustained. This makes Webb's strong resistance to release of the report even more puzzling, in the opinion of this writer. One can only speculate that Webb felt that publication of the many deficiencies in North American's performance would undermine confidence in NASA's ability to administer the space program, plus the ability and competence of North American as a prime contractor. There were also suggestions that Webb was still sensitive about the fact that North American had not been given the highest rating by the Source Evaluation Board in bidding for the first big Apollo contract.

On the closing day of hearings, Fuqua touched on a problem which concerned every member of the committee, in these remarks to Webb:

I think the committee has gone out of its way to cooperate with NASA in every way. I am getting the feeling that maybe you haven't really cooperated with us in not providing us with the information about some of these management problems that you have with the various contractors. I would certainly hope in the future, with both of us sharing some of the blame, that we can try to work more closely together, and the committee can be more closely informed about the problems.

Unable to obtain any satisfaction through the committee or from NASA, Ryan decided to proceed on his own. From a source he would not disclose, he received a copy of the Phillips report. With the help of Washington Evening Star Reporter William Hines, Ryan ran off duplicates of the report at the newspaper office and then called a press conference to distribute them publicly. Despite the fact that Ryan had the text of the Phillips report printed in the Congressional Record of May 1, 1967, Webb would never concede either the accuracy or completeness of the Ryan version. The issue still troubled Webb many years later, as he wrote in his foreword to NASA's Administrative History, 1963–69:

> One of the difficult matters which faced NASA during my term as Administrator was the demand, in the context of the congressional investigations of the Apollo 204 fire, for the public release of what became known as the Phillips report. This was a collection of contractor evaluations generated by a group under Maj. Gen. Samuel C. Phillips, Apollo Program Director, about a year previous to the fire. NASA's response to the requests of individual legislators to produce these evaluations for release to the public was based upon a strong need not to destroy the system which had been carefully worked out over the years whereby contractors and their key personnel cooperated to the fullest extent in assessing inadequacies in performance of both in-house and out-of-house organizations and equipment. This system was designed to assist in overcoming the inadequacies rather than to fix blame.

Although Ryan, in letters to both Teague and Chairman Miller, asked that the committee seek to obtain a fully authorized copy of the Phillips report and incorporate it into the official record of the fire hearings, no action was taken. Following an executive session with the Senate committee at which Webb furnished a copy of the report to the Senate committee, Webb offered to do the same for the House committee. But Chairman Miller advised Webb on May 17:

> I appreciate your furnishing me with the information on your agreement with Senator Anderson relative to making certain details of the original Phillips notes available for staff study.
>
> Before we make any arrangements in this committee, I would want to assure myself that all of the members of the committee are available to receive such information which can be presented by NASA. It so happens that in the next 2 or 3 weeks a number of our members have pressing engagements outside the city. For example, you will recall that the Paris Air Show is scheduled to take place shortly and we have several members and staff planning to attend.

There was no further action.

Another report, made by a North American employee, Thomas R. Baron, was not only made available to the subcommittee, but Teague also invited Baron to testify at special field hearings of the subcommittee at Cape Kennedy on April 21. Baron, a "preflight inspector," had listed a number of incidents and deficiencies which he had observed, and reported, only to be dismissed from his job for his pains. A large percentage of these deficiencies North American acknowledged were accurate and were being corrected, while others were denied or rejected. The fact that Baron, a subordinate employee, was allowed to testify and that Teague also invited anyone else who wished to testify to step forward, added credibility to the Teague investigation.

On May 10, Webb, Seamans, and Mueller returned to the subcommittee and presented their recommendations. Taking up each of the recommendations of the Review Board, NASA indicated that drastic measures were being instituted to eliminate combustible materials from the spacecraft, to design a new hatch which would enable escape from within in a few seconds rather than the 90 seconds previously required, space suits were being redesigned to make them fireproof, materials in the spacecraft were being fireproofed, and many other measures were being taken to prevent leakage at metal joints and otherwise recondition the spacecraft to guarantee the safety of the astronauts. A nitrogen/oxygen mixture was substituted for pure oxygen at ground level, going toward pure oxygen for use in space.

EFFECT OF THE TEAGUE COMMITTEE HEARINGS

Despite the committee's somewhat ambiguous handling of the Phillips report and the North American contract, the Teague subcommittee hearings were impressive in their thoroughness. Astronaut Frank Borman, as a member of the review board, bore a heavy burden of the testimony after the fire, and also personally assisted the subcommittee members in their excruciatingly personal examination of every phase of the Cape Kennedy details. Borman had these conclusions:

> My own particular association with the committee was most frequent during the investigation into the Apollo 204 fire. The investigation was tough, impartial, and a positive factor in the ultimate success of the Apollo program. Had the committee been so inclined, it is conceivable the lunar program could have been delayed or abandoned at that point. Instead, it proceeded with renewed vigor and determination. I am confident that the maturity of the chairman and senior members of the committee had a great deal to do with its independent weighing of the facts. * * * Congressman Teague and the committee members contributed immeasurably to the final success of the program.

Col. Rocco Petrone also observed concerning Teague:

> To me, it was his actions during the fire that kept us at NASA alive. He very coolly and smoothly played his role in oversight to make sure all things came out, and at the same time he kept us together—because there was a political opportunity to make NASA a scapegoat.

Although Representative H. R. Gross (Republican of Iowa) repeatedly called for Webb's resignation as a result of the fire, the committee rallied to the support of both Webb and the Apollo program. Wydler, a strong NASA critic, stated as the hearings ended:

I want to say this to you, Mr. Webb. Over the past few years * * * I probably have been one of the most critical members on this committee of NASA * * * It appeared to me * * * that you have had it too easy for your own good from this committee. This is not a criticism being directed inwardly at the Congress and this committee. I feel right now that you got less criticism than you deserved (in the past, but now) you are getting more criticism than you deserve. I don't intend to add to it for that reason.

In awarding Teague NASA's Distinguished Public Service Medal on October 3, 1978, NASA Administrator Robert A. Frosch commented:

The single episode which best epitomizes Mr. Teague's profound faith in the space effort, was the leadership he demonstrated at the time of the Apollo fire in early 1967. The space program was in severe jeopardy because of the tragic deaths of the Apollo crew; many influential Americans questioned the wisdom of proceeding with the lunar landing program; the basic concept of the space effort was challenged; and many potentially disrupting actions were being proposed. The dynamic leadership of Chairman Teague spurred a prompt identification of the issues and a clear-cut course of action to resolve them. Undoubtedly, more than any other single individual, Chairman Teague "saved" the program and redirected our energies in a direction which resulted in the successful lunar landing within the decade of the 1960's.

Chairman Teague and members of his investigating committee inspect materials recovered from the Apollo spacecraft after the fire. From left, Teague, Representative Guy A. Vander Jagt (Republican of Michigan), Astronaut Frank Borman, Representative Ken Hechler (Democrat of West Virginia).

Five astronauts testifying April 17, 1967, before Science and Astronautics Committee, following the Apollo fire: From left, Frank Borman, James A. McDivitt, Donald K. (Deke) Slayton.

During the week following the release of the Apollo Review Board report, when Teague was holding daily Oversight Subcommittee hearings morning, afternoon and evening, he spent a good deal of time during committee recesses working behind the scenes to help repair the shattered morale of the NASA Apollo team. Teague was busy on the telephone with NASA and contractor personnel, handling inquiries from the press, appearing on television programs like NBC's "Today" show and ABC's "Issues and Answers" with Astronaut Frank Borman. Tough, thorough, fair and exacting while he presided over the Oversight Subcommittee in the glare of publicity in the huge and imposing committee room in 2318 Rayburn Building, Teague usually asked the witnesses to stop by for a relaxing chat in his friendly office around the corner from the forbidding hearing room.

Five astronauts who appeared before the committee—Borman, Shepard, Slayton, McDivitt, and Schirra—spent considerable time in Teague's office. He wanted to be sure that they understood that the purpose of the searching hearings was not to find scapegoats through a witch hunt, but to get to the bottom of what really happened and what needed to be done to correct deficiencies, not only to protect the astronauts, but to inform the public and restore confidence in the program.

Perhaps it took a medical doctor really to understand what Teague was trying to do behind the scenes during the week of the hearings. Dr. Charles A. Berry, Director of Medical Research and Operations at the Manned Spacecraft Center (the personal physician for the astronauts), wrote to Teague on May 1, 1967:

> I want to express my deep personal gratitude for your many kindnesses during the hectic week of hearings in Washington. The very effective professional manner in which you chaired these hearings should be made known to every American and should indeed make one proud of our Congress. It was certainly a morale-booster for all of us in the program.
>
> The great understanding and friendship shown to me by you and your wife and the two wonderful Texas barbequed steak dinners in your office will never be forgotten. You made a week, which could have been unbearable, into a memorable experience.

On trips to Texas, Teague also went out of his way to call on the widows and families of Gus Grissom, Ed White, and Roger Chaffee, in an attempt to console them. On April 18, 1967, a hand-written note came in from retired Maj. Gen. Edward H. White:

> DEAR MR. TEAGUE: Mrs. White and I wish to commend and congratulate you for the calm and dignified manner in which you conducted the hearing on the Apollo tragedy.
>
> The exact cause of the accident may never be known, and I am convinced that it was a freakish coincidence that would never occur again, and that all reasonable precaution could not have prevented.
>
> The conquest of outer space must be pushed aggressively if our Nation is to retain its technological leadership. Astronaut Ed White would have insisted on it.
> Sincerely,
>
> EDWARD H. WHITE,
> *Major General, USAF (ret.)*

Teague responded to General White on May 12, 1967:

> I wish you could know how very much I appreciate your letter of about a month ago. It has been a most difficult task for me to conduct the hearings on the Apollo tragedy.
>
> It was not my desire to protect anyone or to persecute anyone; but to paint a clear picture for the American people of the space program. However, with the press interested mostly in headlines, it was rather difficult. I think it is all over now and I hope we have been fair to everyone concerned, including your wonderful son, Ed White.

After the fire, the Oversight Subcommittee rode herd on both the technical and administrative changes which were carried out. Boeing was assigned a technical integration and evaluation contract, which enabled NASA to have an extra watchdog. There was a wholesale personnel shakeup at North American Aviation. Harold Finger was promoted from NASA's Director of Nuclear Systems and Space Power to Associate Administrator for Organization and Management. As a result of the fire and the Oversight Subcommittee hearings, and the initiative of Representative Donald Rumsfeld (Republican of Illinois), the authorization bill passed in 1967 included an Aerospace Safety Advisory Panel to report on and make suggestions regarding facilities and operations. The bill ran into stormy seas both in the committee and on the House floor. There were 36 pages of various minority views out of the 194-page committee report.

Fulton caught his colleagues by surprise with a motion to recommit the authorization bill with provisions for cuts of about $170 million below NASA's budget request, and for the establishment of the Aerospace Safety Advisory Panel. An even greater surprise occurred when Fulton's recommittal motion passed by a rollcall vote of 239 to 157. One Democrat (Ryan) and seven Republicans (Fulton, Roudebush, Pelly, Rumsfeld, Wydler, Winn, and Hunt) on the committee voted to recommit the bill. It then passed as amended by the recommittal motion by a 342 to 53 vote. The final conference committee version passed in 1967 authorized $4.865 billion— or $235 million below what NASA had requested. A big slash was made in the Apollo applications program, which was clipped down to $347 million in contrast to the $454 million originally asked by NASA.

Once again, Gross was the most outspoken critic of the manned lunar landing, proclaiming:

I live in fear of the day when, if ever, we plant a man on the Moon because if we find a single, living human being on the Moon, this Government will start a whole new multibillion-dollar foreign giveaway program—a whole new foreign aid program.

With the encouragement and full support of the Science Committee, NASA made a brilliant recovery from the catastrophe on pad 34. Apollo 4, the first unmanned Saturn V, was launched in November 1967. Teague characterized it as "the free world's largest and most complex space vehicle." In April 1968, the Saturn was again successfully tested in near-Earth orbit, and driven back into the atmosphere at the 25,000-mile-an-hour speed of a return trip from the Moon. Despite another successful attack on the Apollo applications program by Fulton in 1968, the committee lines held to preserve support for the Apollo program, and the authorization bill survived by a 262 to 106 vote. A conservative-liberal coalition cut across party lines to mount opposition to the bill because of an unbalanced budget, the Vietnam war, and the pressure of social programs.

Although the fire probably delayed the lunar landing by about a year, and was a severe blow to the morale of all concerned, by 1968, there was a new air of optimism in the committee and NASA about the chances for success in 1969. At the beginning of 1968, the following was the lineup of members of the Manned Space Flight Subcommittee:

Democrats	Republicans
Olin E. Teague, Texas, *Chairman*	James G. Fulton, Pennsylvania
Emilio Q. Daddario, Connecticut	Richard L. Roudebush, Indiana
Joe D. Waggonner, Jr., Louisiana	Alphonzo Bell, California
Don Fuqua, Florida	Edward J. Gurney, Florida
William J. Green, Pennsylvania	Donald Rumsfeld, Illinois
Earle Cabell, Texas	
Robert O. Tiernan, Rhode Island	

Several changes in the composition of the committee, mainly as a result of the 1968 elections, produced the following roster of the subcommittee in 1969:

Democrats	Republicans
Olin E. Teague, Texas, *Chairman*	James G. Fulton, Pennsylvania
Emilio Q. Daddario, Connecticut	Richard L. Roudebush, Indiana
Joe D. Waggonner, Jr., Louisiana	Alphonzo Bell, California
Don Fuqua, Florida	Donald Rumsfeld, Illinois
Earle Cabell, Texas	Larry Winn, Jr., Kansas
Bertram L. Podell, New York	
Wayne N. Aspinall, Colorado	

The first manned Saturn flight of Apollo 7, a perfect textbook mission, was completed in October, followed by the famous circumlunar voyage of Borman, Anders, and Lovell, at Christmastime 1968.

True to his custom, Teague took his subcommittee on its annual whirlwind tour of NASA installations and key Apollo contractors prior to the 1969 hearings, which he opened with these comments:

This year is perhaps the most crucial year in our national space program. Apollo 9 is still in orbit, and Astronauts McDivitt, Scott and Schweickart are performing with distinction. The lunar module on the flight which will be completed this Thursday has justified our faith in the ability of the NASA-industry team to accomplish our national objective of a lunar landing in this decade.

A NEW ADMINISTRATOR: DR. THOMAS O. PAINE

With only a few months to go before realizing the goal of the decade, NASA was represented before the committee in 1969 by a new Administrator, Thomas O. Paine. Webb, who had fought so hard and successfully through the 1960's, was only a spectator and no longer the man in charge when the committee and NASA jointly moved toward that golden moment when Neil Armstrong first set foot on the Moon on July 20, 1969.

On September 16, 1968, Webb had announced at a White House news conference that he would resign effective October 7, his 62d birthday. President Johnson soon thereafter appointed Paine as Acting Administrator, and following President Nixon's nomination, Paine was confirmed by the Senate as Administrator on March 5, 1969.

Webb's departure 9 months before the manned lunar landing represented the end of an era, the close of 8 extraordinarily successful years during which his relations with the Science Committee had been close and generally very cordial. When Webb resigned, Teague commented:

Jim Webb has met the test of great responsibility and the demand of leadership. His abilities as a manager and a leader will be sorely missed. But he can leave NASA with a realization that he had established the greatest technological team that the world has ever known—a team well capable of reaching the goals which have been set forth.

Mosher put it this way:

It was very fortunate that a fellow with Jim Webb's genius headed NASA at just the right time to communicate with a bunch of people like we are. He knew government inside and out, and he was a political animal, he knew politics and how to deal with politicians. He was a born salesman, just a terrific salesman. I could have seen where some terrifically competent engineer or scientist might have been chosen to head NASA, and he would have been a disaster in terms of talking to us.

At one of the night hearings on the Apollo fire, a question was posed to Webb on the adequacy of North American Aviation's work on the Apollo contract, and Webb's response was so wide ranging and expansive as to prompt Teague to observe:

I like Mr. Webb. He has a wonderful reputation but it is not for short answers.

One NASA official commented:

Trying to make conversation with Jim Webb is like trying to drink out of a fire hydrant.

Webb was a genius at organizing the vast, multibillion-dollar enterprise which relied on thousands of private contractors and subcontractors, employing over 400,000 people throughout the Nation. The Science Committee marveled at his ability to present a very complex budget every year with the enthusiasm of a true believer, and the detailed knowledge of a man who had done his homework thoroughly. Congress respected Webb, and the members of the Science Committee regretted seeing him leave.

On the eve of the Moon landing, the Science Committee faced a tough fight in Congress over the perennial issue of declining funds for

NASA. The high-water mark for NASA appropriations had been reached in calendar year 1964 when NASA was furnished with $5.25 billion, and ever since then the funds had dwindled each year. At the beginning of 1969, NASA officials publicly indicated to the committee, for the first time, that they would offer three alternative budgets, one of which was officially approved by the Bureau of the Budget. First, there was a barebones budget of $4.2 billion which deferred new projects and was designed as a "minimum program for continuing ongoing programs." At the same time, NASA submitted an optional budget of $4.7 billion as the amount "required to maintain world leadership in space." The Bureau of the Budget responded by including only $3.76 billion for NASA—$1 billion short of the optimum, and half a billion dollars below the minimum. These figures furnished a clear-cut challenge to the Science Committee, which responded with a recommendation that some $250 million should be added to the rockbottom budget, about $200 million of which was earmarked for manned space flight and Apollo Applications. However, the Senate stuck to the budgeted figure and persuaded the House in conference to conform to the budgeted figure of $3.76 billion.

Fuqua helped stave off some of the opposition which had been grumbling about duplication between Apollo and the Air Force's Manned Orbiting Laboratory program. He pointed out that the Air Force was abandoning the MOL, leaving the manned space field exclusively to NASA. And Representative Bob Casey (Democrat of Texas) helped spice the debate with this gem:

> If Queen Isabella, after she pawned her jewelry to send Columbus on his adventurous trip to the New World, had had to stand for reelection, she would have probably been beaten for taking that gamble * * *. Let us show the strength and the fortitude and the leadership that we need to keep this country first in space.

After spirited debate, the authorization bill passed by a 330 to 52 margin on June 10, 1969.

Administrator Paine had a mission which was indeed painful: he came to NASA to preside over the dissolution of much of NASA's former power, as a result of severe budgetary constraints. Largely through his personal leadership, however, there was inaugurated a new spirit of scientific cooperation with the Soviet Union. From Paine's initial contacts with M. V. Keldysh, President of the Soviet Academy of Sciences, grew the Apollo-Soyuz linkup of American astronauts and Soviet cosmonauts in 1975.   After eight years of solid foundations which were laid by Webb, Paine had the honor of serving as NASA Administrator when the giant leap for mankind was recorded on July 20, 1969.

Well over 200 House and Senate Members, 19 Governors, 60 Ambassadors, countless mayors, about one million visitors, plus millions more television viewers throughout the world watched in awe on the morning of July 16, 1969, as Armstrong, Aldrin, and Collins successfully were launched moonward from pad 39–A at Kennedy Space Center. Once the astronauts had successfully landed on the Moon and returned to splashdown in the Pacific Ocean on July 24, Bob Gilruth, the Director of the Manned Spacecraft Center in Houston, took time out to send a little note to his old friend, Tiger Teague:

> Through all the stress and turmoil, the good days and the tough ones, you have stood with us—a tower of strength and an inspiration.
>
> My friend, I salute you!

Dr. and Mrs. Robert R. Gilruth greet Congressman Teague on one of his many visits to the Manned Spacecraft Center in Houston, Tex.

Representative Joseph E. Karth (Democrat of Minnesota), left, inspects a model of the Mariner spacecraft with NASA Administrator James E. Webb.

CHAPTER VII

Space Science, Applications, and Advanced Research, 1963–69

Congressman Joe Karth of Minnesota dominated every subcommittee session he chaired. To a far greater extent than most chairmen, he not only led the questioning, but his insistently probing mind and meticulously thorough preparation enabled him to set the tone of all his subcommittee activities and dominate the public hearings. Joe Karth was no committee dictator, and his subcommittee members always were given free rein and plenty of opportunity to participate. He built up respect among his subcommittee members on both sides of the aisle, winning and retaining that respect through sheer force of personality and knowledge of the subject matter. But there was never any doubt who was boss.

The Karth subcommittee in 1963–64: Seated, from left, Representatives William J. Randall (Democrat of Missouri), Thomas G. Morris (Democrat of New Mexico), Chairman Karth, Full Committee Chairman Miller, J. Edgar Chenoweth (Republican of Colorado). Standing, Representatives Neil Staebler (Democrat of Michigan), Thomas N. Downing (Democrat of Virginia), William K. Van Pelt (Republican of Wisconsin), Charles A. Mosher (Republican of Ohio), and James D. Weaver (Republican of Pennsylvania).

As noted in chapter IV, Karth became the first of the eight freshman Democratic charter members who joined the Science Committee in 1959 to chair a subcommittee. Third in seniority among the eight new members, Karth moved up to become a subcommittee chairman early in 1961 following the death of Representative David Hall (Democrat of North Carolina) and the 1960 election defeat of Representative Leonard Wolf (Democrat of Iowa).

After some reshuffling of the subcommittee jurisdictions to conform with internal NASA reorganizations, Karth wound up in 1963 chairing a subcommittee with the formidable title "Subcommittee on Space Science and Advanced Research and Technology," with the following Members:

Democrats	*Republicans*
Joseph E. Karth, Minnesota, *Chairman*	J. Edgar Chenoweth, Colorado
Thomas G. Morris, New Mexico	William K. Van Pelt, Wisconsin
William J. Randall, Missouri	Charles A. Mosher, Ohio
Thomas N. Downing, Virginia	James D. Weaver, Pennsylvania
Neil Staebler, Michigan	

The lion's share of the time and effort of the Karth subcommittee was devoted to wrestling with the annual NASA authorization bill. This meant poking and probing, trying to measure and weigh the arguments advanced by some outstanding scientific talent on how much should be spent for a bewildering variety of scientific experiments. In 30 separate public sessions and additional no-holds-barred, off-the-record or executive meetings, the Karth subcommittee carefully quizzed NASA witnesses to establish whether their money requests were fully justified, whether they were needed in 1963, what they would contribute toward future programs, what would happen if they were canceled or deferred and whether they were important enough to receive support in the rest of the Congress and the Nation.

Karth confessed to his colleagues in presenting the Space Science and Advanced Research portion of the NASA budget on the floor on August 1, 1963:

> There is nothing really exciting or glamorous about basic research and technology and I might add there is nothing really glamorous about space sciences, either.

He pointed out that his subcommittee had labored "under very trying circumstances, with very little, if any, fanfare, on a most tedious and most difficult job." Karth was well aware, as were other members of the subcommittee, of the pulse-throbbing public excitement right next door involving the astronauts in the Mercury program, and all the glamor associated with the race to get to the Moon. Meanwhile, the

Karth subcommittee was dealing with "more than 50 highly technical, highly scientific, and grossly difficult to understand programs" which he generally described to his colleagues in the House as follows:

> Just to give you a feel for the diversity of projects before the committee, let me call to your attention that they range through energetic particle explorers, ionospheric monitors, physical and astronomical observatories to propulsion systems of all types, chemical, nuclear, and electric, research grants and facilities to universities and colleges, space programs of all kinds, human-factor systems, the supersonic transport, international satellites, and so on.

The Karth subcommittee operated somewhat differently from the Teague Manned Space Flight Subcommittee. Some field trips were made to NASA and contractor installations, but for the most part the annual authorizations were hammered out in very intensive, exhaustive discussions in Washington between the subcommittee members on one side of the table and the responsible NASA officials on the other. There were very few agency briefings outside of the formal hearings, and no regular visits to NASA headquarters such as Teague scheduled. Karth's philosophy was that the agency should be kept at arm's length, that there should be a frank but adversary relationship, and in no event should the subcommittee develop into a kind of appendage or apologist for the agency—as he felt wrongfully occurred in some other congressional committees and subcommittees.

Karth's relations with Chairman Miller developed along an interesting pattern. Miller often used the expression "you don't get yourself a watch-dog and then do your own barking." So from the standpoint of his own philosophy as well as his reaction against the Brooks practice of not delegating much authority to the subcommittees, Miller was inclined to let Karth and the other subcommittee chairmen pursue their own courses of action. On the other hand, as a team player who believed in party regularity and was strongly inclined to go along with NASA, Miller sometimes clashed with Karth and the other subcommittee chairmen in full committee meetings. In Karth's words:

> He had a fuse that was three-quarters of an inch long. That means a very low boiling point and a very hot temper. He was so pro-agency that if you asked a question that sounded in his judgment to be a negative-type question, to be one that wasn't necessarily laudatory toward the agency, he himself would become incensed and he'd rap the gavel on you. I remember that very specifically. He'd almost want to rule you out of order, and at times would rap the gavel much before your 5 minutes was up, just because he was displeased with the line of questioning you were pursuing.

In addition to the annual authorization bills, the Karth subcommittee was extremely active in oversight investigations to insure that appropriations were properly spent, that management and cost factors were being properly observed, and that the taxpayers were really getting their money's worth in effective and efficient administration. These oversight investigations were carried on independent of the authorization hearings. Oversight investigations of Project Ranger (an instrumented, hard-landing probe to the Moon), Surveyor (soft landing spacecraft to the Moon, with lunar experiments) and Advent (military communications satellite) were successfully completed under Karth's leadership. Other early investigations by the Karth Subcommittee of Projects Centaur (launch vehicle) and Anna (geodetic satellite) are discussed in chapter IV.

One of the most nagging and difficult problems which the Karth subcommittee faced was how to justify spending some of the limited funds when you couldn't see the end result. Karth described it graphically this way:

> Our scientific and technological developments which we take for granted today have their roots deep in the basic research of yesterday. I do not care if it is the automobile or the airplane or the telephone or the radio or TV * * * first came the tedious, expensive and unglamorous basic research which provided the technological breakthroughs.

On the other hand, there were some instances beyond the intangibles of basic research, where the subcommittee had to judge whether research was necessary for a mission which had not yet been clearly pinpointed for the future. A case in point was the following discussion in an executive session on May 23, 1963, on the feasibility of studying how to cope with the extreme heat of reentering the Earth's atmosphere:

> Mr. CHENOWETH. What is going to be the speed of the Gemini?
>
> Dr. BISPLINGHOFF. (Dr. Raymond Bisplinghoff, NASA Director of Advanced Research and Technology.) 25,000 feet per second.
>
> Mr. CHENOWETH. Now the Apollo is going to be how much?
>
> Dr. BISPLINGHOFF. It will be about 36,000 feet per second.
>
> Mr. CHENOWETH. You could go ahead with the Apollo now?
>
> Dr. BISPLINGHOFF. Yes.
>
> Mr. CHENOWETH. I don't understand why you need this program.
>
> Dr. BISPLINGHOFF. We need this program to develop the knowledge beyond the Apollo speed.
>
> Mr. CHENOWETH. If you were sitting in our places, do you think you'd be justified in voting for something problematical and something in the future that may never come to pass?
>
> Dr. BISPLINGHOFF. Yes, sir, I do. I think we should invest a small part of our resources into looking in the future. * * * If we come to 1970, and you ask us to reenter from one of these planets, and we have not done (the research), we are going to be in a bad way.
>
> Mr. CHENOWETH. You think there would be any great jeopardy in postponing this one year?
>
> Dr. BISPLINGHOFF. Our movement toward preeminence in space would be jeopardized——
>
> Mr. MOSHER. It really postpones it. It doesn't jeopardize it.

Dr. BISPLINGHOFF. Well, I don't know quite what the word "jeopardize" means, but it certainly postpones it, and what I fear most of all is starting—giving a company a contract to develop something before the technology is ready. If there is anything that has cost this country money, it's that kind of money, and money invested in research ahead of time is well worth the effort.

WEATHER SATELLITES

On November 1, 1963, NASA's internal reorganization established the Office of Space Science and Applications, as well as the Office of Advanced Research and Technology and the Office of Manned Space Flight. This signaled an immediate reallocation of jurisdictions between the Karth subcommittee (which was renamed the Subcommittee on Space Science and Applications) and the Hechler subcommittee, which became the Subcommittee on Advanced Research and Technology. Prior to 1963, weather satellites were under the jurisdiction of the Hechler subcommittee which was at first called Applications and Tracking and Data Acquisition.

Ken Hechler, a Columbia University Ph. D., author of the combat story of the first Rhine crossing in World War II, *The Bridge at Remagen*, one-time speechwriter and researcher for President Truman and Adlai Stevenson, had been elected first in 1958 to represent the coal-rich area of southern West Virginia. A charter member of the committee, he became a subcommittee chairman during his second term in 1962.

At seven hearing sessions during August and September 1962, the Hechler subcommittee held successful hearings on weather satellites, communications satellites, and radio astronomy. For the weather satellite hearings, witnesses were called from the Weather Bureau, NASA, Department of Defense, as well as the National Science Foundation, the Department of State, and U.S. Information Agency. Chairman Hechler praised NASA for six consecutive successful launches of the Tiros (Television Infrared Observation Satellite) weather satellite, the last of which was launched while the hearings were in progress. When the hearings developed differing testimony from NASA (responsible for R. & D. on Tiros), the Weather Bureau and the Department of Defense (as operational users), Chairman Hechler called top representatives of all three agencies around the table to work out better coordination. In remarks to Deputy Assistant Secretary of State Richard N. Gardner, Chairman Hechler termed weather satellites as "tremendous weapons of freedom (which could) fire the imagination of the people throughout the world," and he urged that greater efforts be pushed forward through the World Meteorological Organization and the United Nations to make the findings from Tiros available to all peoples. In his report to Congress on weather satellites on August 1, 1963, Chairman Hechler told his colleagues:

A total of over a quarter of a million photographs have been sent back to Earth. Tiros has done a good, sturdy job in the past 3 years in identifying many hurricanes and typhoons and relaying advance warnings. Improvements in weather predictions in the future carry vast implications for farmers and businessmen. These improvements will be of prime importance in underdeveloped areas of the world.

When the Karth subcommittee was assigned jurisdiction over weather satellites at the beginning of 1964, strong support was given to the program. However, the subcommittee was disturbed to learn that a radical decision had been made by the Weather Bureau. NASA had developed and launched Tiros, a relatively simple spin-stabilized craft. At the same time, NASA was developing for Weather Bureau operational use a more complex three-axis stabilized satellite called Nimbus to incorporate more instrumentation, for weather prediction.

In September 1963, however, the Weather Bureau decided that Nimbus was too expensive and "too rich for its blood." So the Weather Bureau rather belatedly notified NASA that because of cost factors Nimbus could not be used in any operational system. To the subcommittee, this seemed to be an unfortunate and inefficient turn of events. But NASA was not to blame, and NASA continued to develop and launch more Nimbus craft to test new instrumentation, even though the Weather Bureau couldn't pay for Nimbus.

In 1965, Karth reported to the House that Tiros had run up a record string of nine straight successes, adding:

The data received from these experiments have opened up new horizons of research into the Earth's atmosphere. Pictures of cloud cover received from Tiros satellites are valuable, but new advanced sensors to measure temperature, wind velocity, and moisture content at various altitudes are now under development.

Several members in 1966 raised questions as to possible duplication among the many weather services in several different Federal agencies. However, Representatives Weston Vivian (Democrat of Michigan), Barber B. Conable Jr. (Republican of New York), and Karth all concluded that from a cost effectiveness standpoint, weather satellites were a sound investment.

In 1967, the Karth subcommittee voted to defer a $5 million item for two additional Nimbus satellites, but the Senate and the conference committee overruled their efforts. General Electric obtained the cost-plus-fixed-fee contracts, and when the new weather satellites were launched, they still proved too expensive for the Weather Bureau to opt to utilize them.

Karth told the House in 1968:

NASA's meteorological satellite projects have been the most successful of all NASA programs. The United States has launched 18 meteorological satellites without a single failure * * *. The Environmental Science Services Administration—ESSA—is now using satellites and sensors developed by NASA for weather prediction on a daily basis.

Mosher also praised the fact that NASA had achieved "an almost unbelievable 100-percent record of success" in launching weather satellites.

At the end of the decade, the support of the Karth subcommittee for weather satellites had helped make this program one of the most popular aspects of the applications program. The effective warnings afforded by weather satellites enabled hundreds of thousands of residents of the coastal areas to evacuate successfully and safely, rather than be overwhelmed, and suffer the fate which residents of the same areas had met in prior hurricanes.

COMMUNICATIONS SATELLITES

Karth and Hechler, who joined the committee together in 1959 as charter members, sat next to each other in full committee meetings and never had a single harsh word with each other even though they frequently were on opposite sides of substantive issues. Hence it was that when both subcommittee chairmen investigated communications satellites in their respective subcommittees in the summer and fall of 1962, no sparks of jurisdictional squabbles were evident.

As noted in chapter IV, Karth's subcommittee had a very productive series of hearings on the Centaur launch vehicle in mid-May of 1962. These hearings revealed serious development problems with the Centaur launch vehicle which the Department of Defense had banked on to boost its communications satellite Advent into synchronous equatorial orbit. In June 1962, control of the Advent program was shifted from the Army to the Air Force. Chairman Miller, concerned at possible duplication between the NASA and Defense Department communications satellite programs, asked Karth's subcommittee to follow up the Centaur investigation with an Advent inquiry in view of the fact the two programs were interrelated. In a report on November 1, 1962, Karth's subcommittee uncovered many management problems in the Advent program. The subcommittee recorded its strong support of NASA's Associate Administrator Seamans' statement that NASA and the Department of Defense—

jointly have a very great responsibility to see that the total research and development that is carried out in the communications field makes sense, that there is not undue duplication, that (DOD and NASA must consider) the total requirements of the Nation, both for commercial and for military purposes.

The Karth subcommittee predicted that DOD would never meet its schedule for Advent; what happened ultimately was that Advent was scrapped in favor of newer technology. When the Hechler subcommittee conducted its investigation of civilian communications satellites, September 18 through October 4, 1962, Representative Hechler opened the hearings with the announcement:

We are pleased to have our colleague, Congressman Karth, sit with the sub-committee. Congressman Karth's subcommittee investigated the military require-ments for communications satellites. It is very useful to have that continuity with your presence, Congressman Karth.

The purpose of the Hechler hearings was to air the hotly com-petitive claims of the commercial developers, particularly Hughes Aircraft Co., which was developing a synchronous satellite, and Bell Telephone Laboratories, which had developed the medium-altitude relay satellite known as Telstar. "There is high expectation that the satellite will perform as designed," the Hechler subcommittee reported concerning the Hughes satellite, Syncom. In 1964 Syncom became the world's first geostationary satellite, maneuvered into synchronous equatorial orbit so it appeared to stay fixed above one spot on Earth.

Meanwhile, Telstar, which had been launched on July 10, 1962, provided a dramatic illustration of its effectiveness, in a timely fashion for the hearings. Hechler opened the October 4, 1962, hearing with this comment:

Yesterday there were millions of people in Europe who shared the thrill of the successful flight of Walter M. Schirra, Jr., by means of viewing television relayed by Telstar satellite.

As the U.S. Information Agency testified: "Communications and television are something which, unlike shooting for the Moon, can touch each person's life personally."

In 1962 and 1963, the following members were assigned to the Hechler subcommittee:

SUBCOMMITTEE ON ADVANCED RESEARCH AND TECHNOLOGY
1962

Democrats	Republicans
Ken Hechler, West Virginia, *Chairman*	Jessica McC. Weis, New York
J. Edward Roush, Indiana	Thomas M. Pelly, Washington
John W. Davis, Georgia	
Joe D. Waggonner, Jr., Louisiana	

1963

Democrats	Republicans
Ken Hechler, West Virginia, *Chairman*	Thomas M. Pelly, Washington
J. Edward Roush, Indiana	Donald Rumsfeld, Illinois
John W. Davis, Georgia	John W. Wydler, New York
William F. Ryan, New York	
Richard H. Fulton, Tennessee	

When the Karth subcommittee picked up jurisdiction over com-munications satellites at the end of 1963, additional emphasis was placed on the value of the NASA research and development in this area. Karth told his House colleagues in 1964:

Telstar, Relay, and Syncom are virtually household words. Hardly an American or Western European has not witnessed the miracle of intercontinental television transmitted by the first experimental communications satellites. A great deal of research still needs to be done, but an economical commercial system now appears to be just over the horizon.

At the beginning of 1965, the membership of the Karth and Hechler subcommittees changed again, producing the following lineups:

SUBCOMMITTEE ON SPACE SCIENCE AND APPLICATIONS

Democrats	*Republicans*
Joseph E. Karth, Minnesota, *Chairman*	Charles A. Mosher, Ohio
Thomas N. Downing, Virginia	Barber B. Conable, Jr., New York
Roy A. Taylor, North Carolina	
Walter H. Moeller, Ohio	
William R. Anderson, Tennessee	
Weston E. Vivian, Michigan	

SUBCOMMITTEE ON ADVANCED RESEARCH AND TECHNOLOGY

Democrats	*Republicans*
Ken Hechler, West Virginia, *Chairman*	Thomas M. Pelly, Washington
John W. Davis, Georgia	John W. Wydler, New York
William F. Ryan, New York	
George E. Brown, Jr., California	
Lester L. Wolff, New York	

NASA's role in the communications satellite area diminished toward the end of the decade as the Communications Satellite Corporation expanded its activities. Also, budgetary limitations were a strong factor in the phasing out of NASA research and development in communications satellites in the early 1970's.

Meanwhile, the Karth subcommittee had taken an increasing interest in the Applications Technology Satellite—an outgrowth of the Syncom program of the Hughes Aircraft Co.—as well as paving the groundwork for the highly successful Earth Resources Technology Satellite. Starting with the world-wide telecasts of the Tokyo Olympic Games in 1964, the communications satellites represented a dramatic illustration to peoples throughout the world of the success of the space program and the interest of the Science Committee in extending the program for the benefit of all mankind.

RANGER

Project Ranger, developed by Dr. William H. Pickering's Jet Propulsion Laboratory of the California Institute of Technology, was designed to crash land on the Moon in the early 1960's to obtain close-up, high resolution television pictures of the Moon's surface during final approach.

Between 1961 and 1964, there were six failures of the Ranger to perform its mission. After the sixth failure on Feburary 2, 1964, NASA established a special review board headed by Earl D. Hilburn, NASA's Deputy Associate Administrator for Industry Affairs, to report on the reasons for the failure and to make recommendations. In a letter to Chairman Miller on March 31, 1964, Webb reported that the Hilburn review board had found that the most likely cause of the failure was "an unscheduled turn-on of the television equipment for 67 seconds" when the booster engine was jettisoned just two minutes after launch. Webb also told Miller that the Hilburn report was classified for military security reasons and because it was an "internal investigatory document" and "since it does not represent NASA's complete judgment and final implementing plans."

Chairman Miller asked the Oversight Subcommittee to conduct an investigation of the Ranger failures. Since the Karth subcommittee had had responsibility for funding Project Ranger, Miller named Karth as acting chairman of the Oversight Subcommittee to conduct the hearings. Teague specifically recommended this move.

The membership of the Oversight Subcommittee which conducted the investigation of Ranger included the following members:

Democrats	*Republicans*
Joseph E. Karth, Minnesota, *Acting Chairman*	James G. Fulton, Pennsylvania
Ken Hechler, West Virginia	R. Walter Riehlman, New York
Emilio Q. Daddario, Connecticut	Richard L. Roudebush, Indiana
Bob Casey, Texas	Alphonzo Bell, California
Joe D. Waggonner, Jr., Louisiana	Edward J. Gurney, Florida
Edward J. Patten, New Jersey	
Don Fuqua, Florida	

Frank R. Hammill, Jr. staffed the Ranger hearings.

Karth immediately asked Webb for the Hilburn report and bristled when it was denied to him. The Karth subcommittee a year earlier had already been highly critical of the management problems in the relationship between NASA and the Jet Propulsion Laboratory (JPL). In 1963, when the Karth subcommittee slashed the authorization for Ranger from $90 million down to $65 million, Karth had observed that there were "grave doubts" about the "adequacy of the management of this project, both by NASA headquarters and the Jet Propulsion Laboratory * * *. The subcommittee feels that in view of the poor record of Ranger to date, Congress should be given reasonable assurance of success before going forward full speed with more spacecraft."

On the eve of the opening of the Karth oversight hearings, both NASA and JPL headquarters were thrown into a turmoil by differing opinions on the severity of the Hilburn report. Oran W. Nicks, NASA's Director of Lunar and Planetary Programs protested that "If the only purpose of the investigation had been to establish a basis for a critical letter to Congress, we in the program office were naively misled initially into supporting it as a constructive endeavor." Nicks wanted to forward to Congress a rebuttal of the Hilburn charges. Personnel at both the Office of Space Science and JPL were infuriated that Webb had sent Chairman Miller a summary of the Hilburn charges without giving them a chance to refute them. In an attempt to soften opposition within his own headquarters, Webb at a news conference confessed there had been an administrative error in signing the March 31 letter to Miller which he thought had been cleared in the Space Science Office.

Webb tried to deal with the revolt within his own headquarters and JPL, as well as to cope with the rising indignation of Karth who was demanding the Hilburn report. The Los Angeles Times ran an angry editorial on April 8, charging that if NASA wanted to separate JPL from Caltech, "it could do this without first resorting to a campaign of defamation, which not only damages JPL but reflects unfavorably on one of the country's very great schools of science and technology."

But Webb did not fully succeed in pacifying Karth. In a letter written for Webb's signature, but actually signed in his absence by Associate Administrator Dryden, Webb reiterated to Karth on April 22 that "the report represents the views of an internal NASA review group, but it is only one working document. It is not a definitive agency position * * *. For these reasons, the Ranger VI Report should not become a basis for either conclusion or action by the Subcommittee on NASA Oversight and should not be made available publicly."

Webb offered to bring the Hilburn report to Karth personally, with the understanding that it would not be left with the committee. In a last-ditch effort to head off or soften the sharp effects of the hearings, Webb added in his letter to Karth:

I hope you will keep in mind that the timing of these subcommittee hearings is unfortunate in that the factors of morale and program execution are both deeply involved and there are very real dangers that both may be seriously affected. Nevertheless, I can assure you that NASA officials will cooperate fully in the hearings and provide the best answers we have to your subcommittee.

When Karth opened the hearings on April 27, he read Webb's letter in its entirety. He took exception to Webb's statement that "the Ranger VI report should not become a basis for either conclusion or action by the subcommittee." Karth commented:

I think that all of the reports, or all of the investigations, regardless of who has conducted the investigation, should be a matter for this subcommittee's consideration, and could become a basis for conclusion or action by the subcommittee.

Karth also resented the statement by Webb that the timing of his subcommittee hearings was "unfortunate." He dealt with this observation with the following public statement:

I would like to point out to the members of the subcommittee and to the NASA people here represented that while Mr. Webb may feel that these subcommittee hearings are unfortunate, the action that precipitated these hearings, in all probability, are the letters addressed to * * * Chairman Miller.

I might further state that, subsequent to the Ranger VI failure, I did have an opportunity of discussing it with Chairman Miller, and that we both recognized that Ranger had had some difficulties in the past and that certain technological difficulties in a program of this magnitude were something that might be expected. For those reasons, we did not expect that the Oversight Committee would be asked to make a review of the program. However, after the Webb letter, it was hardly reasonable to expect that, with the kind of criticism contained in the letter, a congressional investigation was not in the best interests of the country and the Congress.

Hechler immediately added:

I would simply like to support the remarks made by the acting chairman of this subcommittee. It seems to me that the timing of these hearings is highly propitious, and I am certain they are going to fulfill a constructive purpose to carry out the responsibilities of Congress and of this committee.

In the comprehensive hearings which followed, the Karth subcommittee probed into relationships between NASA and JPL, and also called RCA and Northrop Corp. representatives who had worked on Ranger. During the hearings, Karth observed to Webb that "NASA is the contracting agency of the Government; (it) should be, in fact, the boss of the program. NASA provides the money, and therefore should have more to say about how this work is to be done, and by whom it should be done." Karth added that NASA, in light of the repeated Ranger failures, should have installed a strong technical team at

Pasadena "to oversee or supervise, not just management practices at JPL, but technical approaches as well."

The Karth subcommittee charged that JPL had failed to establish rigid and uniform testing and fabrication standards for the Ranger spacecraft. NASA was faulted for regarding JPL more as a field center than a contractor. The final report, which was unanimously approved by the full committee, recommended tighter NASA supervision over JPL "to manage such complex in-house projects such as Ranger and Mariner." Effective August 1, retired Air Force Maj. Gen. Alvin R. Luedecke, the General Manager of the Atomic Energy Commission, was installed as Deputy Director at JPL. General Luedecke was given responsibility for the day-to-day technical and administrative activities at JPL.

Despite the roughness of the questioning by the Karth Oversight Subcommittee, Webb was generally pleased with the outcome. In a letter to Chairman Miller dated May 4, 1964, Webb confessed to Miller that he was happy that Karth had resisted the pressure to "look for scapegoats." He also remarked that he was happy that Karth had recognized "we are dealing with an extremely delicate situation, much like walking down Fifth Avenue in your BVD's."

Rigorous testing and checkout followed the Karth investigative hearings, and there were major changes in circuitry design and hardware. In July 1964, and subsequently during February and March 1965, three highly successful Ranger missions were flown. The pictures taken just before impact resolved details of the Moon's surface less than two feet apart.

Miller and Karth both used the occasion of the Ranger successes to cement support for the program in Congress. Karth assembled foreign editorial reaction to the Ranger successes for reprinting in the Congressional Record. Chairman Miller exulted: "I want to make it crystal clear that the Jet Propulsion Laboratory is doing a splendid job."

Miller also arranged for a special briefing for Members of the House of Representatives on August 4, 1964, at which he termed Ranger "one of the greatest accomplishments that NASA has ever made."

LUNAR ORBITER AND SURVEYOR

When NASA first began to formulate its plans to investigate the Moon and the planets, primary responsibility was assigned to the Jet Propulsion Laboratory. In the early 1960's, the Surveyor was designed as a soft lander on the Moon, with one version termed a "Surveyor Orbiter." The committee closely followed the development of the Surveyor project from the time in April 1961 that NASA reported to the committee:

In January 1961, after intensive competitive design studies by four major companies, Hughes Aircraft Co. was selected to build the Surveyor spacecraft.

By 1963, NASA became concerned that JPL had its hands full trying to develop the Surveyor Lander along with Ranger and the probes to Venus and Mars being done by Project Mariner. With encouragement from the committee, the Surveyor Orbiter was transferred in mid-1963 from JPL to Langley Research Center. There it was redesigned and renamed "Lunar Orbiter" and contracted out to Boeing.

The Karth subcommittee members sharply challenged both the timing and overlap between a lunar orbiter and lunar lander during the 1963 hearings. In its committee report, sustained by the full House, the observation was made that "funds made available previously were transferred to other projects considered by NASA to have higher priority, and virtually no money has been spent on Surveyor Orbiter to date."

As was customary, NASA appealed to the Senate to get $28.2 million restored and the conference yielded to allow $20 million after impassioned pleas by both NASA and the scientific community. The effect of the House action was to spur NASA to define and clarify the orbiter mission and to give it the management support which had hitherto been sorely lacking.

In repeated hearings, the Karth subcommittee questioned the relationship among the three unmanned lunar missions—Ranger, Surveyor, and Lunar Orbiter, forcing NASA to pinpoint what it really planned to do, when and why. Also, the subcommittee through its rigorous questioning brought out the extent to which these programs were being funded for their scientific value, as against providing data which could assist in insuring the success of the Apollo program. The cost-conscious subcommittee also probed into issues pertaining to the Lunar Orbiter contract with Boeing, which exceeded by $20 million the next highest bidder.

During 1966 and 1967, five Lunar Orbiters were launched and all five were successful. As a direct result of the Lunar Orbiter successes, five Apollo landing sites were certified.

In contrast to the Lunar Orbiter, the Surveyor soft lander program ran into deep trouble. The original Hughes contract called for seven flights and the price tag was $67 million, somewhat above the announcement in a 1961 NASA press release that the Surveyor project "is expected to cost upward of $50 million." In October 1965, when the Oversight Subcommittee submitted its report, it was concluded:

Surveyor already represents an investment by the American taxpayer of almost one-half billion dollars for the first 10 spacecraft plus launch vehicles; the ultimate cost to completion of just this first part of the project is estimated to be approximately $725 million.

By 1965, the first launch of Surveyor had already slipped by 2½ years. The first flight actually occurred on May 30, 1966.

Once again, as with the Ranger probe, Chairman Miller authorized Oversight Subcommittee Chairman Teague to allow Karth to chair the hearings to investigate Surveyor. Karth took his subcommittee on a two-day inspection tour of the Jet Propulsion Laboratory at Pasadena, and then on to the Hughes Aircraft Co. plant in Culver City, Calif., on September 2 and 3, 1965.

The Surveyor spacecraft weighed 2,250 pounds, stood 10 feet high, and had a triangular frame with a landing leg on each of the three corners. A solid propellant retrorocket engine was designed to slow it down as it approached the Moon's surface. A television camera and a surface sampler were among the experiments aboard, enabling a measurement of the physical and chemical properties of the lunar surface to a two-foot depth under the eyes of the television camera.

The Karth hearings reviewed the stormy history of the project, involving many design modifications and bitter disputes among NASA, JPL, and Hughes. The committee found that the repeated technical difficulties had been compounded by poor management and supervision all along the line. In 1964, after NASA reviewed the shortcomings at JPL and Hughes, NASA had recommended that JPL assign more personnel to monitor Hughes. The result was an increase of JPL personnel supervising Hughes from 100 to 500, in what can only be described as "intensive surveillance" which further strained the JPL-Hughes relationship. While NASA was trying to pressure JPL to take a more aggressive supervisory role over Hughes, the Hughes organization resented the "new ideas" which slowed down their work.

The committee uncovered the fact that two "drop test" failures contributed to escalating costs. The following colloquy points up some of the multiple problems involved in the tests, the contractual relationships and the responsibility:

Mr. MOSHER. When you were talking about these Surveyor drop tests, it seemed to me there was an implied criticism when you said you found that they were not using flight quality hardware. Now, who were you criticizing at that point? Whose fault was this?

Dr. NEWELL. I think the Hughes contractor agrees that these tests weren't prepared for or conducted properly——

Mr. MOSHER. Is there any penalty here? In our contract with Hughes, is there any comeback that the Government has with Hughes in this respect?

Dr. NEWELL. We didn't have an incentive contract at that time; no.

Mr. MOSHER. This was a cost-plus arrangement?

Dr. NEWELL. Yes, cost-plus fixed fee.

Mr. CONABLE. But you are moving more toward incentive contracts, aren't you?

Dr. NEWELL. We are moving toward incentive contracts. In fact, the Office of Space Science and Applications has the largest number of such contracts in the agency, as just a means for trying to avoid this sort of problem, for getting the attention of the contractor to these things, since it affects him in the pocketbook.

Mr. MOSHER. The mistake on Hughes' part is something you necessarily write off as experience from which we can benefit next time; is that right?

Dr. NEWELL. Yes.

Mr. KARTH. Now, I assume that some of the JPL management team was also working on the problems of the drop test, or was this an exclusive thing on the part of Hughes? I am trying to pinpoint some responsibility.

Dr. NEWELL. Well, it is a responsibility we all have to share, because if we had penetrated properly into the testing program, we should have spotted this too.

The Karth subcommittee concluded that the first deficiency was NASA's failure to require sufficient preliminary design work before hardware development. Second, NASA should have stepped in and exerted firmer control over JPL sooner than it did. Third, JPL was concentrating so heavily on Ranger and Mariner that it neglected to supervise Hughes until late in the game. Of course, needless to say, Hughes top management was equally to blame.

The Surveyor investigative report ended on an optimistic note, encouraging NASA to "continue their present high level of attention to the Surveyor project."

A pleasant aftermath of the Karth hearings was the fact that from 1966 through 1968, five of the seven Surveyor shots landed successfully and performed their assigned experiments. The data from the experiments were important to the successful manned lunar landings because they substantiated the fact that the lunar surface would support landings by the Apollo astronauts.

MARINER, MARS, AND VENUS

On February 26, 1963, Dr. Homer E. Newell, NASA's Director of Space Science and Applications, briefed the committee in an informal session on the scientific results of Mariner's 36-million-mile trip to fly-by the planet Venus. Mariner was adjudged by the scientific world as a success in revealing new data on the mass, temperature, and nature of Venus. During the early 1960's, it seemed strange to committee members that Ranger had failed six times in a row to complete successful experiments a quarter of a million miles away, while a 36-million-mile shot to Venus was successful. As has been noted, Ranger snapped out of it and scored several later successes after the early failures.

Just when it seemed that Mariner's luck was going against it by a failure in 1964, Mariner IV buoyed the hope of the scientists by flying within 6,200 miles of Mars on July 14, 1965. JPL, NASA, and committee members shared the glory of a special White House ceremony in

1965, and the Karth subcommittee gave full support in 1965 to an ambitious new planetary program called Voyager, to make an instrumented landing on Mars.

Shortly after 10 o'clock on the morning of March 7, 1966, Karth assembled his subcommittee and staff in the smaller of the two main committee rooms, room 2325 of the Rayburn Building. It was one of those lengthy brainstorming sessions during which the committee members chewed and digested the testimony they had elicited in public hearings, and were now down to the hard decisionmaking process when they were airing their opinions in free-wheeling, off-the-cuff discussion in executive session. About an hour into the discussion, the following exchange occurred:

Mr. KARTH. Could we talk about Venus for a minute?

Mr. MOSHER. De Milo?

Mr. KARTH. De Milo—I think the question arises on Venus, whether or not we feel the only Venus shot which is scheduled between the middle or late 1970's is a reasonably decent investment for $30 million?

Mr. MOSHER. Are you suggesting we might just as well leave Venus to the Russians for a while, and let them do the job, and work on that, and we could just ignore it for a while?

Mr. KARTH. I am not suggesting anything, except maybe we discuss this thing——

Mr. CONABLE. Well, the imponderable here is the prestige element, I guess. You raised this implication in your opening remarks about Venus, Mr. Chairman. There is a serious question whether we want to put ourselves in the position of simply saying, "We have no interest in Venus," and the Russians are likely to be talking about it a good deal——

Mr. VIVIAN. I have a feeling the scientific community really put that Venus shot in there * * *. My feeling is if we are going to save anything, any significant fraction of that money, it has to be saved reasonably soon with a positive decision * * *. What you are saying is you would rather put enough eggs in the basket on Mars with the hopes of really doing a job on it, feeling that the peripheral data we are going to pick up on Venus is not going to be worth much.

Mr. KARTH. I am saying I would like to put those extra eggs into the Mars basket without touching any other program.

Same time, same place, the same cast of characters assembled the next day, March 8, to mull over the same issue. Karth again raised the issue of Venus, and the consensus in the committee began to develop, with Mosher observing:

Well, I certainly think, Mr. Chairman, that the Venus program is one that is most expendable. It is the one we can do away with and hurt less than anyplace else.

Mr. CONABLE. It certainly sounds as if we ought to make a serious effort to try to fund Voyager more heavily.

Karth also raised the question whether several European countries might be interested in cosponsoring the Venus shot, adding:

I had Bill Wells, my assistant, yesterday checking around to see whether or not we could make some effort to talk to the European counterparts about the possibility of their undertaking a program like this on a cooperative basis with the United

States. We came to the same stone wall that you usually come to with the State Department when you are talking about doing something in a hurry. They, I think, would be willing to explore this, but it would take them 6 months to properly explore it so they could get an answer. I don't have 6 months.

Gradually, but decisively, the subcommittee moved toward a unanimous decision. Karth personally was not disturbed that NASA and the Space Science Board of the National Academy of Sciences wanted to go ahead with the Venus probe, rather than shifting the concentration to the Mars probe. To this line of argument Karth responded:

I do not agree that we ought to leave all decisions bearing on science to the scientists, or that all political decisions should be made by politicians.

The full committee and the House supported the decision of the Karth subcommittee to concentrate on the Voyager probe to Mars rather than the Venus shot. But as so frequently happened, the decision became untracked when the scientific community mobilized behind NASA to appeal the action in the Senate, where they not only won but also reversed the House action in the conference committee. Reflecting on the turn of events, Luther J. Carter wrote in Science magazine that Karth "by general agreement is an intelligent and unusually hard-working committee chairman" who "has worked diligently at understanding the programs entrusted to his review." He added:

His experience illustrates the classic frustration of Congress in an era of deep Government involvement in science and technology. How does it pass judgment on highly technical programs without being either a rubberstamp or an incompetent intruder upon the affairs of experts?

Usually, the Karth subcommittee managed to cope with that dilemma extremely well, to the benefit of the taxpayers and the Nation.

The Venus fly-by took place in 1967, passing 2,600 miles from the planet, refining the temperature and atmosphere measurements made by the 1962 Mariner flight. The ambitious Voyager program was fostered and encouraged by the Karth subcommittee, but fell victim to the budget woes caused by the Vietnam war, and was mercifully put to sleep by the Senate and House Appropriations Committees in 1967. The decade ended with two Mariner shots which flew within 2,000 miles of the Martian surface and took many excellent close-up pictures of the planet.

EARTH RESOURCES TECHNOLOGY SATELLITES

In 1967, the Karth subcommittee noted in its report:

The Members uniformly support the objectives of the various space applications projects. These efforts are expected to result in tangible and measurable economic benefits to the Nation and to the world in the foreseeable future. Great strides have already been made in space meteorological and communications systems because the Congress has given generous support to these projects in the past years.

Members of the Karth subcommittee took a consistently strong position of support for the Earth resources program, which was originally named the "Earth resources survey" program.

In 1968, Karth urged NASA to be more aggressive in order to produce an early operational Earth resources satellite:

Mr. KARTH. I don't think we have to start way back at point zero with these application satellite programs as we did with Syncom, for example, because with Syncom we started without having done any previous research in an area that was applicable. I think that is not true today by virtue of the fact that we have done a great deal of research in those areas where there is direct applicability. I would think that today the time period could be shortened quite considerably if we really had an aggressive Earth resources program evolving from the agency. Would you agree?

Dr. NAUGLE. I think we should be working to shorten the time period between research and the development of the operational system, certainly. * * *

Mr. MOSHER. Then you should press forward toward it just as the chairman says.

Dr. NAUGLE. Yes.

At first, NASA witnesses balked a little when Karth suggested that the Department of the Interior and the Department of Agriculture would be pleased to receive future Earth resources data in their programs: "Haven't they brought to your attention programs that they feel would be extremely useful and save billions of dollars annually for the American people?" Dr. Naugle wondered whether the economic value of such data had been analyzed, to which Karth replied that in programs like physics and astronomy, NASA had never applied such a yardstick before proceeding with a program. With some exasperation, Karth observed:

I can't for the life of me understand how NASA, with all the brains they have, and indeed I have great respect for the intellectual capability of the people who work for NASA, is having such a hard time finding out if there is any cost effectiveness related to the Earth resources satellite program. Cost effectiveness has never been applied to any one of the other programs that I know of, and I think the most glaring example is the Apollo program itself * * *. I am just not sure I understand what is going on, but I can tell you one thing: as far as I am concerned the subcommittee is going to find out, and if there is a make-work program for the manned space flight people, chickens are going to come home to roost, if I have anything to say about it. I couldn't be any less interested in make-work or more interested in economic benefits.

Appearing before the Karth subcommittee on behalf of speedier progress by NASA in the Earth resources satellite area, Representative Fulton observed: "I believe rumor hath it that you are also dissatisfied with the progress of the program," to which Karth responded:

I think the subcommittee is more interested in the rapid development of an Earth resources satellite program than in any other program in the Office of Space Science and Applications. We feel that here is an area of immediate and widespread economic benefit which, in the long run, can do more to sell an overall space program to the public than any other program.

In arguing for passage of the NASA authorization bill in its committee report, the full committee stated:

> The committee strongly believes that the prospects for economic benefits being achieved in the near future by an Earth resources satellite system are so bright as to justify increased effort in research in this area. Accordingly, NASA is urged to emphasize research and advanced studies pointing toward an early operational Earth resources satellite system.

Various Members also added emphasis in their remarks on the House floor during the May 2, 1968, debate on the authorization bill. For example, Mosher termed the Earth resources survey system "the most exciting new project on the horizon." He added:

> It is expected that remote sensors in space will, in just a few years, provide valuable data on the status of our agriculture and forests, and on the location and availability of mineral and water resources; such a system will contribute to the management, utilization, and conservation of all our natural resources.

In 1969, the House, at the urging of the Karth subcommittee and the full committee, added $10 million to the authorization for the Earth resources technology satellite program. Karth, Mosher, Symington, and other members of the Karth subcommittee led the charge to bring home to the Congress as well as the forgotten "man in the street" that this was a program which had practical applications and potential returns for the taxpayers. The subcommittee solicited support through testimony by Departments of the Interior and Agriculture officials to bolster their case. In addition, the contractors were brought in to furnish additional evidence of their ability to move forward faster if given additional support.

When Karth picked up a copy of Space Business Daily of March 12, 1969, he was angered to read the following note:

> The head of NASA's manned space flight program said this week that the agency's unmanned Earth resources satellite program will not be rushed into development due to technical considerations, and suggested that man may play a major role in the ERS project.

Waving the article at the March 12 subcommittee hearings, Karth commented to NASA witnesses that he felt NASA's Dr. George E. Mueller was trying to hold back the Earth resources satellite development "until such time as we can use these very interesting and very desirable experiments on manned spacecraft to assist in justifying certain other manned flights." Karth, who often threatened drastic action as a means of getting fuller attention, exploded that if NASA couldn't justify its manned space flight program without this ploy, "then I am going to start saying loudly and clearly around here that we don't need a manned space flight program." He added:

This really bothers me. I can't help but feel that some place along the line we have been had, and I would hope an intelligent man like Dr. Mueller would not be so devoid of practical considerations that he would suggest what he reportedly suggested yesterday.

Karth's feeling about the manned space flight program erupted into the strong opposition he voiced against the Space Shuttle during the 1970 consideration of the NASA authorization bill. At that time, Karth led a fight to attempt to bring better balance between the manned and unmanned portions of the space program.

APPLICATIONS TECHNOLOGY SATELLITES

Historically, the applications technology satellite grew out of a 1962 advanced Syncom study project at Hughes Aircraft Co. Following the establishment of the Communications Satellite Corp. in the early 1960's, NASA decided to reorient the advanced Syncom program away from communications research and development, in line with the efforts to turn such developments over to private enterprise. Instead, NASA developed a satellite with the broader ability to carry experiments in several different areas of technology. This led to the "advanced technological satellite" program which eventually was renamed "applications technology satellite."

From the start, the Karth subcommittee strongly supported more aggressive work in this area. As Karth told the House on May 6, 1965:

It is the development of applications satellites—spacecraft which perform meteorological, communications and navigation services—where the United States has its greatest opportunity for continuing leadership in space technology. I believe Congress should fully support this effort.

In 1968, as the ATS program began to bloom with several successful launches, Karth remarked to his colleagues:

I want to mention one other aspect of the unmanned program which I consider to be the most significant—the so-called applications satellite project * * *. NASA is continuing its important research and development of equipment for use in future communications systems with the ATS program. Closely related to this work in communications, research in navigation and traffic control techniques and equipment has already indicated that satellites can assist over-ocean aircraft and ships at sea to obtain more precise position information under all weather conditions and will some day aid in air-sea traffic control, and in coordination of emergency rescue operations.

In the 1969 hearings, the Karth subcommittee discussed what was to become the highly successful use of the ATS–6 satellite over India for communications purposes in the furtherance of education, agriculture, medicine and many other forms of communication for the assistance of the people in villages throughout India.

OTHER PROJECTS IN SPACE SCIENCE AND APPLICATIONS

In addition to the programs discussed above, the Karth subcommittee dealt with a myriad of subject-matter issues during the 1960's. The subcommittee generally supported, but rigorously examined and exercised careful oversight over the following programs:

Observatories—astronomical, solar and geophysical.

Launch vehicle procurement (Scout, Delta, Atlas-Agena, Thor-Agena, and Atlas-Centaur).

Explorers—small satellites for Earth, solar and interplanetary scientific experiments.

Sounding rockets and balloons—for vertical soundings of the atmosphere and ionosphere.

Geodetic satellites—launching and procurement of data from GEOS class satellites for geophysics and oceanography.

Biosatellites—spacecraft with recoverable capsules, to investigate biological effects of weightlessness and cosmic radiation on small animals and primates, as well as plants; also ground-based research in bioscience, and search for extraterrestrial life.

SOLIDS VERSUS LIQUIDS

Throughout the decade of the 1960's, a majority of the committee repeatedly insisted that NASA was not devoting enough effort or resources toward developing a solid rocket motor. Aerospace contractors like Thiokol and Aerojet General pressed their claims on behalf of solids. The committee was not critical of NASA's major decision to go for the use of liquids, but emphasized that NASA was overlooking a good bet by not pushing research and development of solid propellants as a parallel, but modestly funded, course of action. The advocates of a modest level of support for solid boosters generally had a majority on the committee, but they ran into a stone wall of opposition in the top management of NASA. The all-out liquid propellant advocates on the committee did not argue as vociferously; they simply didn't have to. They knew that time and NASA were on their side.

The battle over use of solids was a classic illustration of how NASA used the tactics of divide and conquer to frustrate the will of a majority of the House committee. There were two important keys to the manner in which liquids won the long battle and remained supreme in the Apollo program; first, the Senate Committee on Aeronautical and Space Sciences was never as enthusiastic in their support for solids as was the House committee; second, Dr. Wernher von Braun, the premier rocket genius of the space program, was also the No. 1 cheerleader for liquid propellants. After all, the V–2's on which von Braun had worked so successfully at Peenemünde were fueled by liquid oxygen and alcohol. And the first rocket which was fired in 1926 by Dr.

Robert H. Goddard, "the father of American rocketry," contained liquid propellants. At the same time, the Air Force and the Navy had considerable experience with solids, which powered the Minuteman and Polaris missiles.

On April 20, 1961, Chairman Brooks indicated at an open session of the committee: "Many of the committee members, as well as myself, feel that we are not taking advantage of the state-of-the-art potential of large segmented solid boosters." To supplement the testimony of advocates of the liquid propellant approach, including NASA, Chairman Brooks summoned several witnesses on behalf of solid propellants.

Representative David S. King (Democrat of Utah), wrote a strong letter to President Kennedy early in 1961, urging more emphasis on solids. The President arranged a special meeting with Webb, Secretary of Defense McNamara and Director of the Budget David Bell, at which King presented his views. King, along with Representatives Anfuso, Karth, Randall and several others, lobbied hard for an increase in committee support for solids, and they amended the authorization bill in committee to allocate $18 million instead of $3 million for solids. Witnesses contended that this would enable the Moon flight to take place two years earlier and for less money.

Representatives George P. Miller (Democrat of California) and Perkins Bass (Republican of New Hampshire) issued "Supplemental Views" challenging the committee majority favoring the $15 million increase in support for solids. Miller and Bass stated:

> The committee heard testimony from several industrial witnesses who represented firms that produce solid propellants and solid-fueled rockets. They testified that a rocket test engine of one million and a half-pound thrust can be produced and flight tested with adequate funding, in about 18 months. While we do not question the sincerity and honest conviction of these witnesses, it is obvious that such statements are highly subjective and are qualified by an evident motivation of self-interest.

The majority of the committee, as expressed in its report, took the position that the $15 million increase—

> will permit a rapidly stepped-up program in the development of solid propulsion fuels, an area which in the committee's judgment requires much new work if the United States is to gain leadership in space exploration.

When President Kennedy addressed the Congress and set the goal of a manned lunar landing within the decade, his May 25, 1961 address also recommended that there be parallel development of a solid booster, as well as a liquid-fueled booster. The Congress down through the years, at the initiative of the Science Committee, annually authorized and appropriated funds beyond what NASA requested, earmarked for the specific and directed purpose of research and development on the 260-inch solid booster. Despite the successful development and test-firing of the solid motor by Thiokol Corp., NASA resisted use of solids in the Apollo program and other parts of its program. After a decade of fruitless attempts to lead the NASA horse to solids, the committee reluctantly concluded that NASA preferred to drink liquids.

In the early years of the committee's operation, Fulton regularly regaled many witnesses, and also at times bored his colleagues, with repetitive soliloquies on the virtues of boron as a propellant for space vehicles. NASA and the committee both conscientiously investigated the claims for boron, and found little evidence to recommend its use. However, these findings did not ever deter Fulton from continuing to advocate boron.

SUBCOMMITTEE MEMBERSHIP

At the start of 1967 and 1969, the following Members were assigned to the subcommittees headed by Karth and Hechler:

SUBCOMMITTEE ON SPACE SCIENCE AND APPLICATIONS, 1967

Democrats	Republicans
Joseph E. Karth, Minnesota, *Chairman*	Charles A. Mosher, Ohio
Thomas N. Downing, Virginia	Guy Vander Jagt, Michigan
Lester L. Wolff, New York	Larry Winn, Jr., Kansas
Jack Brinkley, Georgia	Jerry L. Pettis, California
Bob Eckhardt, Texas	

SUBCOMMITTEE ON SPACE SCIENCE AND APPLICATIONS, 1969

Democrats	Republicans
Joseph E. Karth, Minnesota, *Chairman*	Charles A. Mosher, Ohio
Thomas N. Downing, Virginia	Guy Vander Jagt, Michigan
Roy A. Taylor, North Carolina	Larry Winn, Jr., Kansas
James W. Symington, Missouri	Lowell P. Weicker, Jr., Connecticut
Edward I. Koch, New York	

SUBCOMMITTEE ON ADVANCED RESEARCH AND TECHNOLOGY, 1967

Democrats	Republicans
Ken Hechler, West Virginia, *Chairman*	Thomas M. Pelly, Washington
J. Edward Roush, Indiana	John W. Wydler, New York
John W. Davis, Georgia	John E. Hunt, New Jersey
William F. Ryan, New York	D. E. (Buz) Lukens, Ohio
George E. Brown, Jr., California	

SUBCOMMITTEE ON ADVANCED RESEARCH AND TECHNOLOGY, 1969

Democrats	*Republicans*
Ken Hechler, West Virginia, *Chairman*	Thomas M. Pelly, Washington
John W. Davis, Georgia	John W. Wydler, New York
George E. Brown, Jr., California	D. E. (Buz) Lukens, Ohio
Henry Helstoski, New Jersey	Louis Frey, Jr., Florida
Mario Biaggi, New York	

NASA's Langley Research Center Dr. Edgar M. Cortright (left) discusses with Subcommittee Chairman Ken Hechler (Democrat of West Virginia) latest developments in aeronautical research.

NERVA AND NUCLEAR POWER

Many battles raged within the Congress over the issue of whether or not to develop nuclear rockets, as well as on-board nuclear space power. The biggest fights occurred over the Nerva (Nuclear Engine for

Rocket Vehicle Applications) and Snap (Systems for Nuclear Auxiliary Power), most particularly Nerva.

Nerva followed a course of development much akin to the Electronics Research Center: Funding was strongly supported at the start by NASA, bitterly opposed by a minority in the Congress and a strong minority in the Science Committee, and finally the rug was pulled out by NASA under the pressure of budgetary restrictions.

In the early sixties, there wasn't too much opposition to the funding of Nerva or nuclear on-board power. The committee generally supported what NASA recommended in these areas. The programs were conducted jointly by NASA and the Atomic Energy Commission, and the coordination was made smoother by a Joint NASA–AEC office called the Space Nuclear Propulsion Office. When Senator Anderson became chairman of the Senate Space Committee in 1963, he regarded all nuclear programs as his "babies" and took good care to protect them and feed them adequately. The location of Los Alamos in New Mexico helped persuade Senator Anderson to support nuclear programs even more strongly.

The first rumbling of discontent and disagreement between NASA and the committee came in 1965, when the Bureau of the Budget decided to discontinue further funding of the Snap-8 on-board nuclear power development. The House committee, supported by the House, disagreed and put in $6 million to continue the development. At the same time, the "nuclear hawks"—Fulton and Bell—issued a separate report urging greater effort in the whole area of nuclear propulsion research in order "to insure our preeminence and security in the space field."

A bipartisan team of Wydler and Ryan led the fight within the subcommittee against any further funding of the Nerva nuclear rocket. Their opposition suddenly erupted in 1967, primarily because it became apparent that Nerva was being planned by NASA for any manned expedition beyond the Moon to Mars. The opponents argued that the $47 million specifically requested for Nerva in 1967—much of which had been sent to Capitol Hill in a late supplement to the President's January budget—would cost $1.5 billion over a 10-year period, plus more later if and when a mission were chosen. Schedule delays, cost overruns and technical problems also fueled the arguments of Nerva's opponents. The proponents of Nerva pointed out that lack of a nuclear rocket in the future would foreclose the picking of missions to the planets or lifting large payloads in Earth orbit.

Chairman Miller, Bell, Hechler, Davis, Fulton, and Pettis spoke for Nerva. The vote was a close one, with Nerva surviving a 121–91 teller vote. But then Fulton offered his surprise motion to recommit.

Despite the fact that he had previously spoken out in favor of Nerva, his recommit motion slashed the nuclear rocket program by some $24 million.

Senator Anderson, that perennial friend of nuclear development, helped the conference committee to restore the full amount of Nerva in 1967. But the Appropriations Subcommittee hit the program on the blind side and forced NASA to scale down its Nerva program from a nuclear rocket with 200,000 pounds thrust to one with only 75,000 pounds thrust.

The same cast of characters marched out to do battle in 1968. This time the Hechler subcommittee, supported by the full committee and the House, refused to fund Nerva, but once again Senator Anderson and his power in the conference committee prevailed. Pelly and Rumsfeld joined the opponents of Nerva. The House and Senate Appropriations Committees, while cutting NASA's total budget, expressed the opinion that there was enough flexibility for NASA to go ahead with the smaller Nerva rocket motor if they really wanted to.

The House had a strange reversal of feeling as the decade drew to a close. In 1969, NASA asked for $36.5 million for nuclear rockets, and this time the House voted $13.5 million more than was requested in order to speed up Nerva. It was argued that Nerva would be cheaper for post-Apollo missions. On the committee, Wydler opposed the $13.5 million increase, but muted his objections to the amount NASA requested. Representative Edward I. Koch (Democrat of New York), later to become mayor of New York, was the only committee member to make an all-out fight against Nerva in 1969.

In 1971, the House and Senate Appropriations Committees finally killed Nerva, by denying further funds. And when Senator Anderson retired from the Senate in 1972, Nerva lost its last big clout on Capitol Hill.

UNIDENTIFIED FLYING OBJECTS

As noted in chapter I, the Select Committee on Astronautics and Space Exploration had conducted a subcommittee hearing on unidentified flying objects. No conclusions were reached, and testimony was confined to an Air Force presentation on material assembled on sightings, plus explanations of phenomena where available.

The successive chairmen of the Science and Astronautics and Science and Technology committees were all reluctant to authorize full-blown inquiries into unidentified flying objects, on the grounds that the jurisdiction of the committee did not warrant coverage of the issue. Perhaps the real reason for the reluctance of the committee to grapple directly with the subject was the feeling that this was a "hot potato" which might consume an inordinate amount of time, plus focusing undue attention on the "UFO buffs" who might unduly divert the committee from more important missions.

Representative J. Edward Roush (Democrat of Indiana) was the most outspoken advocate on the committee who supported the need for public hearings. Chairman Brooks, who sanctioned committee inquiries on a wide variety of subjects, drew the line against any investigation of UFO's because he feared that such a hearing would bring public ridicule against the committee. Chairman Miller also declined to sanction any UFO inquiry on the grounds that the subject properly belonged within the jurisdiction of the Air Force and the Armed Services Committee. Congressman Roush bided his time, bringing up the issue casually on a number of occasions, realizing that gentle prodding and compromise worked better with Chairman Miller than direct confrontations. Finally in 1968 Roush worked out a formula which met Miller's approval: Roush offered to chair a one-man "Symposium" which would appear to be something less than a formal committee hearing. Roush agreed to limit the meeting to one day, to allow only bona fide scientists to testify, not to set up a special subcommittee for the purpose, and not to issue any kind of official report of the proceedings other than the text of the recorded symposium itself.

Six participants all accepted invitations and appeared at the symposium on July 29, 1968: Dr. James E. McDonald, Institute of Atmospheric Physics, University of Arizona; Dr. J. Allen Hynek, head of Department of Astronomy, Northwestern University; Dr. Robert L. Hall, head of Department of Sociology, University of Illinois at Chicago; Dr. Robert M. L. Baker, Jr., senior scientist, Computer Sciences Corp.; Dr. James A. Harder, associate professor of civil engineering, University of California at Berkeley; and Dr. Carl Sagan, Department of Astronomy, Cornell University. In addition, prepared papers were presented by Dr. Donald H. Menzel, Harvard College Observatory; Dr. R. Leo Sprinkle, Division of Counseling and Testing, University of Wyoming; Dr. Garry C. Henderson, senior research scientist, space sciences, General Dynamics, Fort Worth, Tex.; Dr. Stanton T. Friedman, Westinghouse Astronuclear Laboratory; Dr. Roger N. Shepard, Department of Psychology, Stanford University; and Dr. Frank B. Salisbury, head, Plant Science Department, Utah State University.

In opening the Symposium, Representative Roush declared:

We approach the question of unidentified flying objects as purely a scientific problem, one of unanswered questions. Certainly the rigid and exacting discipline of science should be marshaled to explore the nature of phenomena which reliable citizens continue to report.

A significant part of the problem has been that the sightings reported have not been accompanied by so-called hardware or materials that could be investigated and analyzed. So we are left with hypotheses about the nature of the UFO's. These hypotheses range from the conclusion that they are purely psychological phenomena, that is, some kind of hallucinatory phenomena; to that of some kind of natural physical phenomena; to that of advanced technological machinery manned by some kind of intelligence, that is, the extraterrestrial hypothesis.

We take no stand on these matters. Indeed, we are here today to listen to their assessment of the nature of the problem; to any tentative conclusions or suggestions they might offer, so that our judgments and our actions might be based on reliable and expert information. We are here to listen and to learn.

Chairman Miller, in welcoming the participants to the symposium, took great pains to underline his apprehension:

I want to point out that your presence here is not a challenge to the work that is being done by the Air Force, a particular agency that has to deal with this subject. * * * I want you to know that we are in no way trying to go into the field that is theirs by law, and thus we are not critical of what the Air Force is doing. We should look at the problem from every angle, and we are here in that respect. I just want to point out we are not here to criticize the actions of the Air Force.

In general, those who testified recommended that UFO sightings merited scientific study, rather than ridicule. One committee member, Representative Jerry L. Pettis (Republican of California), an experienced pilot, indicated that a number of his fellow pilots had observed unusual phenomena caused by "UFO's" which they had been reluctant to report for fear of being exposed to ridicule.

At one point, Representative Roush asked Dr. Sagan whether he believed in extraterrestrial life, and Dr. Sagan responded:

Congressman Roush, I have enough difficulty trying to determine if there is intelligent life on Earth, to be sure if there is intelligent life anywhere else.

One witness, Dr. Baker, stated his preference for the term "anomalistic observational phenomena" rather than "unidentified flying objects." When Roush protested that his Hoosier constituents might not cotton to the lengthy new characterization, and would prefer the term "UFO," Dr. Baker insisted that his new phrase "comes trippingly off the tongue" and the phenomena could be labeled "AOP's."

The symposium continued until after 4:30 p.m. on July 29 before adjourning. As indicated, no report or conclusions were issued on behalf of the committee, and no further action was taken on the subject.

TRACKING AND DATA ACQUISITION

In the period from the creation of NASA to the end of the 1960 decade, close to $1 billion was spent on building and operating the tracking networks and acquiring the almost endless flow of data which spewed forth from manned and unmanned missions, and many far out projects tra cked through the deep space network. Spacecraft tracking, data telemetry, spacecraft command, voice communication, and television all figured in these efforts.

Early in the development of the tracking networks, the committee encouraged closer cooperation with the tracking efforts of the Department of Defense, as well as the establishment of cooperative relationships with foreign countries and tracking facilities with the know-how and reputation of Jodrell Bank in England. A visit by several committee members to Jodrell Bank early in the program helped cement these relationships, and avoid some of the duplication which otherwise would have arisen.

A good example of how the committee influenced NASA policy, improved coordination with the Department of Defense, saved money for the taxpayers, and instilled a greater measure of common sense into a program occurred in 1963. For several days, the Hechler subcommittee meticulously examined, dollar by dollar, the requested expenditures for tracking and data acquisition. Chairman Miller, visiting the hearings, made some observations on the process:

Chairman MILLER. Mr. Chairman, may I say to the committee on those things, we are meeting here, authorizing money which is the upper limit that can be spent. Before any of this money is spent, it has to be appropriated. This is one of the functions of the Committee on Appropriations to determine whether or not $3 million or $1½ million or $24 million or $5 million will be immediately spent for this. This is the upper limit for the thing. Let's not confuse the fact that we are an authorizing committee and not an appropriations committee.

Mr. ROUSH. Thank you, Mr. Chairman.

Mr. RUMSFELD. I think that is a good point. However, recently the House considered an authorization bill, and each item was approved by the legislative committee exactly as requested with, in my opinion, insufficient discussion. I think it would be of great value to the members of the appropriations committee if these matters were gone over in some detail by the authorizing committee.

Chairman MILLER. I may say to the gentleman that I think this committee has gone over these items from time to time since it has received this authority with much greater care and skill than any other authorizing committee and there are only two in the Congress. We still have the matter of appropriating the money for these.

Now no one wants to cut you off or interfere with your right to investigate any of these items, but I just point this out to you, that in this field particularly and in the whole field in which we deal, it is almost highly impossible to predicate today what you are going to spend 18 months or a year from now or to get definitive information. * * *

Mr. HECHLER. Mr. Chairman, I have never seen as energetic or broad a committee chairman as would come to take such an interest in what a subcommittee is doing. We certainly appreciate it, Mr. Chairman.

Chairman MILLER. Whenever I say that I want the new members to know that it is in no criticism. I want them to satisfy themselves because that is the only way we can get the basis on which the committee operates.

Mr. HECHLER. Let me complete my thought. I appreciate serving with a subcommittee here which is as energetic and thorough in its work, also.

Mr. RUMSFELD. Well, I would say that this has certainly been the chairman's policy, as well as the chairman of the subcommittee, since I have been on the committee, which amounts to just a few weeks now, and I certainly appreciate it.

The following day, NASA was besieged with a barrage of questions concerning $90 million NASA asked to modify three ships for NASA's exclusive use in the Apollo program. Roush, Wydler, Hechler, Rumsfeld, Fulton of Tennessee, and other members of the subcommittee raised the warning flag that they objected to the concept that NASA seemed to want its own Navy. NASA insisted they had negotiated over a year with the Department of Defense, and could not conclude an arrangement which would give them the instrumentation ships on short notice when they needed them for the critical tracking missions required by the Apollo program. The subcommittee directed NASA to send a new letter to the Department of Defense, with specifications for the ships, and then called Assistant Secretary of Defense John H. Rubel before the subcommittee to clarify the fact that Defense could offer a satisfactory arrangement to NASA for $10 million less than NASA was proposing. Also, for $80 million, the Defense Department indicated it could supply five ships instead of the three initially requested by NASA, at an annual operating cost of $4.5 million less than NASA estimated.

The House conferees on the authorization bill then threw the whole problem into the laps of the conference committee which met in the waning days of August 1963. NASA persuaded the Senate conferees to give them a little more than $80 million, but the House conferees won the fight to require some tough language on coordination. In exchange for raising the authorization to $83.3 million, the House conferees persuaded the conference to stipulate that none of the funds could be obligated until a joint NASA-DOD study had been completed by January 1964, "that would result in a pooling of tracking ship resources." The conference backed up the subcommittee position 100 percent, and further required that priority for the use of the ships by NASA should be given to NASA, but that DOD should have responsibility for navigating and operating the ships under regulations jointly negotiated by NASA and DOD.

Following up the victory by the subcommittee, Chairman Miller dispatched a letter to Vice President Johnson asking him to take the initiative to crack some heads together, through the Space Council, to get some coordination. The fur began to fly. The Council rode herd on NASA and DOD to get them to give a high priority to the joint study. Within NASA, Retired Navy Adm. W. Fred Boone, head of the NASA Office of Defense Affairs, took central responsibility to move the study forward.

Admiral Boone reported that the negotiations "were long and at times contentious." But an agreed-upon solution met the congressional deadline of January 1964, and coordination was achieved. The committee decided to stay out of the argument over what to name the new ships. The Secretary of the Navy wanted to name them after cities which had little to do with the space program. Admiral Boone suggested that since two ships had already been named after Air Force generals Arnold and Vandenberg that it would be appropriate to name the new instrumentation ships the James E. Webb, Hugh L. Dryden, and Robert C. Seamans who, according to Admiral Boone were "the three men most responsible for the Apollo program." This was promptly vetoed by Webb, and the ships were named instead the Redstone, Vanguard, and Mercury, three names prominently associated with the early days of the space program.

Coordination, cost saving, and rigorous oversight were high on the list of committee priorities.

This philosophy dominated the efforts of the Hechler subcommittee to insure that building and equipping the network were accomplished at the lowest possible cost. As Representative Roush told the House in 1964:

> The committee has emphasized to NASA that, insofar as possible, the equipment and facilities authorized for the tracking network must serve all users to the maximum feasible extent. The placement of tracking stations should include consideration of the future space network requirements of the Department of Defense as well as NASA.

In 1965, the committee discovered that NASA wanted to purchase 40 acres of land for a tracking station at Antigua at a cost of $5,000 per acre. The committee asked the Corps of Engineers to examine the availability of other land, and directed that NASA look into working out cooperative arrangements with the Air Force or negotiate for land owned by the British crown. Roush initiated an inquiry which established the fact that because of agreements with Great Britain, crown lands could be obtained rent free. This information was forwarded to NASA, and NASA proceeded to select a new site on crown land, thus saving the taxpayers the purchase price of private land on Antigua. Furthermore, cooperative arrangements were worked out so that the Air Force provided ground support for the Antigua tracking station.

Working with the tracking and data acquisition program was a very complex business for members of the committee. Those in the business had an esoteric language which was difficult to comprehend. Even when committee members made inspection trips to view and ask questions about the tracking network, it was a different world where computers and tapes whirred and mathematical formulae seemed so complex as to defy any layman. Addressing tracking network personnel on March 18, 1972, Neil Armstrong said:

To those of you out there on the network who made all of the electrons go to the right places, at the right time—and not only during Apollo XI—I would like to say thank you.

Teague and Mosher both put it more succinctly. In separate statements at different times, they commented that those involved in the tracking and data acquisition program were "the unsung heroes of our space program."

FOR THE BENEFIT OF ALL MANKIND

If there were one theme which dominated the committee's thinking in relation to the entire space program, it was the strong determination that practical applications growing out of NASA's work should be made available quickly and effectively to American industry and consumers. The very first sentence in the National Aeronautics and Space Act of 1958 stipulates:

> The Congress hereby declares that it is the policy of the United States that activities in space should be devoted to peaceful purposes for the benefit of all mankind.

On July 5, 1960, the committee published its first report on "The Practical Values of Space Exploration." It became one of the most popular publications the committee produced, and it was reproduced by the thousands under different titles such as "For the Benefit of All Mankind."

Organizationally, NASA seemed almost determined to hide or down-grade many of its efforts on behalf of the average man in the street. The Space Act also provided, and NASA was directed to—

> provide for the widest practicable and appropriate dissemination of information concerning its activities and the results thereof.

To the distress of committee members, NASA seemed to overlook some of the obvious opportunities to tell the world that going to the Moon meant something far more to the American people than bringing back a load of Moon rocks.

Press releases and brochures were available through the NASA Public Information Office to send to those who asked, and by far the greatest number of inquiries came from school pupils. But it took the initiative of the committee to produce and distribute the publication on "The Practical Values of Space Exploration," a chore which NASA shunned for many years.

When Morton J. Stoller, Director of the Office of Applications, testified before the Hechler subcommittee on March 5, 1962, the bulk of his testimony was devoted to weather and communications satellites—two areas which clearly were devoted to practical applications, and which throughout the committee's history received strong support from the Congress. Included in a statement which ran 32 printed pages, which Hechler said had "broken all records for a massive statement before the committee," Stoller devoted a fleeting quarter of a page to what he termed "industrial applications." In terms which could hardly be termed ringingly enthusiastic, Stoller mumbled:

> Many industrial firms normally will not be exposed to the new developments in space technology in the course of their routine operations. However, NASA's program will be generating much in the way of new technical capability which all commercial organizations should have an opportunity to evaluate and use.

Many committee members continued to pressure NASA to take a more aggressive role in making space benefits available to both industrial and other users. As a result, in 1963, Dr. Seamans announced to the subcommittee that Dr. George L. Simpson had been named Assistant Administrator for Technology Utilization and Policy Planning with a responsibility for public information, data storage and retrieval, educational programs and industrial applications. Simpson himself was a good public relations man, and he pleased the committee with his opening statement:

> NASA is committed to a hard-driving effort to transfer the useful fruits of our research and development effort to the private sector of the economy in as quick and as useful a way as possible.

Although the subcommittee and full committee, supported by the House, annually attempted to raise the authorization for technology utilization, Congress never seemed to be able to instill in NASA the same enthusiasm which the committee felt for the value of the program. At a time in the midsixties when NASA's total expenditures soared over $5 billion, NASA was still budgeting only $5 million— less than one-tenth of 1 percent of the total budget—to technology utilization. The committee also found it very distressing that the average Administrator of Technology Utilization stayed in office about one year.

Despite these handicaps, the technology utilization program began turning out a vast number of "Tech Briefs" to alert industrial users of available products developed through the space program—products like aluminized mylar, developed originally as reflectors for satellites, used for jackets, parkas, blankets and sleeping bags; a lightweight fireman's air tank and breathing system based on technology developed for astronauts' equipment; tiny television transmitters which could be swallowed in a capsule and used to examine the stomach; and exotic lubricants developed to withstand extreme temperatures on the Moon.

With the reorganization of NASA and the subcommittees in 1963, the Karth subcommittee took over weather and communications satellites, as well as applications technology and other forms of satellites, while the Hechler subcommittee retained authorization authority over technology utilization. As noted, the Karth subcommittee continued to stress the superior investment opportunities in applications such as the Earth resources technology satellites. Chairman Miller and the subcommittee chairmen led the fight to persuade NASA to place greater emphasis in areas understood, appreciated and utilized by the public.

At the close of the decade, Representative Lou Frey, Jr. (Republican of Florida) began a renewed campaign, supported by all members of the committee, to focus more attention on the practical benefits of the space program. As Frey stated in his views appended to the 1969 committee authorization report:

First, increased steps must be taken by NASA to insure that a "payoff" orientation is present in all NASA planning for the future. Second, greater efforts must be made by NASA to transfer the scientific knowledge and technology from the space program to other phases of our life. Third, the citizens of this country who pay hard-earned dollars for this program must be shown by example and through non-technical language that they are receiving their money's worth, which they certainly deserve.

The committee pointed out countless other examples of practical benefits first developed in the space program, from the use of lasers in eye surgery to the home use of fuel cells, fire-resistant clothing and home furnishings, and the grooving of highways to prevent hydroplaning accidents. The electronic pacer, rechargeable from the outside, need now be implanted in the chest only once to give a new heart to the afflicted. The remarkable "sight switch," developed for activating switches in a spacecraft by a mere movement of the astronaut's eye, has been adapted to aid paralyzed people, and has been demonstrated before the Science Committee. Just as space scientists have used digital computers to clarify pictures televised from spacecraft, so is the same technique used to clarify and sharpen medical X-rays. Railroad tank cars, weighing half as much as steel cars, are being produced from the light-weight plastics developed for NASA for use in its rockets.

As the committee reached the end of the decade of the 1960's, the immediate goal of the lunar landing had been realized. Yet the corrosive influence of the Vietnamese war, as it did in every phase of American life, deeply affected the future of the committee's work.

In the 1970's, new blood energized the committee and its outlook. In 1975, a totally new challenge faced the committee as the Congress expanded its jurisdiction to cover research and development in nonnuclear energy and many other nonmilitary areas, including civil aviation, environmental research and development and the National Weather Service. In 1977, following the termination of the Atomic Energy Commission and the Joint Committee on Atomic Energy, the Science Committee also was given jurisdiction over nuclear energy research and development.

These and other challenges, and how the Science Committee met them, are discussed in subsequent chapters.

Dr. William H. Pickering (center), Director of NASA's Jet Propulsion Laboratory, with Representatives James D. Weaver (Republican of Pennsylvania), left, and Joseph E. Karth (Democrat of Minnesota), right.

Committee personalities of the 1960s':
From left, Representatives Guy Vander Jagt
(Republican of Michigan), Bob Eckhardt
(Democrat of Texas), and Jerry L. Pettis
(Republican of California).

Representative Thomas M. Pelly (Republican of Washington).

Representative William J. Randall (Democrat of Missouri).

Representative William Fitts Ryan (Democrat of New York).

Representative John E. Hunt (Republican of New Jersey).

Artist's conception of the first Shuttle flight.

CHAPTER VIII

Decision on the Space Shuttle

As the 1970's dawned, it was the worst of times for the space program.

The high drama of the first landing on the Moon was over. The players and stagehands stood around waiting for more curtain calls, but the audience drifted away. Five more successful Apollo flights to the Moon brought back valuable data far beyond the original expectations of the scientific community. But to many voters and taxpayers they were anticlimactic. The Nation sweated out the safe return of the Apollo 13 astronauts after an oxygen tank ruptured and aborted their mission, yet the brief and emotional concentration on the accident in space did not rally broad-based national support for expansion of the space program.

The bloody carnage in Vietnam, the plight of the cities, the revolt on the campuses, the monetary woes of budget deficits and inflation, plus a widespread determination to reorder priorities pushed the manned space effort lower in national support.

SHOULD WE LAND ON MARS?

Ignoring the storm signals, Vice President Spiro T. Agnew tried to copy what President Kennedy had accomplished in 1961. In a nationwide television interview at Cape Kennedy just before the launch of Apollo 11 to the Moon, Agnew called for "a manned flight to Mars by the end of this century." Unlike the enthusiastic response to the Moon goal by the Committee on Science and Astronautics in 1961, the idea of a Mars mission was greeted by a cold shoulder in Congress.

Chairman Miller, in an address to the House on August 11, 1969, bluntly stated: "I do not at this time wish to commit ourselves to a specific time period for setting sail for Mars." Teague, as Chairman of the Manned Space Flight Subcommittee, also shied away from supporting a manned flight to Mars. When Vice President Agnew, as Chairman of a Presidentially appointed Space Task Group charged

241

with outlining future goals, asked Teague for his views, Teague responded:

> I know of one major contribution that can be made. That is the development of space vehicles that can be used repeatedly, with basic characteristics in common with transport aircraft. In view of the potential in this area, I believe the reusable space transport should stand very high on our list of priorities.

It was not a new idea for Teague. As early as 1966, in a report of the Subcommittee on NASA Oversight which he chaired, entitled "Future National Space Objectives," Teague had included this recommendation:

> Immediate planning for a new generation of spacecraft capable of recovery at low cost and which are ground recoverable is a requisite to attaining lower total mission cost.

SPACE TASK GROUP RECOMMENDATIONS

The Space Task Group, chaired by the Vice President, included Dr. Lee A. DuBridge, Science Adviser to the President; NASA Administrator Thomas O. Paine; and Secretary of the Air Force and former NASA Deputy Administrator Robert C. Seamans, Jr. Reporting to the President in September 1969, the Space Task Group presented a smorgasbord of manned and unmanned space projects, with a series of options which virtually afforded the opportunity to move in almost any direction at varying speeds. Throwing in everything but the kitchen sink, the Space Task Group did mention that through concentrated effort, a manned mission to Mars would be possible by 1981.

Without congressional or public support for such a mission, the Mars project appeared doomed. Some committee members, notably Representative Thomas N. Downing (Democrat of Virginia) mentioned that if the unmanned probes to Mars confirmed the possible existence of life on the red planet, "then it was an entirely new ball game" and a manned mission would receive strong support. As no such evidence developed, congressional support for a manned Mars mission collapsed and attention was directed toward other areas.

The Space Task Group included among its multiple recommendations "a reusable chemically fueled shuttle operating between the surface of the Earth and low earth orbit in an airline-type mode." Other optimistic suggestions for "a space tug, or vehicle for moving men and equipment to different earth orbits", plus a 6–12-man "space station" and 50–100 man "space base" were recommended but never fully implemented because of budgetary considerations.

COMMITTEE REACTION TO THE SPACE SHUTTLE

Powerful support for the Space Shuttle came from Chairman Miller, Manned Space Flight Subcommittee Chairman Teague, Ranking Republican Member Fulton, as well as articulate senior members such as Waggonner, Fuqua, and Roudebush. The opposition was led by Karth, the third-ranked member of the committee, who was supported by Hechler, the fourth-ranked member, as well as Mosher and Pelly, who ranked second and fifth on the Republican side.

Chairman Miller may have unwittingly added some fuel to the flames of opposition within his own committee by the fashion he structured the 1970 committee hearings on the NASA authorization. In a move which harked back to the days when Chairman Overton Brooks centralized power in the full committee at the expense of the subcommittees, Chairman Miller dispatched a memorandum on February 11, 1970, announcing his intention of holding all NASA authorization hearings in 1970 within the full committee. This contrasted sharply with the practice in other years when the subcommittees had an opportunity to probe more deeply into NASA's projected program through more detailed public hearings.

Chairman Miller made one compromise toward subcommittee delegation in his memorandum:

It is my intention, this year, to take all testimony before the full committee, and thereafter have the subcommittees mark up the bill and make recommendations to the full committee.

When he opened the public hearings on February 17, 1970, Chairman Miller made another pointed observation which was received by some committee members in a quiet spirit of resistance:

I hope we will be able to accomplish this matter in two weeks. I will ask that your questions be short and to the point because a number of the questions which are posed to the witnesses can be answered by reading the backup books which have been sent to your offices. As I indicated in my memorandum, when we have conducted these hearings, the bill will be sent to the subcommittees for markup. I will ask the subcommittees to report back to the full committee within three days with their recommendations.

Subsequently, a bruising battle occurred within the committee on whether to support the Space Shuttle, which rapidly developed as the centerpiece of NASA's program for the 1970's. Added to this fight was a projection of the furor arising throughout the Nation over the issue of whether manned space flight deserved so large a slice of the national budget pie. Even though the budget presented for NASA in

1970 was the lowest since 1962—$3.3 billion—this issue erupted into heated debate within the committee, the Congress and the Nation.

COMMITTEE HEARINGS ON THE SPACE SHUTTLE

In six morning hearings totalling 12 hours, the committee rushed through the entire NASA budget between February 17 and 26, 1970. Most of the time was taken up by NASA presentations. When the committee members had an opportunity to question, their queries centered on the Space Shuttle and Space Station, with the major questions being directed toward cost and feasibility. Dale D. Myers, NASA's Associate Administrator for Manned Space Flight, pointed out that "we can carry to the Space Station people that are not trained as astronauts. We can carry chemists, metallurgists, physicians, astronomers, photographers—and I have added, since yesterday, Congressmen and Congresswomen." This prompted Karth to observe:

Do you have any candidates in mind? I will send you a list that I have.

With Chairman Miller's full knowledge, Teague took his Manned Space Flight Subcommittee out on his annual field trip to visit and interrogate contractors, and inspect NASA installations, just before Chairman Miller issued his edict against subcommittee hearings. As a result, early in February Teague's subcommittee visited and quizzed the officials of the Martin Marietta Corp. in Denver; North American Rockwell Corp. in Downey, California; McDonnell Douglas Astronautics Co. in Huntington Beach, California; and a joint meeting of the Boeing Company and Lockheed Missile and Space Co., in Sunnyvale, California.

Subsequent to the full committee hearings, Teague took his subcommittee to the Space Division, Chrysler Corp., at New Orleans, La.; Grumman Aerospace Corp. at Bethpage, N.Y.; and received reports from the Kennedy, Marshall and Manned Spacecraft Centers.

SUBCOMMITTEE MARKUP OF NASA AUTHORIZATION IN 1970

The Manned Space Flight Subcommittee was in a runaway mood in 1970. There was a unanimity of feeling, expressed by Chairman Teague, Ranking Republican Member Fulton, and members on both the majority and minority sides, that NASA's request for funds should be sharply increased. Fulton put it this way:

I believe that this is the year that we should move forward on manned space flight because we have had budgetary restrictions in the past two years. In light of that policy, I would recommend that we move up to the $4.2 billion level for the manned flight operations. So that would move the figure of $1,651,100,000 request of NASA for Apollo to $1,777,500,000.

In addition to stepped-up support for the Apollo program, the Manned Space Flight Subcommittee took a strongly bullish approach to the Space Shuttle and Space Station programs, urging an $80 million increase in these areas. All in all, the subcommittee opted for an increase of $297 million over the budget, in order to keep pace with the recommendations of the Space Task Group and to purchase long lead-time items necessary to keep the manned space flight program on schedule.

During the markup by the Karth Subcommittee on Space Science and Applications there erupted the first confrontation between supporters and opponents of the Space Shuttle. In the presence of Chairman Miller, who was attending as an ex officio member of the Karth subcommittee, Karth began to criticize NASA's cost estimates on Project Viking (the unmanned probe to Mars) as "atrociously inaccurate." Karth went on to suggest:

Here we are going into contracts on the Shuttle which for all practical purposes is a new program, not even a year old, and we haven't done the basic research necessary in the laboratory to determine just how this Shuttle vehicle ought to be built. * * *

I will predict on the record right here that program will cost at least three times what NASA today is saying it is going to cost just on the basis of our experience here in these other programs.

Chairman Miller sprang to an immediate defense of the Shuttle:

You are going to have something eventually that has to go out and visit the synchronous satellites that are going to be in space and be used for all time hence. We can't afford to orbit these things and have them go to pieces in six months to a year without being able to go out and recover them or perhaps fix them in space. That is what part of it is. You have got to have the Shuttle.

During the second day of the Karth subcommittee markup, Staff Director Frank Hammill observed: "I have heard that Mr. Teague is going to propose to the full committee that the manned space flight authorization be increased." Karth exploded:

I think that is ridiculous in a budget year like this year, particularly since the budget is within an eyelash of being unbalanced.

This stimulated the following interchange:

Mr. MOSHER. What does Tiger want to do with the extra money?

Mr. HAMMILL. Mr. Teague asked Mr. Myers to tell him what he would do if the manned space flight budget were increased. * * *

Mr. KARTH. The manned space flight people think they are bad off, but 50 percent of the budget is going to manned space flight.

Mr. HAMMILL. A little more, 54 percent, I think.

Mr. MOSHER. We could try all sorts of things worth doing if we wanted to add some money. Every direction I look I see crucial needs.

Mr. SYMINGTON. It is sort of Parkinson's law of budgeting.

For New York Congressman Ed Koch, the budgetary problem was even more serious than trying to decide on whether to support the Space Shuttle. Koch zeroed in against the Viking-Mars project in the Karth subcommittee, where he told his colleagues:

> I just for the life of me can't see voting for monies to find out whether or not there is some microbe on Mars, when in fact I know there are rats in the Harlem apartments.

KARTH BLASTS SHUTTLE

Even before the final committee markup of the NASA authorization bill in 1970, Karth went publicly on record with a scathing attack on the Space Shuttle. In an address on March 3, 1970 at a meeting of the American Institute of Aeronautics and Astronautics in Annapolis, Md., Karth labelled the President's Space Task Group Report and the NASA program which was based on the Report as "totally unrealistic." He exploded:

> Based upon my experience with Ranger, Centaur, Surveyor, Mariner, Viking and even Explorer, NASA's projected cost estimates are asinine. * * * NASA must consider the Members of the Congress a bunch of stupid idiots. Worse yet, they may believe their own estimates—and then we really are in bad shape.

Chairman Miller and Teague bristled at Karth's opposition. Miller did not want to add fuel to the flames by denouncing Karth publicly, but Teague was not at all bashful about expressing himself. In his long and successful service as Chairman of the Veterans' Affairs Committee, Teague looked on it as an unprecedented breach of Congressional courtesy and practice for the chairman of another subcommittee, who had not attended the hearings and field investigations, to take a strongly critical position against the findings and recommendations of a subcommittee of which he was not a member. Teague labelled Karth's public attack on the Shuttle as "just plain stupid," adding:

> Karth could have had much more influence had he worked within the committee, but instead he went out and made a bunch of speeches and got nowhere.

Teague was so angry at Karth's public opposition to the findings of the Manned Space Flight Subcommittee, that he swore that as long as he lived Karth would not become chairman of the full committee.

FULL COMMITTEE AND NASA AUTHORIZATION IN 1970

By the time the subcommittee reports reached the full committee, the battle lines were tightly drawn.

The Teague subcommittee report, adding $80 million for the Space Shuttle and Space Station, called for "more extensive and inclusive trade-off analyses and additional engineering studies" as well as "advanced prototype effort for testing and verification of preliminary designs." The report expressed the opinion that the additional funds would assist NASA and the Congress in future years to reach sounder decisions on the progress and timing of Shuttle and Space Station development. Despite the increase, plus other increases totalling close to $300 million over the budget, the report noted that this was the lowest construction request for manned space flight since the inception of the program.

The sparks began to fly in the full committee the minute the Teague subcommittee report had been completed. Referring to Fulton's long-time, repeated effort to get NASA to use boron as a launching fuel, Mosher's first crack was:

When I look at these proposed increases on page 3 and page 5, I get the impression that you guys have been drinking some of Jim Fulton's boron juice.

Teague countered by reminding the committee that the President's Space Task Group had urged a billion dollars more than the Bureau of the Budget recommended, and all the subcommittee was doing was restoring less than half that amount. When Karth got the floor, he said he was willing to vote for increases in the latter stages of the Apollo program, but as for the Shuttle, "for goodness sake, wait until that Phase B study is completed, so that we have at least some grasp of the magnitude of that program, so that we have some grasp of its potential, and so that we have some grasp of precisely what we are going to do with it, and the cost effectiveness of the program. * * * Now, I say, Mr. Chairman, if I didn't have so much respect for you I would probably be shouting and waving my arms and emphasizing my points with some profanity."

Fuqua jumped into the fray to demonstrate that the increases for the Shuttle and Space Station had been well thought out:

I don't think that we are proceeding on a crash-type basis, or in any manner other than a prudent manner, in trying to get the best for our space dollar.

Hechler appalled Karth with the suggestion that a rift was developing in the committee, and the following colloquy ensued:

Mr. HECHLER. All I can say is that when the fire bell of rebellion sounds, this old fire horse has great difficulty in not joining with Mr. Karth now that there is some indication that there is a spirit of rebellion.

Mr. KARTH. Mr. Chairman, I object to being held to the word "rebellion".

Mr. DOWNING. Mr. Chairman, I certainly don't think that this is a rebellion * * *. But I do think we are making a big mistake here to increase this budget by $300 million. Here we are going on the floor of Congress and advocating an additional

$300 million above the budget on a program that is losing romance with the American people.

Teague was angry at the criticism levelled at his subcommittee, and shot back:

> I think that our subcommittee—Lou Frey, Bob Price, Joe Waggonner, and Don Fuqua, and the group that got out and talked and worked from daylight to dark, have got a good feel for this thing, and I don't think we have to rubber-stamp something the Bureau of the Budget does. We are going along with the people halfway, going along with the people who are supposed to know something. That was the President's Task Group. What should we do, just sit back on our cans and let the Bureau of the Budget dictate every damn thing we do? * * * We are right, and we know we are right, and we know more about it than they do, and I bet you this subcommittee of mine knows more about this program than the Bureau of the Budget does.

When the full committee report came out, Teague's position was voted as the majority position. Karth's minority views on the Shuttle were endorsed by three Democrats—Hechler, Downing and Biaggi; and three Republicans—Pelly, Vander Jagt and Pettis in a written minority report. Mosher, in "Additional Views", also criticized the budget-busting recommendations of the Manned Space Flight Subcommittee.

THE SPLIT AMONG COMMITTEE REPUBLICANS

Just before the battle over the Shuttle on the House floor, it became apparent that in a close vote the position of the Republicans on the committee represented a crucial swing element. Fulton, as the long-time ranking Republican on both the full committee and the Manned Space Flight Subcommittee, had usually been an exasperating thorn in the side of both Miller and Teague, but now in 1970 his support of the Shuttle suddenly became of towering importance. Fulton's hospitalization as he recovered from a heart attack loomed as an important factor in the outcome of the vote on the House floor, which was expected to be close.

Teague had done his usual workmanlike job of lining up Republican support on his subcommittee. He could count on active help from Roudebush, Winn, Frey and Price, and certainly from Fulton on the Shuttle. It was an open question how many of the other Republicans would join Mosher in opposition.

On April 10, Mosher sent the following memo to all Republican members of the committee:

> Due to Mr. Fulton's illness, and expecting him not to be here for the NASA authorization bill debate on the floor next week (probably Thursday), I am preparing to manage the bill for our side.

As you know, I will take action to reduce the authorization to the level presented to us by NASA. I have talked to Jerry Ford in regards to how we can best handle this on the Floor and believe it is imperative that we have a meeting prior to the floor debate.

Several unexpected developments produced marked changes in the scenario. The rescheduling of the House debate until the later date of April 23 enabled Fulton to make a dramatic appearance on the House floor and manage the bill for the Republican side. Mosher's meeting of the minority members of the committee did not produce a consensus. Finally, Republican Leader Ford, who initially was sympathetic to Mosher's position, eventually wound up as a supporter of the committee position rather than joining the ranks of the Karth-Mosher opposition.

THE SHUTTLE FIGHT IN THE HOUSE

On April 23, 1970, Chairman Miller led off the debate on the Space Shuttle and Space Station with these comments:

The key to the success of this Nation's future space effort lies in the development of a low cost, recoverable, and reusable space transportation system. The reusable Space Shuttle will drastically reduce the cost of putting people and cargo into space In particular, the Shuttle will facilitate construction of a manned orbiting Space Station that will open up new areas of scientific and technological activity in the near neighborhood of earth.

"Frankly, I have hesitated to grab this tiger by the tail," Mosher told his colleagues in firing the opening gun of the opposition to "Tiger" Teague's efforts to increase the manned space flight authorization. Mosher argued:

We must put relatively greater emphasis on those aspects of the space program (where) the practical returns are the greatest * * * to the human beings right here on earth.

Mosher contended that at a time when the budget constraints were the most severe, it simply did not make sense to spend nearly $300 million above the budget for manned projects while holding the more practical applications of unmanned experiments at the lowest possible level.

Karth insisted that his purpose was "not to kill the project, but simply to establish a realistic pace for development." He added:

Before the Space Shuttle can be a reality, many difficult technological advances must be made in such areas as configuration and aerodynamics, heat protection, guidance and control, and propulsion. * * * As a matter of fact, NASA officials are divided on the fundamental questions of whether the Space Shuttle should be a fully reusable, two-stage vehicle, or simply a recoverable orbital stage launched by an expendable first stage.

KARTH AMENDMENT AGAINST SHUTTLE

Karth's amendment to the authorization bill in 1970 cut $240 million from the entire bill, eliminating the $80 million increase voted by the committee, cutting back an additional $110 million asked by NASA in its budget, and lopping another $50 million from manned space programs.

Representative H. R. Gross (Republican of Iowa) poked fun at the Karth Amendment by contending it did not go far enough. Gross proposed instead to slash $1.5 billion from the $3.6 billion authorization bill for this reason:

> There has been much talk about austerity here today. Well, anyone would have to have moon rocks in his head to believe there is any austerity in this program. As a matter of fact, with the amendment offered by the gentleman from Minnesota (Mr. Karth), it is still above President Nixon's budget. And that is austerity?

When Roudebush pointed out to Gross that the bill was less than spent the previous year, the caustic Gross shot back: "Well, so what? It was far too much last year." The Gross amendment was voted down, 67–19.

In the closing minutes of the debate, Karth made a strategic error in suggesting that the Space Shuttle would necessarily lead to a $50 billion to $100 billion manned landing on Mars. Karth stated that this "back door" approach to a Mars landing "is something I think we ought to debate loud and clear."

"There is no money in here for a manned trip to Mars," countered Fuqua. Roudebush added: "I am puzzled by the statement that the Shuttle is in some way mixed up with the Mars landing, when nothing is further from the truth." Chairman Miller also authoritatively persuaded his colleagues that there was no relation between the Space Shuttle and a manned Mars trip.

When the roll was called, many Members had left the floor and the results were in doubt. Miller and Teague had lined up their troops to stay at their posts, but the opposition was strong also. In a teller vote, as Members passed down the center aisle and were counted by Miller and Karth, it was obvious that the result was going to be a close one. Representative Louis Frey, Jr. (Republican of Florida) recalls:

> I'd lobbied pretty hard with the freshmen, and after the first rush of people went through, one of the freshmen from Maryland came rushing in from a meeting and went through the line on our side. He was followed by another Maryland Congressman. The gavel came down, it was announced to be a tie vote, and so the Shuttle stayed in. The second Maryland Congressman said 'Blankety-blank it! I went through the wrong way!' As I look at the Shuttle now, I often wonder what would have happened if he'd walked through the right way.

Just before the vote was announced, several people asked Chairman Miller, who was one of the tellers: "Have you voted yet? You had better vote!" Miller quickly went through the line supporting the Shuttle, although there was some question whether he had cast his vote at the same time as Karth had, when the teller vote commenced. In any case, both Miller and Karth announced their total votes as 53, which meant that on a tie vote the Karth amendment failed by an eyelash.

FULTON'S RECOMMIT MOTION

Now occurred another crisis affecting the fate of the Shuttle. Mosher, who opposed additional funds for the Shuttle, would have been the senior minority member authorized to offer the motion to recommit the bill but for the fact that Fulton, the senior Republican on the committee, had left his sick bed to be present for the Shuttle debate. Had Mosher been allowed to present his recommit motion, it was his intention to include the Karth-Mosher amendment in that motion. There was some question whether Fulton's strength would allow him physically to remain until the end of the debate, but Fulton was a stubborn man.

By offering the recommittal motion, Fulton saved the Shuttle, since Fulton's motion reduced the total authorization bill by only $30 million in the Apollo and space flight operations areas. Chairman Miller and Minority Leader Ford startled the House by announcing that they both were going to support the modest cut contained in the motion to recommit.

Ford then paid his compliments to Karth and Mosher for their opposition to the Shuttle, indicated he had conferred with both opponents several weeks prior to the debate, and "I must admit that many of their arguments were persuasive." Ford pointed out that the recommit motion "will do no harm to the program and yet will not hamstring the agency as to any new decisions for the future." He gave the distinct impression that the White House would not object to a net increase of $270 million over the President's budget, and certainly would be happier if the Karth-Mosher attack on the Shuttle were rebuffed.

The recommit motion breezed through on a voice vote, and then a surprising amount of opposition arose on final passage of the authorization, which survived by a vote of 229–105. Three committee members—Karth, Mosher and Koch—voted against the final passage of the NASA authorization bill in 1970.

WHERE DO WE GO FROM HERE?

The President's Space Task Group Report in 1969, and the first efforts in 1970 by the committee to finance the Space Shuttle, failed to answer all the questions about the future goals of the space program. There was a big let-down after the first manned landing on the Moon. Near-panic struck the aerospace industry as employment sagged from well over 400,000 to scarcely over 100,000 by 1970 among NASA-supported contractors. From January 1969 through July 1970, 74,000 people employed in manufacturing in the five-county Los Angeles area were thrown out of work; 57,600 of these were aerospace workers.

Technically trained engineers and scientists were pumping gasoline, or drawing unemployment or welfare checks, while entire divisions of aerospace corporations were being phased out. In the late summer of 1970, Chairman Teague's Oversight Subcommittee planned to hold September and October hearings on the present and future of the space program. An extended session of Congress and uncertainty over NASA's appropriation legislation forced the cancellation of the formal hearings, but the testimony submitted by NASA and industry officials was published as a special committee print. After outlining the sad state of the aerospace industry, the testimony agreed that "a revitalized space program, given strong direction and adequate funding, is needed for the United States to retain its technological preeminence in the decades ahead."

COMMITTEE LEADERSHIP ON FUTURE OF SPACE

Teague had no patience with those who were contending that this country had too many internal problems to afford a high level of spending for space. As Teague put it:

If Columbus had waited until Europe had no more internal problems, he would still be waiting, but the opening of the new world did more to revive European culture and economy than any internal actions could possibly have done.

Chairman Miller, who frequently stressed the need for a more personalized, less formal dialogue than the forum of a committee hearing would allow, decided late in 1970 to plan "a small gathering of the senior members of this Committee and leaders of the major aerospace companies." In a private note to a few committee members, Miller stated: "It is urgent that we share in an exchange of ideas on what the Congress and industry can do toward assuring a vigorous and continuing space program through the 1970's." Cocktails, dinner and an extended after-dinner confab took place at the Federal City Club, Sheraton-Carlton Hotel, in Washington on the evening of

January 28, 1971. Among the aerospace leaders who attended were
Thomas G. Pownall, President, Aerospace Group, Martin Marietta
Corp.; L. J. Evans, President, Grumman Corp.; Allen E. Puckett,
Executive Vice President, Hughes Aircraft Co.; and D. J. Haughton,
Chairman of the Board, Lockheed Aircraft Corp. Teague used the
opportunity to indicate that the aerospace companies were not doing
as much as they should to publicize and "sell" the necessity of main-
taining a vigorous space program. Pownall, in a letter to Teague on
February 4, 1971, acknowledged:

> Please be assured that we did get a message and that we will make an effort to
> improve our usefulness in some of the ways suggested.

Yet the evening's discussion failed to produce a firm consensus
among all concerned as to how to recapture the spirit and vigor of the
space program of the 1960's. As observed by Puckett in a letter to Chairman Miller on February 3:

> It seemed to me that our discussion gave evidence once again that even among
> quite knowledgeable people in this field there is a considerable diversity of views as
> to where our space program should go, and what should be its rationale.

TEAGUE ACCENTUATES THE POSITIVE

Early in 1971, President Nixon decided to appoint Dr. James C.
Fletcher as NASA Administrator to succeed Dr. Thomas O. Paine, who
had resigned September 15, 1970. Dr. Fletcher remained during the
Nixon and Ford administrations until 1977, winning the respect of the
committee for his candor and leadership.

Even before Dr. Fletcher took office, he received a jolting reminder
from Capitol Hill that underscored the intense interest and concern
which the committee had for the future of the space program. On
Sunday, February 28, 1971, Tiger Teague picked up his copy of the
Washington Sunday Star, and did a slow boil as he read an Associated
Press interview with Dr. Fletcher, based on a press conference in Salt
Lake City. "We may tend to reduce manned space flights in favor of
unmanned flights. It would be very exciting for man to go beyond the
Moon but I suspect that's beyond the country's budget. We will go
beyond the Moon but probably with unmanned flights," Dr. Fletcher
correctly predicted. What really caught Teague's eye, and angered him
was the line reading: "Fletcher said public interest is waning in the
space program 'and it's going to be up to us to have more exciting
things to rekindle that interest.'"

When Teague reached his office on Monday morning, March 1,
he was really fuming. He got on the phone to Jim Fulton, who was

equally upset, and the two fired off a 3-page telegram to Dr. Fletcher which really sizzled:

> I completely disagree with your view that public interest is waning in our national space program. 41 million people in the United States in the last year have looked at the lunar rock samples that have been returned from the Moon. Another 2 million have visited the facilities of the National Aeronautics and Space Administration across the United States. 14 million people alone viewed lunar samples at Expo '70 in Osaka, Japan. * * * The largest number of visitors in 1970 were at two of the NASA Centers most closely associated with manned space flight. Well over one million people visited the Manned Spacecraft Center in Houston, Texas. At the Kennedy Space Center in Florida, approximately 1,200,000 visitors toured the Center because of their interest in our national space program. I can personally attest to the fact that the Apollo 14 launch attracted more visitors to the Cape Kennedy area than ever before. * * *

> I hope this will help you recognize that there is great support on the part of the people of the United States in the manned space flight program and the NASA program in total. It seems to me that the most important job of the new Administration of NASA is to harness this grass roots support and to encourage a similar enthusiasm within the executive and legislative branches for our national space program. I regret that you come to a very positive agency with negative statements.

As Dr. Fletcher correctly notes today, the real waning of interest took place several years prior to 1971 with the escalation of the war in Vietnam and after the first landing on the Moon. The telegram itself stirred Dr. Fletcher to pay a personal visit to Teague. There Teague reiterated what he had told Dr. Mueller (see page 158), underlining the fact that the committee and Teague personally would back him up as long as he fought for the program.

The impulsive telegram did not constitute Teague's finest hour. Despite the examples he cited, any objective observer pretty well had to conclude that interest in and support for the space program had certainly declined since the glory days of the 1960's when Congress and the Nation were solidly and enthusiastically behind the Apollo program. As a matter of fact, Teague's own Subcommittee on Oversight, in a December 10, 1970 report—less than three months prior to the chiding of Dr. Fletcher—included this sentence in its Foreword:

> And despite truly remarkable successes, public enthusiasm for the NASA program seems to have waned.

One of Teague's best friends in the House, Representative Bob Casey of Houston, Tex., who was on the Science and Astronautics Committee from 1961 to 1965 before moving to the Appropriations Committee, brought no argument or response from Teague when he made this statement during House debate on June 3, 1971:

> Mr. Chairman, I want to state that in these days, when the interest of the public is waning in the space program, and when many people feel that we have

done all that we can or should do in the space program, it takes dedication and hard work on the part of this great committee—which I once had the pleasure of serving upon—to generate interest and to keep our great space program going.

Fortunately, Dr. Fletcher understood precisely what Teague was driving at in his telegraphic blast. At the University of Utah, where he had served as President, he was described by a fellow administrator as "devoid of vanity and willing to talk candidly about any issue, even those which might be embarrassing to either the university or himself." Teague got his point across forcefully: that at a time when the budget squeeze was on, the space program needed leadership and a minimum of negative soul-searching. Teague helped instill a fighting spirit into NASA, and never ceased to insist that the space program and its allies must accentuate the positive.

At the beginning of the 92nd Congress in 1971, the committee included the following members:

Democrats	*Republicans*
George P. Miller, California, *Chairman*	James G. Fulton, Pennsylvania
Olin E. Teague, Texas	Charles A. Mosher, Ohio
Joseph E. Karth, Minnesota	Alphonzo Bell, California
Ken Hechler, West Virginia	Thomas M. Pelly, Washington
John W. Davis, Georgia	John W. Wydler, New York
Thomas N. Downing, Virginia	Larry Winn, Jr., Kansas
Don Fuqua, Florida	Robert Price, Texas
Earle Cabell, Texas	Louis Frey, Jr., Florida
James W. Symington, Missouri	Barry M. Goldwater, Jr., California
Richard T. Hanna, California	Marvin L. Esch, Michigan
Walter Flowers, Alabama	R. Lawrence Coughlin, Pennsylvania
Robert A. Roe, New Jersey	John N. Happy Camp, Oklahoma
John F. Seiberling, Jr., Ohio	
William R. Cotter, Connecticut	
Charles B. Rangel, New York	
Morgan F. Murphy, Illinois	
Mike McCormack, Washington	

The Subcommittee on Manned Space Flight in 1971 was assigned the following members:

Democrats	*Republicans*
Olin E. Teague, Texas, *Chairman*	James G. Fulton, Pennsylvania
Don Fuqua, Florida	Alphonzo Bell, California
Earle Cabell, Texas	Larry Winn, Jr., Kansas
Richard T. Hanna, California	Robert Price, Texas
Walter Flowers, Alabama	Louis Frey, Jr., Florida
Robert A. Roe, New Jersey	

WINNING KARTH'S SUPPORT FOR THE SHUTTLE

In June of 1971, Joe Karth penned a little note to Tiger Teague about the space program. Enclosing an article from the St. Paul, Minn., Pioneer Press which depicted Karth's support for both the Space Shuttle and other space expenditures, Karth's note read:

DEAR TIGER: I just wanted you to know that while I don't agree with every d—— thing the Agency has for sale, I support the program even back home where my poll showed constituent support 5 to 1—in the *negative!*

Karth, who along with Mosher and a sizable group of committee rebels had led a nearly successful fight against overfunding the Space Shuttle in 1970, came around to supporting the Shuttle in 1971. He even went so far as to come out publicly for the Shuttle in these terms:

If we're going to have a space program, we're eventually going to have to develop a new transportation system, there's no question about it. We can't afford to build the short-launch vehicles that cost $5 million to $15 million which are treated like skyrockets.

What led Karth and the other rebels of 1970 to reverse their position in 1971? The answer lies in an interesting bit of parliamentary maneuvering within the committee.

In its 1971 presentation to the committee, NASA quietly dropped all references to a Space Station in their discussion of the Space Shuttle. The Space Station was a victim of chloroforming by the Office of Management and Budget. OMB also slashed the NASA request for funding the Shuttle from $190 million down to $100 million.

Following extensive hearings and their customary series of field trips, the Teague Manned Space Flight Subcommittee recommended increasing the authorization for the Space Shuttle from $100 million to $135 million. Additional increases in the manned space flight area made the subcommittee total $90 million above the budget. Since the Subcommittee on Advanced Research and Technology was advocating a $71.4 million increase in calendar year 1971 over the NASA budget request, and the Subcommittee on Space Science and Applications was holding its increases to $2.5 million, it almost seemed as though the stage was set for a repeat of the 1970 fight by Karth and his subcommittee colleagues.

When the full committee met on March 30, 1971, to consider the subcommittee reports, the mood appeared to be less combative than in 1970. Mosher, who had helped lead the Republican side of the fight against the Shuttle, was more subdued in his 1971 criticisms. He told the full committee in its executive session on March 30 that he was "somewhat unhappy and full of doubt" about the manned space flight increases. He explained his position this way:

> I certainly respect the subcommittee's judgment that this money could be well used. I think my doubts hinge around the matters of political expediency more than anything else. I just have to raise the question with the mood of Congress and the mood of the people the way it is today, is it good for the space program to offer such substantial increases over and above the budget on the floor of the House? Don't we just invite resistance, invite the House to cut back even further than they might if we didn't so dramatically draw their attention to these increases?
>
> With these various doubts in mind, I think I am going to have to reserve judgment and probably vote "present" so far as this committee report is concerned.

Teague responded to Mosher by pointing out that when the space program was being slashed a billion dollars a year, $4 billion was being voted for welfare. Bell mused that "one of the very reasons we made the increase was the fear they would be cut back and sometimes it is better to ask for a little more than we expect to get."

Wydler then expressed a common sentiment on both sides of the aisle when he observed:

> I have been going along with these cuts year to year. I really feel we have reached the point where we should stand up and say "enough." I think we have allowed the space program to be treated as a form of foreign aid in the public's mind, that we are just spending money, and getting us nothing. I think we better start redirecting the public's attention to the fact they are spent to hire American people, to do American productive work and to try to save that money a little bit that we are talking about particularly, and say that we are going to then have some kind of associate program to take care of some of these people that are thrown out of this work.

Karth then announced his opposition to the $35 million increase for the Space Shuttle:

> I personally have said I guess maybe on fifty occasions that I support the orderly, well-defined and well-engineered Shuttle program, but I do think that a $35 million increase at this point when we do not yet have the Phase B studies completed, and when we are embarking upon a major nine or ten billion dollar program * * * is probably the kind of an increase that will attract attention.

Karth termed the 35 percent increase desired by the Manned Space Flight Subcommittee "a little exorbitant." Chairman Miller and Teague both stood firm in their insistence on a $35 million increase. Teague was the first to give the glimmer of a possible compromise, when he said: "As far as I am personally concerned, $35 million is no magic figure at all. It was a figure we came up with after going back to NASA."

Karth then indicated he would offer an amendment to limit the increase to $20 million. In his last markup session prior to his death from a heart attack on October 6, 1971, Fulton finally made the suggestion that there should be a compromise because "I don't want a floor fight with this committee." Karth picked up the concept by suggesting that "I am trying to find some area where we can go to the House with a solid front." Wydler clinched the decision to move toward a compromise by a very timely statement:

> I just want to say to Mr. Karth I look at this as a practical matter now. * * * I think it is much more important on the authorization bill we act as one man. I think that is the significant and important thing, and quite frankly Karth help on this bill is of very great importance to the way the bill will move. * * * I think it is terribly important that we have a unified committee, and if that is the price, let's pay it, it is well worth it in my judgment.

Fulton's compromise amendment to limit the recommended increase in the Space Shuttle authorization to $25 million won support of the full committee. The important point is that once the compromise was adopted, Karth was locked in to support it, and his colleagues who had rebelled against the Shuttle in 1970 felt obligated to support the increase also. Mosher, who had threatened to vote "Present", wound up supporting the $125 million for the Shuttle and also the balance of the program.

FLOOR DEBATE ON NASA AUTHORIZATION IN 1971

The floor debate on the NASA authorization bill found the committee members unanimous in their support of the Shuttle. Fuqua and Frey, joined by nine other Members, issued "Additional Views" in the committee report which outlined the strongest arguments for proceeding with the Shuttle. The Fuqua-Frey statement ended with this assertion:

> The development of the Space Shuttle is essential if this Nation is to maintain its preeminence in space. We should proceed without delay. The technology necessary for the Space Shuttle development is at hand. What is required is the will to do it.

Clearly, the corner had been turned. Committee members who had opposed the Shuttle now joined a united committee front in its support. Instead of a separate report by opponents, as had occurred in 1970, there was now a separate report by Fuqua, Frey and their allies, on the positive side. The new display of unity effectively gunned down opposition to the Shuttle in the House authorization debate. And when Representative Bella S. Abzug (Democrat of New York) introduced an amendment to remove the $125 million Shuttle authorization from the bill, the committee members moved in with a whoop and a holler and obliterated her effort by a crushing voice vote.

THE COMMITTEE AND THE 1972 SHUTTLE DECISION

One of the Shuttle's early critics, Congressman Karth, left the committee in October 1971 to assume a position on the House Ways and Means Committee. The year 1972 also marked the last year in which Chairman Miller, defeated in the California primary, served on the committee. Both Miller and Karth had been charter members who had served since the creation of the Science and Astronautics Committee in 1959.

As support for the Shuttle rose in the House of Representatives, thanks to the leadership of the committee, the opposition mounted in the Senate, where Senators Walter F. Mondale (Democrat of Minnesota) and William Proxmire (Democrat of Wisconsin) led the criticism. NASA, and to an even greater extent the Nixon Office of Management and Budget, kept a wary eye on Congress in attempting to cost out the economics of the Shuttle.

"We did not think we could sell a 10 to 15-billion-dollar program to the Congress right then", recalls Dr. Fletcher in looking back in 1979 on the decision in 1972 to reduce the size and expense of the Shuttle. Clearly, Teague, Fuqua, Frey and the strongest boosters of the Shuttle felt that the correct course of action was to press forward with the original program for a completely reusable Shuttle and Space Station at a higher cost. To Teague's consternation, the President appeared to be leaning strongly toward his budget advisers instead of choosing the bold solution. Teague publicly denounced President Nixon for failing to support the Shuttle and the space program while the big debate on the Shuttle's future was going on during 1971:

He isn't even following the advice of his own Space Task Group. They told him and us that anything below a $4-billion budget for NASA was a "going out of business budget", but he's allowed those damned pencil-pushers in the Budget Bureau to set policy instead of following the experts' recommendations.

While the debate was going on during 1971 over the size and configuration of the Shuttle, the political cross-currents were already swirling over where the Shuttle was to be launched. Fuqua and Frey were the most articulate leaders to keep the launch facilities at Cape Kennedy, to protect NASA's billion-dollar construction investment at the Cape. The Chairman of the Senate Aeronautical and Space Sciences Committee, Senator Clinton P. Anderson (Democrat of New Mexico), terribly upset by termination of his favorite Nerva project, insisted that the Shuttle would be in trouble unless it were launched from White Sands Missile Range in New Mexico. Meanwhile, Chairman Miller mobilized the Californians to press for the use of Vandenberg or Edwards Air Force Base. NASA responded with a somewhat tongue-in-cheek letter pointing out that the Roush amendment to the authorization bill required that "consideration be given to the geographical distribution of Federal research funds whenever feasible", a provision which had rarely been used to govern decisions. In a letter to George Low of NASA, Teague seemed to lean toward the Florida site:

> Unless I am convinced that NASA is making maximum use of existing facilities, I intend to oppose any money for the Shuttle in every way, form or fashion * * *. It is not "pork barrel" as far as I am concerned.

NASA made a Solomon's choice in 1972 by concluding that Cape Kennedy and Vandenberg on the Atlantic and Pacific coasts should share the Shuttle's launching facilities.

On January 5, 1972, President Nixon announced his decision:

> I have decided today that the United States should proceed at once with the development of an entirely new type of space transportation system designed to help transform the space frontier of the 1970's into familiar territory, easily accessible for human endeavor in the 1980's and 1990's.

The decision involved a pared-down version of the Shuttle. Instead of a fully reusable system with a larger, manned crew, the final selection favored a two-man, recoverable orbiter which would still glide in for an Earth landing on return, but there would be an unmanned and recoverable booster and expendable fuel tanks. The smaller version was estimated to cost about $5.15 billion to the end of the decade of the 1970's. On behalf of NASA, Dr. Fletcher issued a public statement that the Shuttle would cost $5.15 billion plus or minus 20 percent. OMB instructed him never to mention this contingency again.

COMMITTEE REACTION TO THE DECISION

Although the committee had pressed for an early, firm decision on the Shuttle, and individual members like Frey had warned that to defer a decision beyond January would strengthen the opposition, Teague and Fuqua were not entirely happy with the cutback in the size of the Shuttle. With some asperity, Fuqua asked Dr. Fletcher when he appeared before the committee on February 8, 1972:

> Is this the final configuration, or later in the year are we going to hear that has been modified, as we did last year.

Frey noted:

> Of course, one of our sales pitches on the recoverable craft was the reduction in cost per pound, but now that you're going back to this concept, it seems to me you're

increasing the cost per pound, and it almost in a sense destroys the sales pitch that we originally conceived for the craft.

Not long after the decision was announced on the new configuration of the Shuttle, Fuqua recalled:

We had just finished defending one configuration on the Floor and then suddenly they announced they were going to change it. Tiger Teague got the top brass from NASA over here and raked them over the coals. We all wanted to know how long they had known they were going to change and how much of this kind of thing was going on behind the committee's back. They explained the reasons behind the changes, and everybody calmed down.

After that, though events moved pretty fast, they did try to keep us reasonably well informed.

For Teague, the decision was still the kind of small solution he wondered was really sound. At the public hearing on February 17, 1972, Teague pondered aloud whether 4 or 5 years hence "we will not look back and be sorry we did not have a more aggressive program".

THE PRESIDENTIAL CAMPAIGN OF 1972

Science Committee members who were disenchanted with President Nixon's "small solution" position on the Space Shuttle were horrified with the outright opposition by the initial 1972 Democratic National ticket—Senators George S. McGovern and Thomas F. Eagleton—as well as Mrs. Jean Westwood, the Chairman of the Democratic National Committee. Two weeks after President Nixon's decision was announced on January 5, Senator McGovern told a Florida campaign audience that if he were President he "wouldn't manufacture a foolish project like the Space Shuttle to provide jobs" and that the Shuttle was "an enormous waste of money". He labelled it "Nixon's boondoggle" and even attacked it in speeches to aerospace workers. Senator Eagleton, before his withdrawal from the Democratic ticket, indicated that the Shuttle "will deprive important social programs of much-needed revenue".

When NASA awarded a $2.6 billion Shuttle contract to the North American Rockwell Corporation, Democratic Chairman Westwood charged it was—

the latest, and perhaps most blatant, example of President Nixon's calculated use of the American taxpayers' dollars for his own reelection purposes.

Teague publicly took exception to Mrs. Westwood's statement, and dismissed her as "uninformed on the space program," which was for Teague an understatement. Fuqua said he—

deeply regretted that McGovern and Eagleton have taken such strong stands because, at least from Sen. McGovern's statements during the Florida primary, it was quite clear that he knew really very little about the Shuttle system that the President had endorsed.

And for Downing, fifth-ranked Democrat on the committee, it was also a shock:

> I don't know how large a bloc they are, but the ticket might as well write off all the voters who are affected either directly or indirectly by the aerospace industry.

A McGovern victory in 1972 would have meant a serious blow to the future hopes of the Shuttle, especially with the rising opposition in the Senate. Senate Majority Leader Mike Mansfield (Democrat of Montana) announced his public opposition to the Shuttle on January 15, 1972—ten days after President Nixon's decisive support. The death of Senate Appropriations Committee Chairman Senator Allen J. Ellender (Democrat of Louisiana) on July 27, 1972, caused NASA to shake in its boots, because it eventually resulted in Senator Proxmire, an avowed opponent of the Shuttle, moving up to the subcommittee chairmanship over NASA appropriations. But the powerful, united support by the House Committee on Science and Astronautics helped save the Shuttle in 1972.

"BELLA, IT IS NICE TO HAVE YOU WITH US"

After hearing NASA witnesses, and visiting the contractors and NASA Centers, the Manned Space Flight Subcommittee threw its hearings open to anyone who wanted to present testimony. At 1:30 p.m. on the afternoon of March 14, 1972, Representative Bella S. Abzug (Democrat of New York) appeared in opposition to the Shuttle. For some 700 pages prior to her appearance, the subcommittee had commended and complimented a parade of Government and industry witnesses describing how necessary the Shuttle was for America's future. Now the subcommittee members eagerly awaited the opportunity to tangle with an opponent.

The hearing opened innocently enough. Chairman Teague started off with a friendly greeting: "Bella, it is nice to have you with us". Mrs. Abzug responded in kind: "It is nice to be here, Mr. Chairman". She then launched into a broadscaled attack on the Shuttle:

> Now that NASA has reached the Moon, it is seeking a new, similarly glamorous toy for its next project and it feels that a Space Shuttle would be just the ticket. * * * I would remind you that the President recently vetoed as fiscally irresponsible a bill that would provide only $2 billion for child care centers, a mundane but urgent issue for the millions of working parents in this country * * *. I favor the use of lightweight, unmanned, instrumented systems which can produce the same results as our manned program at a fraction of its cost.

Representative Larry Winn, Jr. (Republican of Kansas) asked Representative Abzug whether she knew that the Shuttle would add $914 million to the gross state product of New York State and produce 12,594 new jobs for aerospace workers in New York. Mrs. Abzug answered that she was aware that this would add temporary jobs, but she favored a full employment program:

I believe we should utilize the energies and strength and the creativity of our own people and our own science to create new ways to improve the quality of life here on earth.

Winn pointed out that it was impossible "to take the money that we would spend on the Space Shuttle and put it into the expenditures for the social ills * * *. We don't have an appropriation program of 'put and take'. Each must stand on its own priorities".

Wydler commented on the plight of the aerospace workers on Long Island:

Thousands of these people are out of work, thousands of their children are suffering, they are suffering, their wives are suffering. They need some relief, some help from us in the Government. It would seem to me we would be helping those people if we were to pass this program. For that reason alone, it would seem a very people-oriented program. This money doesn't go overseas as foreign aid. It goes into our own country, to our own people, working and struggling to keep our Nation first in space and ahead in technology.

Fuqua called attention to the mobile cancer detection units in operation in the Bedford-Stuyvesant area in New York City, adding:

I think we are getting many things from our space program that are being applied to solve many of the human needs we have. I cannot think of a better program than a cancer detection unit in such areas as Bedford-Stuyvesant.

Mrs. Abzug countered:

We are not in disagreement on that. It is that I am inclined to favor the use of lightweight, unmanned instrumented systems which I understand can produce the same results as our manned programs, at a fraction of the cost.

Fuqua responded:

The cost would be reduced in half, even in the unmanned area, by having a Shuttle in order to place it in proper orbit, and the fact that they can return vehicles.

Chairman Miller entered the hearing room, and he and Representative Richard T. Hanna (Democrat of California) tried in vain to get Representative Abzug to admit that the Shuttle was really beneficial. Other committee members gave Mrs. Abzug a hazing for daring to oppose and presuming to know more than they did about the worth of the Shuttle. It almost seemed at times that they were attempting to accomplish a rite of exorcism for the heretical beliefs she espoused. Teague, presiding over the hearing, who was ordinarily the toughest-talking and bluntest member of the committee, ended the confrontation on a note as soft as he had opened it:

> Mrs. Abzug, thank you for your testimony. I think I have a point we can all agree on, that it would be great to be a Member of Congress, if we just knew the right answers.

COMMITTEE SUPPORT IN 1972 DEBATE

Although the committee report strongly endorsed the Shuttle once again in 1972, Congressmen Fuqua and Frey rounded up eleven additional committee members to sign "Additional Views" containing even stronger support. The "Additional Views" labelled the Shuttle a "national necessity", and were signed by eight Republicans (Frey, Bell, Pelly, Wydler, Price, Esch, Coughlin and Camp) and five Democrats (Fuqua, Davis, Cabell, Hanna and Flowers).

Committee members inspect progress on Space Shuttle at Marshall Space Flight Center. From left, Representatives Olin E. Teague (Democrat of Texas), Walter Flowers (Democrat of Alabama) and Don Fuqua (Democrat of Florida).

The committee recommended $200 million for NASA in 1972 to develop the Shuttle—precisely the amount NASA had requested from the Office of Management and Budget and the same amount budgeted by the President. Mosher, taking the floor for the first time as the ranking Republican on the committee, recalled the fight that he and

Karth had led against the Shuttle in 1970. He contrasted the situation in 1972, when the committee supported "a Shuttle program very much diminished and simplified, much less costly than was contemplated two years ago." Mosher added:

> It is the product of a tremendous amount of careful reconsideration and better preparation. It does represent the "go slow" policy which we urged two years ago. So, I stand before you today, confessing that I once was very much a skeptic concerning the Shuttle plans. Now, I have changed my mind and I believe it is for entirely valid reasons * * *. The Space Shuttle program as now proposed is considerably different from that first recommended. The total development costs for the Shuttle have been reduced from $13 billion to $5 billion, to be spread over some 6 years.

The floor attack on the Shuttle was led by Representative Les Aspin (Democrat of Wisconsin), who predicted costly over-runs in the development of the new program. Fuqua predicted that NASA would pretty much stick to the ball park estimates of $5.15 billion. NASA came close, but needed more funding in 1979 and later years.

Aspin introduced an amendment on the floor to eliminate the $200 million authorization for the Shuttle, and have the National Academy of Sciences conduct a one-year study on whether further funds should be spent on the Shuttle. Wydler pointed out that "there is no earthly use in sending to the National Academy of Sciences for a study * * *. We have that information available now. There is nothing more to study regarding it. This body can make up its mind whether it wants a Shuttle or not."

Both Majority Leader Hale Boggs (Democrat of Louisiana) and Minority Leader Gerald R. Ford (Republican of Michigan) spoke against the Aspin amendment. Boggs warned that "if we delay for another year, we will never regain the momentum the space program now has." Ford used the analogy of military weapons programs which had resulted in cost over-runs due to starting and stopping, and he urged that the Shuttle be carried through to its conclusion "on the schedule that has been announced."

The Aspin amendment was clobbered on a division (standing) vote by 103–11. Then the NASA authorization bill was passed by the comfortable margin of 277–60. The committee was victorious in keeping the $200 million authorization at that level throughout the legislative process, including the appropriation by both the House and Senate.

THE COMMITTEE AND SHUTTLE CONTRACTS

The committee, its individual members, and the Manned Space Flight and Oversight Subcommittees took a vigorous and continuing interest in how NASA awarded and administered the contracts for the Space Shuttle. The biggest controversy erupted over the initial contract for the main engine. The chief competitors were North American Rockwell Corp. and Pratt & Whitney Aircraft Division of United Aircraft.

When NASA announced July 13, 1971 that the Rocketdyne Division of North American Rockwell had been awarded a contract of half a billion dollars for the development of 35 Space Shuttle main engines by 1978, Representative William R. Cotter (Democrat of Connecticut) protested the award and he also requested an investigation by the General Accounting Office. Congressman Cotter and the committee, as well as the G.A.O., asked NASA for the reports of their Source Evaluation Board. It is interesting to note the different fashion in which NASA Administrator Dr. James C. Fletcher responded, as contrasted with two of his predecessors, Administrators Dr. T. Keith Glennan and James E. Webb. Dr. Glennan and Webb had adamantly insisted that any release of Source Evaluation Board data to the committee would compromise the confidentiality of their private-source assessments of the companies that were bidding. Dr. Fletcher blithely went ahead and furnished the committee with the requested information, including the detailed analysis showing the point totals of the competing companies in various categories, and precisely how the decision was reached to award the contract to the Rocketdyne Division. Interestingly enough, the dire predictions made by Dr. Fletcher's predecessors about destroying the confidentiality of the evaluation system never materialized. And the G.A.O. on March 31, 1972 concluded that the NASA award "was consistent with applicable law and regulations." Subsequently, on July 26, 1972, North American's Rocketdyne Division was awarded a $2.6 billion, 6-year contract for the Shuttle.

Understandably, all members of the Science Committee fought like tigers (no pun intended) for their states and congressional districts when it came to the awarding of contracts. Congressman Wydler was an effective advocate for Grumman Aerospace Corp., and joined the Manned Space Flight Subcommittee in order better to represent his district's interest in the Long Island firm's role in developing the Space Shuttle. Fuqua and Frey worked long and hard to help stress the preferable location of Cape Canaveral as a launching and recovery site for the Space Shuttle. Chairman Miller, despite his frequent admonitions that "the space program is no WPA program," unabashedly touted the superior advantages of California aerospace concerns. In a letter to Dr. Fletcher dated January 28, 1972, Chairman Miller urged NASA to get on with the final award of the main engine-Space Shuttle contract:

I trust the decision will be forthcoming soon to continue with the contract as awarded to Rocketdyne for the good of the Space Shuttle program and the welfare of the country.

An innocent little note attached to the carbon copy of Chairman Miller's letter included the typed notation: "Attached letter drafted by North American-Rockwell."

Even the have-nots of the committee tried to get into the act. At the request of Congressman Hechler, NASA staged an all-day conference and expended considerable Federal funds for an elaborate briefing of West Virginia manufacturers and small businessmen, advising them how to get a fairer share of space contracts.

The oversight activities of the committee were searching, analytical and thorough. Under the leadership of Chairman Teague, while he chaired the Manned Space Flight and Oversight Subcommittees and later the full committee, and ably followed by his successors—Fuqua and Downing—the annual visits to the contractors and installations, plus the insistence on close committee and staff contact, produced excellent results. The committee and its leadership effectively carried out the dictum of Chairman Teague: "We don't want any scandals in NASA." And there were no major scandals, either.

To be sure, there were instances of waste. There were instances of mismanagement, fostered by poor administration. There were losses of human life, and incredible errors which resulted in the loss of valuable Federal investments in spacecraft. Yet the space program was remarkably free of the kind of criminal activity resulting in the enrichment of private or Federal officials as sometimes seemed to occur in other Federal programs. The alertness of the committee oversight deserves credit for this result.

A NEW CHAIRMAN FOR MANNED SPACE FLIGHT SUBCOMMITTEE

When Tiger Teague succeeded George Miller as chairman of the full committee in 1973, the game of musical chairs for subcommittee chairmanships began. Throughout the 1960's and early '70's, Teague had been the guiding force of the Manned Space Flight Subcommittee which he chaired. Starting in 1963, Don Fuqua had joined Teague's subcommittee and the two worked well in tandem, ably supported as the years went on by Republican veterans like Winn, Wydler and Frey.

Fresh from two terms in the Florida state legislature, where he had been named in 1961 as one of the most valuable members, Fuqua was the youngest Democrat in the U.S. Congress when he was first elected at the age of 29 in 1962. The former state president of the

Florida Future Farmers of America was a farmer by both trade and training, and it would have been natural for him to be assigned to the House Committee on Agriculture, but another Floridian, Representative D. B. "Billy" Matthews, already occupied a slot on that committee. Up to 1963, Florida had never had a member on the Science and Astronautics Committee, and in that year both parties corrected the oversight by assigning Republican Edward J. Gurney and Democrat Fuqua to the Science Committee.

Fuqua's district includes the University of Florida, Florida State University and Florida A. & M. University, as well as five community colleges. The 1960 census had resulted in the creation of his district—which at first included 13 counties sprawled across the middle of Florida's panhandle. Another redistricting in 1966 spelled danger for Fuqua, pitting him against the popular veteran Congressman Matthews. The two Democratic incumbents fought it out and Fuqua survived the biggest political challenge of his career.

On the Science Committee, Fuqua rose through the ranks and in 1971 was named the first chairman of the new Subcommittee on International Cooperation in Science and Space. In January, 1972, he moved up to take over the chairmanship of the important Subcommittee on NASA Oversight of which Teague himself had been the first chairman.

As Chairman of the Manned Space Flight Subcommittee, Fuqua presided over the big development decisions relating to NASA's Space Shuttle, the final phases of the manned orbiting Skylab, and the successful Apollo-Soyuz link-up in space between American astronauts and Soviet cosmonauts. The Manned Space Flight Subcommittee merged with and was renamed the Subcommittee on Space Science and Applications after the reorganization and expansion of the full committee's jurisdiction in 1975. Fuqua then chaired in the following years not only the further development of the Shuttle, but also all other activities of the National Aeronautics and Space Administration with the exception of aeronautics.

OVERSIGHT ON THE SHUTTLE PROGRAM

In its oversight activities, the committee closely followed NASA's procedures throughout the Shuttle program, with emphasis on safety, scheduling, costs, manpower, facilities, and reliability. As initially planned, the Shuttle was to cost $5.15 billion in 1971, at which time the first manned orbital flight was scheduled for 1978. At the committee review on October 18, 1979, NASA indicated that the cost would probably amount to 20 percent over the estimate in 1971 dollars, while the launching of the first manned orbital flight was scheduled by the middle of 1980.

Attempting to avoid delays, the committee spent considerable effort weighing the amounts of money needed each year to accomplish the announced objectives. This meant insuring that sufficient manpower, both in NASA installations and industrial contractors, was consistently available to do the job and carry out the program effectively.

The committee recognized that it was crucial to provide sufficient funds for getting the Shuttle program designed, and to procure the necessary long-lead items essential to carry out the schedule. Through their questioning in the 1973 hearings, for example, Fuqua and Frey ascertained that the tight cost ceilings imposed by the Office of Management and Budget were holding up the employment of manpower on critical subcontracts. As a result, $25 million was added by the committee to the $475 million which NASA originally was allowed in the President's budget. Still, the program was underfunded.

THE OPPOSITION IN 1973

The indefatigable Representative Bella S. Abzug challenged the Shuttle authorization in 1973, through an amendment to eliminate all funds for that purpose. She noted during the debate:

It has been argued the Space Shuttle would enable us to leave the Earth when it becomes too crowded or too polluted for existence here. I can understand people wanting to leave the planet, especially—at this time—some people at the White House. I think the Space Shuttle will be so stuffed with armaments that there may be no room for people.

Returning to the House for the first time since a hospital stay, Teague led the fight against the Abzug amendment in 1973. In addition to the usual arguments favoring a low-cost transportation system in space, Teague brought out the benefits which would accrue from the numerous payloads which the Shuttle could carry—which could result in "medical and health care, materials and manufacturing processes, and earth resource exploration." The Abzug amendment was defeated on May 23, 1973 by a division (standing) vote of 95–20.

THE MAIN ENGINE PROBLEMS IN 1974

From the start of the Shuttle program, it was clear that the development and qualification of the main engine was the principal pacing factor. In an attempt to avoid time slippages, NASA requested $889 million from the Office of Management and Budget in 1974, and was granted only $800 million. This cut critically affected the schedule on the main engine. William A. Anders stated on September 19, 1979:

The Nixon administration did not live up to agreements of initial funding and subsequent budget levels nor was the contingency recommended by NASA allowed.

During the authorization hearings, Fuqua had this colloquy with Dale Myers, Associate Administrator for Manned Space Flight in 1974:

Mr. FUQUA. The $89 million that was reduced by OMB in the request for the Shuttle, if you had that money, where would you put it?

Mr. MYERS. We would put it, I think, in the area of the orbiter subsystems, mostly into the subcontracts on the orbiter and into the main engine.

Mr. FUQUA. How much in the main engine would you think you'd need?

Mr. MYERS. I think it would be about $20 million of that $89 million that would go into the main engines.

As a result of this and other inquiries, primarily in the field, the committee decided to add $20 million to the NASA authorization to enable the work on the main engine to get back on schedule. In 1974, Representative Abzug confined her opposition to critical questions and neither she nor any other Member introduced an amendment to cut Shuttle funding in 1974.

The $20 million which the House added, however, was pared down to a mere $5 million increase when the issue was resolved in the conference committee. This prompted Mosher to observe when the conference report returned to the House floor:

The compromise reached was an increase of $5 million, or $15 million less than the House had sought. This compromise is a signal that the Congress is looking to NASA to hold the Space Shuttle program to original NASA estimates; we will be very reluctant to provide supplementary funding for every minor program perturbation encountered.

As on other occasions, the committee was deeply concerned with the problem of continuity of trained technicians and general manpower problems in 1974. When Rockwell International President J. P. McNamara was briefing the committee on minority hiring, Representative John N. Happy Camp (Republican of Oklahoma) wondered: "Is that minority you're talking about Republican?"

THE SHUTTLE AND THE AUTOMOBILE

"The Shuttle will do for the exploitation of space what the automobile did for interstate travel," Winn told his colleagues during the 1975 debate on the NASA authorization bill. Frey, another outspoken advocate of the Shuttle, indicated that in 1975 there were almost 31,000 contractor employees in 47 states working on the Shuttle, an employment figure which was due to rise to 34,000 in 1976 and 50,000 by 1977.

By 1975, the serious, organized opposition to the Shuttle was winding down. Instead of attacking the Shuttle, Representative Bella S. Abzug directed her fire at a NASA-drafted section of the authorization bill which empowered the NASA Administrator to prohibit the disclosure of technical information if it "contains ideas, concepts or designs which have been submitted in confidence to the Administration by any person, firm or institution" or might result in release of information to foreign competitors. Mrs. Abzug pointed out that the Freedom of Information Act fully protects both trade secrets and the national security, and she submitted an amendment to delete the offending section. Chairman Teague, after consultation with the House Government Operations Committee which administers the Freedom of Information Act, accepted the Abzug amendment.

Representative Herman Badillo (Democrat of New York) noted that "President Ford has asked Congress to hold the line on spending, exercise fiscal restraint, and enact no new social welfare programs, yet would provide more than $1 billion for research and development of a Space Shuttle." But neither Badillo nor any other Member of the House offered any amendment in 1975 to cut the Shuttle authorization, a signal that the Shuttle expenditure was generally supported and well justified through the committee's leadership.

AUSTERITY HITS THE SHUTTLE

The effect of funding cuts in 1974 and 1975 produced a 15-month slippage in the Shuttle schedule, pushing the first planned orbital flight farther down the road. The stretch-out in schedule also increased the projected total research and development cost from $5.15 billion to $5.22 billion, in 1971 dollars. By 1979, cost was over $6 billion.

In 1976, NASA's budget request was cut $182.6 million by the Office of Management and Budget. Most of this cut was sustained by the research and development area, and of course had its effect in the development of the Space Shuttle as well. One of the challenges faced by the committee was how to insure that safety, reliability, good management and cost controls could be achieved with a minimum adverse effect on the scheduling. Fuqua remarked during the floor debate on the NASA authorization bill in 1976:

> In the past year, detailed reviews were made to assure us that our program was both technically sound and cost effective. As a result, some tests were deleted and deferred * * *. Too much testing can be costly * * *. In the case of the Shuttle, NASA deleted some large module testing. In these cases, the tests were found to be redundant, or alternate verification methods were defined which were more cost effective.

Winn, the ranking Republican on the subcommittee, observed in 1976:

> NASA's Space Shuttle program remains within cost and schedule despite budgetary constraints and past deferrals. During our series of field hearings, I became convinced that the morale of NASA personnel and their principal contractors is quite high in the face of these pressures, and they are doing an excellent job on the Shuttle.

According to Frey:

It appears to me that NASA and the major Shuttle contractors have responded very well to budget austerity and are still within cost and schedule on this very challenging program. NASA has had "to do more with less" and will suffer the loss of an additional 500 civil servants (next year) * * *. I should remind my colleagues also there is no more room for stretching out the Shuttle and meeting cost and schedule commitments.

THE AIR FORCE AND THE SHUTTLE

Ever since the designation of Vandenberg Air Force Base as a western Shuttle launch and recovery site, the Air Force participated with NASA in the development of the Shuttle program. Hearings by the subcommittee in 1977, chaired by freshman Representative Albert Gore, Jr. (Democrat of Tennessee) explored the nature of the Air Force activities. As budget difficulties forced the scaling down of the Space Tug, originally planned to boost Shuttle payloads from Earth orbit into geosynchronous orbit or deep space, the Air Force proceeded with development of an "inertial/upper stage" (non-recoverable).

At the same time, private industry was developing spin stabilized upper stages with the use of Atlas/Centaur and Delta expendable launch vehicles. In his testimony before the committee on January 26, 1978, Lt. Gen. Thomas F. Stafford, Commander of the Air Force Flight Test Center and former Gemini and Apollo astronaut, praised the cooperative relationship between NASA and the Air Force on the development of the Shuttle:

I just cannot agree with the critics that NASA and the Air Force are not getting along well, since I have been in it deeply for 13 years.

THE FIFTH SPACE SHUTTLE ORBITER

In 1978, members of the committee were horrified to find out that NASA's budget had been clipped again, to cut off funds for a proposed fifth Space Shuttle orbiter. Representative Jim Lloyd (Democrat of California) was the first to raise the flag on behalf of the fifth orbiter in this question to General Stafford:

I notice that we have cut out one of the Space Shuttles in this year's budget. Within the realm of practicality, from your own point or view and wearing the uniform that you are, could you comment on that? Is that the right thing to do or should we go forward a little stronger?

General Stafford went as far as he could and still remain supportive of the President's budget:

Looking at the Department of Defense's requirements for satellite payloads, and what NASA has before it, it would be a very tight schedule, particularly if there is any delay. But as long as the production line is open to procure the fifth one at this time, I certainly go along with it.

Congressman Winn was in a determined mood when NASA Administrator Dr. Robert A. Frosch appeared before the Subcommittee on Space Science and Applications on February 23, 1978. He laid it on the line:

The decision to delay the procurement of the fifth orbiter is very depressing. It's my understanding that in the next two years the plan is to save $57 million by delaying the procurement. But if the fifth orbiter is procured it will then cost an additional $235 million. I can only draw one conclusion from this: that the Administration must feel that the odds are really stacked against the fifth orbiter to make that wild a gamble.

Representative Jim Lloyd (Democrat of California)—second from left—examines model of the Space Shuttle with Representative Doug Walgren (Democrat of Pennsylvania)—far right. Astronauts Charles G. Fullerton (left) and Fred W. Haise, Jr. (second from right) participated in the approach and landing tests of the Space Shuttle at the Dryden Flight Center in California.

Lloyd and Winn received unanimous support from the subcommittee and full committee, as well as the House of Representatives, in adding $4 million to the NASA authorization in 1978 to restore the fifth Space Shuttle orbiter which had been budgeted out. Wydler, the ranking Republican on the full committee, added his support for the fifth orbiter during the floor debate. Strong support for the Shuttle program also came from another Republican committee member, Representative Robert K. Dornan of California. The committee-recommended increases for purchase of long-lead items and to keep the option for a fifth orbiter also passed the Senate, thereby insuring its inclusion in the final conference committee action. In addition, the committee persuaded the Appropriations Committee to include the $4 million increase to lock it in.

THE SHUTTLE IN PERSPECTIVE

Because millions of people throughout the Nation did not share in the same commitment to the Space Shuttle as they had to the Apollo program when it first started, it is remarkable that the committee successfully managed to push the program through the Congress throughout the decade of the 1970's. It is true that in areas closer to Vandenberg Air Force Base, Kennedy Space Center, and the various contractors, people flocked by the thousands to see and support this new development in space transportation. Yet the committee shouldered a major educational burden in convincing the Congress that over $5 billion should be invested in a bird which would not fly until after 1979. In the early 1970's the argument was used that tremendous savings would result from bringing down the launch costs. As time went on, the justification for the Shuttle continued to emphasize savings, and also stressed the versatility of missions which the Shuttle could perform.

The leadership of Tiger Teague, first as chairman of the Subcommittee on Manned Space Flight, then as chairman of the full committee from 1973 through 1978, was a major factor in the successful progress of the Shuttle. Teague initiated the practice of annual visits to the contractors and space installations, which was carried on by Don Fuqua when he succeeded to the subcommittee chairmanship and to an even greater extent as full committee chairman in the 96th Congress. The active participation and support of leading Republican Members like Jack Wydler, Larry Winn, and Lou Frey were vital in paving the way toward smoother progress for the Shuttle. Aggressive oversight by the committee, through repeated hearings, field trips, queries and published reports, also was an important feature of the legislative process. The annual reports which the committee developed on the Shuttle, supplementing the formal hearings record, show how this incredibly complex mechanism was developed and pushed forward despite many obstacles and delays.

Looking back on the many milestones of the past decade, the day after he was chosen as the new Chairman of the Committee on Science and Technology, Congressman Don Fuqua reflected that one of the Shuttle's biggest technical and administrative problems had been the development of its main engines. Fuqua observed that management problems dominated the early phases of the Shuttle development, and gradually as time went on they were intermixed with engineering problems. The decision to use existing facilities, on which the committee insisted, saved millions of dollars and as many headaches for the program.

SUPPLEMENTAL FOR SHUTTLE

On March 28, 1979, Fuqua addressed the House on the first of two big monetary deficits which plagued the Shuttle in 1979. He told the House that unless NASA received an $185 million supplemental shot in the arm, there would be a 4–6 months' delay in the first orbital launch and more than $1 billion of additional costs to NASA. The tight budget restrictions which had kept NASA's spending within 10 percent of the original estimates "have required pushing testing to late in the program and consequently difficult technical problems have been encountered and are being overcome later in the development cycle", Fuqua told his colleagues. Winn added that "nearly half of the 42,000 contractor personnel would be laid off" if the $185 million supplemental were not enacted. He made this observation on the management of the program:

> The very nature of this program has the potential of many serious impairments and setbacks. In spite of this huge potential, however, this program is within 6 to 12 months of the original 1971 schedule and 10 percent of the original cost. I submit to you that there cannot be any mismanagement when a program of this magnitude and complexity is as close as it is to the original plan.

Wydler also stressed the bipartisan nature of the support for the increased funding. Nelson and Flippo also spoke for authorizing the additional $185 million, which passed by the topheavy vote of 354–39.

ADDITIONAL FUNDS REQUIRED

Not until after the passage of the supplemental authorization by the House did NASA notify the committee that several hundred million dollars more would be needed in 1979 and 1980 to meet unforeseen problems. The 1979 funds were subsequently reprogramed from production funds. Committee members were understandably angered that NASA officials had assured the committee in January and February that the supplemental would be enough to keep the Shuttle on schedule. As soon as he learned the shocking news, Fuqua on May 4, 1979 ordered a review of NASA's operating procedures and management practices and scheduled Shuttle hearings for June 28, 1979.

The atmosphere was tense when Dr. Frosch and the contractor representatives assembled in 2318 Rayburn on June 28. Fuqua and Winn reiterated their strong and continuing support for the Shuttle, but made no bones about their displeasure with the failure of NASA to communicate the problem. Winn put it most sharply:

> The apparent cost over-runs which have been incurred could have profound effects on the entire space program, not just the Space Shuttle. The political controversies that will occur because of these over-runs will continue for some time and may do irreparable damage to the integrity of NASA as a mission-oriented agency. * * *
>
> After spending all of these years traveling from one briefing on Shuttle status to the next, I feel like I have totally wasted my time. The visits gave me the confidence to go before my colleagues in the House of Representatives and fight for the necessary support to move this program along. I can see now that it was a false sense of confidence.

Dr. Frosch explained simply that "it has been necessary for us to spend more resources to accomplish the development program than we had planned", requiring an additional $220 million as a budget amendment to the regular authorization bill passed in 1979. Even with these additional funds, NASA estimated that the first manned orbiting Shuttle flight would be delayed from its projected November 1979 date until 1980. Dr. Frosch added:

> Early in the Space Shuttle program NASA established a philosophy of maintaining an austere budget environment. Budgetary reserves were maintained at Headquarters and only utilized after review by the highest levels of management. This was a different philosophy than used in Apollo, in which reserves were approved and maintained at lower levels of management.

The unforeseen developments raised the total cost of the Shuttle to over $6 billion in 1971 dollars, which was about 20 percent above initial estimates. These events resulted in a tightening of NASA's management control, as well as a much closer oversight by the committee through its visits to NASA centers and more frequent and franker communication with both headquarters and field personnel, as well as contractors.

Although NASA had a reserve fund known as "Allowance for Program Adjustments" (APA), Dr. John Yardley, NASA's Associate Administrator for Space Transportation explained it this way in his colloquy with Fuqua:

> We also, I will have to confess, thought we were getting a little pressure from Marshall and Kennedy to get in and get some of the APA before Johnson and Rockwell used it all up, if you want me to be brutally frank. So we were somewhat suspicious of the inputs at this time. They were pretty fuzzy.
>
> Mr. FUQUA. There was a raid on the cookie jar?
>
> Mr. YARDLEY. Right.

Winn added this graphic comment:

> It seems to me that you guys were drowning, but you didn't really know you were drowning so you didn't yell for help.

To this, Dr. Frosch responded:

It is always a question of difficult judgment as to when you cross the line between crying wolf because you think something might happen and informing people because you're pretty sure something might happen.

On August 30, 1979, Chairman Fuqua released a report of the Subcommittee on Space Science and Applications which took both NASA and Space Shuttle contractor management to task for their shortcomings in assessing Shuttle budget requirements. In releasing the report, Fuqua noted:

The Space Shuttle program has been austere from the very beginning and program reserves have been inadequate to cope with cost growths and schedule delays, which have resulted from work deferrals from one year to the next throughout the life of the program.

Winn stated:

I have been very displeased with the financial planning that has taken place in the past year. There is no doubt in my mind that NASA has the capability to effectively manage the Shuttle program and develop realistic financial estimates. However, this recent cost overrun is a drastic mistake in these times of fiscal austerity. I hope NASA will draw upon their capability to provide more realistic cost estimates in the future.

On a more positive note, the subcommittee report expressed confidence in the integrity of the system design of the Space Shuttle program. Fuqua warned that "NASA must demonstrate and reestablish its credibility with regard to controlling cost growth and forecasting budget requirements." The report recommended that an annual financial assessment of the Space Shuttle program be conducted above the level of NASA's Office of Space Transportation Systems.

At the subcommittee's fall program review of the Space Shuttle program on October 18, 1979, Wydler remarked:

I am deeply worried about what is happening to the Space Shuttle. * * * it could well mean serious difficulties for our national space program in the years ahead. My feeling is that we haven't got this program under control, that we really don't know when we are going to be ready to fly, that the cost overruns are well in the neighborhood of about a billion dollars.

The final chapter has not yet been written on the success or failure of committee oversight on the Shuttle, as the first manned orbital flight has, as of this writing, not yet occurred. But the record of the 1970's is an instructive and revealing account of how a congressional committee grappled with a totally new program in a highly technical field where the targets were always moving at incredible speeds.

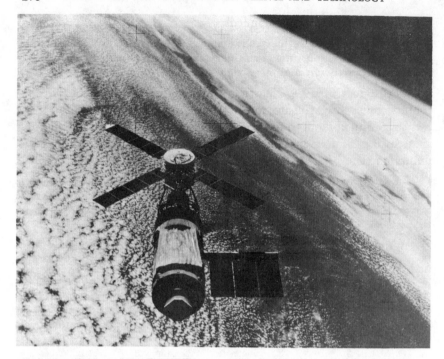

Overhead view of the Skylab space station taken by Skylab 4 crew which included Astronauts Gerald P. Carr, Edward G. Gibson, and William R. Pogue.

Representative Bob Bergland (Democrat of Minnesota), later Secretary of Agriculture, a committee member from 1972 to 1975.

Representative Charles B. Rangel (Democrat of New York), who fought against racism among NASA employees at Johannesburg (see pages 345–50).

CHAPTER IX

Space Science and Applications in the 1970's

On the first anniversary of the first manned landing on the Moon, Chairman Miller on July 20, 1970, made this observation in a public address before the Engineering Foundation Conference in Deerfield, Mass.:

> The Apollo program met a very real national need. * * * The Committee on Science and Astronautics of the House of Representatives identified that need nearly a year before President Kennedy made his appeal to the Nation to launch the Apollo program. The President could hardly have set such a bold and challenging goal for the Nation in the sixties without knowing that many key Members of Congress were already behind him. * * * From a Congressman's point of view, I can say only that it is of great value in the annual battle for funds to have a firm commitment to completing the job and a schedule that must be met.

APOLLO APPLICATIONS BECOMES SKYLAB

The committee enthusiastically supported funding of the Apollo Applications program during the nineteen sixties, although a majority of the committee became miffed at Fulton for slashing the program through several recommittal motions. Early in 1970, NASA announced that "Apollo Applications" had been redesignated as "Skylab." With the cancellation of the Air Force's Manned Orbiting Laboratory (MOL), NASA had the manned orbiting workshop to itself, a decision which the committee encouraged and helped to fund.

Somewhat overshadowed in the hectic debate over funding the Space Shuttle in 1970 was the strong committee support for adding $75 million for the manned Skylab in the authorization bill passed by the House in 1970. Roudebush, during the debate in the House, reminded his colleagues:

> Personally, I am a great believer in the authorization committees of this House. I do not suppose there is any group of men more familiar with our space program than the House Committee on Science and Astronautics. The committee's considered judgment was that a portion at least of the original recommendations by NASA to the Bureau of the Budget should be restored. * * * $75 million has been added to augment the spacecraft and subsystems for a low Earth orbiting laboratory called Skylab. This additional funding would give emphasis to Earth resources and medical experiments, and would permit work to commence in the field of design for a second orbital workshop.

279

RENEWED SUPPORT FOR SKYLAB

In the authorization hearings in 1971, Congressman Winn first raised the question of a possible second series of Skylab flights, noting:

> After the Skylab missions in 1973, we face at least four years in which there will be no U.S. manned space flight.

The committee in 1971 recommended adding $30 million to examine the possibility of a second Skylab set of missions in 1974. The House approved this addition. Also, the committee recommended, and the House endorsed adding $15 million for Skylab "for a rescue capability for the most probable failure situations."

The $15 million add-on was a personal victory for Manned Space Flight Subcommittee Chairman Teague. Safety in space had been one of Teague's highest priorities from the start of the space program. The loss of three personal friends and the investigative hearings on the Apollo fire seared the issue even more deeply on his mind. In a brief and pointed letter to the President on November 5, 1969, Teague noted:

> One portion of this future effort continues to concern me and that is the ability to provide space rescue and to react adequately to space flight emergency. Both the Space Task Group Report and the NASA report discussed in substantial detail future significant directions necessary for a well-balanced space program. However, no discussion or consideration is provided in this report to react to space flight emergencies and to provide for space rescue capability. The programs proposed fail to provide the focus and impetus necessary to assure the adequate planning for a true space rescue capability.

As their initial response to the Teague letter, NASA appointed a space station safety adviser and also established a Shuttle safety advisory panel. As a subsequent followup, during 1970 the NASA centers at Cape Canaveral, Huntsville, and Houston made feasibility assessments of providing a crew rescue capability for Skylab. This resulted in a decision to modify a command and service module by removing the astronauts' stowage lockers so as to accommodate a five-man crew instead of three; the modified craft would then be launched with two astronauts if necessary to rescue the three astronauts in the Skylab orbital workshop.

The committee solidly supported Skylab, and some members expressed their feelings even more strongly. In 1971, Bell, supported by Goldwater and Fulton, had this to say in "Additional Views" attached to the committee report:

> There is equipment in inventory which would permit follow-on Skylab activity at a minimal additional investment. Furthermore, there are numerous productive experiments which could be flown, particularly in the area of applications.

Similar views, with strong minority support for both extending Skylab missions and speeding up the Shuttle, were expressed by Winn and Price, and endorsed by Frey, Goldwater, Camp, and Fulton.

The enthusiasm of the House committee for a second series of Skylab missions, plus progress toward a space station, did not meet the same response in the Senate. The "other body", as House Members by tradition and courtesy referred to the Senate during formal debate, simply declined to make any changes whatsoever in either the budget or substance of the Skylab program. The Senate report in 1971 bluntly stated:

> Your committee does not agree with the position taken by the House of increasing funds (for Skylab). NASA has testified that they have no intention of going forward with a second Skylab. Therefore, your committee feels that the additional $45 million is unnecessary.

When the conference committee met, the Senate conferees stood firm in their opposition to a second Skylab series. It was all the House conferees could do to get the Senate to agree to adding $15 million for Skylab rescue capability. The action of the conference committee doomed the Skylab series to end after the 1973 flights.

Following the conference committee meeting, Teague, as chairman of the Manned Space Flight Subcommittee asked the Subcommittee on NASA Oversight in October 1971 to do a review and status report on both Skylab and the Space Shuttle. The report was based on extensive visits to contractors and installations in the field. The optimism expressed in the report concerning costs, performance and scheduling proved fully justified by the actual results achieved when the Skylab missions were flown in 1973. At the time of the report, which was completed in January 1972, the committee was still holding out the option that it might somehow be possible to have a second series of Skylab missions. These hopes were dashed with the realization—a familiar story—that there simply wasn't enough money available for anything extra.

The Subcommittee on NASA Oversight which made the Skylab report included the following:

Democrats	Republicans
Thomas N. Downing, Virginia, *Chairman*	John W. Wydler, New York
Olin E. Teague, Texas	Robert Price, Texas
Ken Hechler, West Virginia	Barry M. Goldwater, Jr., California
Walter Flowers, Alabama	John N. Happy Camp, Oklahoma
Charles B. Rangel, New York	

"One of the most significant benefit-oriented programs of the space age" was the characterization applied to Skylab by NASA's Dale D. Myers in his 1972 testimony before Teague's Manned Space Flight Subcommittee.

1973—THE YEAR OF SKYLAB

Three successive and successful manned flights of 28, 59, and 84 days were followed very closely by the committee in 1973 and 1974 as the members assessed the achievements of Skylab. The three crews of Skylab astronauts orbited the Earth 2,475 times and traveled over 61 million nautical miles in space. On May 14, 1973, shortly after the unmanned Skylab workshop was launched, the meteoroid shield was torn away, and two solar cell arrays were lost. This meant that the valuable workshop was overheated and underpowered.

Eleven days after the first flight, astronauts Charles Conrad, Joseph Kerwin and Paul Weitz were launched to rendezvous with and repair the workshop. They demonstrated the ability of human beings to perform difficult repair and construction work in space. A portable sunshade was deployed and one of the solar arrays was freed.

On July 17, 1973, the first crew of Skylab astronauts appeared before a joint meeting of the House Committee on Science and Astronautics and the Senate Committee on Aeronautical and Space Sciences, with Chairman Teague presiding. The 1973 lineups of the full committee and the Subcommittee on Manned Space Flight were as follows:

Democrats	*Republicans*
Olin E. Teague, Texas, *Chairman*	Charles A. Mosher, Ohio
Ken Hechler, West Virginia	Alphonzo Bell, California
John W. Davis, Georgia	John W. Wydler, New York
Thomas N. Downing, Virginia	Larry Winn, Jr., Kansas
Don Fuqua, Florida	Louis Frey, Jr., Florida
James W. Symington, Missouri	Barry M. Goldwater, Jr., California
Richard T. Hanna, California	Marvin L. Esch, Michigan
Walter Flowers, Alabama	John N. Happy Camp, Oklahoma
Robert A. Roe, New Jersey	John B. Conlan, Arizona
William R. Cotter, Connecticut	Stanford E. Parris, Virginia
Mike McCormack, Washington	Paul W. Cronin, Massachusetts
Bob Bergland, Minnesota	James G. Martin, North Carolina
J. J. Pickle, Texas	
George E. Brown, Jr., California	
Dale Milford, Texas	
Ray Thornton, Arkansas	
Bill Gunter, Florida	

SUBCOMMITTEE ON MANNED SPACE FLIGHT

Democrats	*Republicans*
Don Fuqua, Florida, *Chairman*	Larry Winn, Jr., Kansas
Walter Flowers, Alabama	Alphonzo Bell, California
William R. Cotter, Connecticut	John W. Wydler, New York
Bob Bergland, Minnesota	Louis Frey, Jr., Florida
Bill Gunter, Florida	John N. Happy Camp, Oklahoma

When Representative Don Fuqua (Democrat of Florida), center, became Chairman of
the Manned Space Flight Subcommittee, he worked closely with Representatives Larry
Winn, Jr. (Republican of Kansas) and Louis Frey, Jr. (Republican of Florida), right. Here
they are shown (right) at NASA's Ames Research Center, California. At left is Martin A.
Knutson of Ames and, second from left, Thomas N. Tate, committee staff.

At the close of the hearing where the first Skylab crew testified,
Congressman Esch brought general agreement with his statement:

> More than anything else, your flight demonstrates the need to have man in space.
> But for the last 12, 13 or 14 years, we have had a strange dichotomy in this com-
> mittee and in NASA between manned and unmanned flights. Isn't it about time we
> get over that dichotomy, admit we need both * * * an integrated system of manned
> and unmanned?

It fell to the Manned Space Flight Subcommittee to hold a public
hearing on the investigation report on what had gone wrong with the
shield and solar arrays in the first Skylab workshop. On August 1, 1973,
Chairman Fuqua opened the hearing by noting that despite the criticisms contained in the investigative report by NASA "it is important,
however, that we not forget the overwhelming success of the first
Skylab mission, accomplished by the dedication and outstanding work
of the Skylab astronauts and the ground team to turn a potential
failure into an outstanding success." The subcommittee concentrated
on the reasons why the meteoroid shield had not been designed to fit
tighter to the tank of the workshop so that aerodynamic pressures

would not have torn it off after liftoff. The committee's objective was to make certain that both in design, test and inspections these errors not be repeated in future programs. The committee brought out the failure of forceful communication between the designer and the contractor who was carrying out the design; Fuqua characterized it as "almost like building a house and failing to hook up the plumbing in the bathroom."

In presenting the NASA program to the House of Representatives in 1974, the committee for the first time was asking no money for Skylab, but had an opportunity to express justifiable pride in the accomplishments of the $2.5 billion program. The last manned Skylab flight of 84 days splashed down on February 8, 1974, conclusively proving that human beings could withstand extended stays in space and perform useful tasks. Reflecting on Skylab's accomplishments, Congressman Bell told his colleagues during the authorization debate in 1974:

> Skylab gathered information on the Earth's resources and environment to help with such problems as air and water pollution, flooding, crop deterioration, and erosion.

Congressman Fuqua in 1974, labeled Skylab an "unprecedented success," adding:

> From its unique vantage point in space—beyond the atmospheric veil of Earth—Skylab's sensors searched out and recorded new and far-reaching information about the solar system, the Sun, the Earth, and man himself.

In a letter to Sam Lindsey of Old Town, Fla. on July 23, 1979, Fuqua explained:

> At the time that the Skylab development and launch was completed in May 1973, the design of the Space Shuttle was underway and planned for first launch in 1978. At that time the orbital life of Skylab was estimated to extend from 1979 to 1983, depending on assumptions as to predicted solar activity. It was also envisioned at the time of Skylab launch that the Space Shuttle would be available to support either a reboost of Skylab during reentry or a controlled deboost of Skylab during reentry into a remote location. However, since the time Skylab was launched, the Space Shuttle first launch schedule was slipped and the Skylab was reentered during the early portion of the previously predicted period.
>
> With the advent of the Space Shuttle next year, we should be in a position to avoid random reentry such as that which occurred with Skylab.

Following some hysterically overblown news media warnings, Skylab fell harmlessly to Earth on July 11, 1979, primarily in the Australian outback, with some pieces falling into the Indian Ocean. Fuqua summed up some of the contributions of Skylab in his letter to Sam Lindsey:

Skylab was a particularly productive scientific program. It demonstrated that such space activities can be of enormous practical value to life on Earth. The program included over 50 scientific, technological, and medical experiments. There were high-resolution astronomical studies of the Sun at short wavelengths not observable from Earth, medical research associated with man's living and working in space for extended periods of time, and investigation and application of remote sensing to the location, measurement, and protection of Earth resources.

All told, the three crews spent 740 hours observing the Sun with telescopes and brought home more than 175,000 solar pictures. Such data are changing longstanding theories of solar physics and could lead to more practical use of the Sun's vast energy on Earth.

More than 46,000 photographs and 40 miles of data tape obtained by Skylab's Earth resources instruments have been used by government and industry for studies ranging from agriculture to zoology.

SCIENTIFIC EXPLORATION OF THE MOON

Between 1970 and 1972, when the Apollo program officially ended after the splashdown of Apollo 17 on December 19, 1972, the committee fought a losing rearguard action to try to extend the number of flights to the Moon. The committee argued that the equipment, personnel and facilities were all available, and therefore they should be used to capitalize on the investment of $23.5 billion over a 11½-year period.

In 1970, the Manned Space Flight Subcommittee added nearly $300 million to the President's budget, including funds for Apollo 18 and 19, which NASA eventually canceled when later budgets got tighter. Congressman Wydler interrupted a long and enthusiastic statement by Congressman Fulton, who was supporting the Moon flight increase, with this question:

I am wondering exactly how do we justify to the people the fact that we would have an Apollo 18 and 19? What are they going to do with them?

Fulton answered:

The technology developed in the space program has changed the lives of every one of us, and we shouldn't ridicule it. It has advanced the boundaries of human knowledge so far that if we begin to limit our horizons and not look ahead, to advance at the edge of the unknown, and if we prejudge it, we will still be sitting looking up at a green cheese Moon.

Wydler was still unconvinced:

I am just wondering what you are going to tell the general public when they say, why do you need Apollo 18 and 19 in addition to all the rest of it?

Fulton's response raised the specter of the Russian threat which had spurred the decision made in the early 1960's:

We have to get going on the planning and the long lead items or we are going to get caught again, just like we were in 1958. And I am simply not going to let the USSR get ahead of us once we beat them out.

Interestingly enough, Congressman Karth, who led the close and bitter fight against the Space Shuttle in 1970, said to Chairman Teague concerning proposed Moon flight increases by his subcommittee:

Now, let me say, Mr. Chairman, that I agree with the principle involved in what your Committee did insofar as it relates to the actions you took on Apollo and Saturn V production. * * * And I say that because here we are talking about a program that, Mr. Chairman, we have spent $25 billion on. And now we want to discontinue it. That doesn't make much sense to me.

Although the House supported the big increases the committee wanted in 1970 to extend the Apollo program, the Senate balked at the increased amounts for manned space flight. A lively conference between the House and Senate resulted in a substantial increase in the Apollo authorization in 1970, but in later years the committee resigned itself to sticking pretty close to the budget.

Chairman Miller's campaign to stir up aerospace workers, scientists, and other space enthusiasts to lobby harder for the manned space program did produce a great deal of activity on Capitol Hill. The scientific community was split; some strongly favored greater emphasis on unmanned missions, while other scientists deplored the cancellation of Apollo 18 and 19. The President felt the additional flights were impossible because of budgetary limitations. In early September 1970, Chairman Miller received a flood of letters from all over the country, urging continuation of the Apollo program. Although sympathetic toward the manned space concept, Chairman Miller sagely observed:

However, the gain from these additional two missions must be balanced with the current NASA fiscal restraints.

He went on to defend the course of action which NASA was taking, in the context of the President's budget decisions, and delivered this parting shot at many of his correspondents:

Had your views on the Apollo program been as forcefully expressed to NASA and the Congress a year or more ago, this situation might have been prevented.

By hindsight, it seems unlikely that even the strongest and most adept mobilization of the supporters of more manned flights to the Moon could have successfully overcome the adverse feeling in the country in the early 1970's. Congress and the Nation could be persuaded to support Skylab, the Space Shuttle, and a modest level of activity by NASA in many other areas. But as the NASA budget was squeezed down to the plateau between $3 billion and $4 billion annually, it became obvious that manned flight would be restricted to Earth orbital activities. Von Braun's dream of a manned flight to Mars was not in the cards for the 20th century, at least.

THE COMMITTEE AND APOLLO IN 1971–72

Opening the hearings on NASA's authorization on March 2, 1971, Chairman Miller began by introducing Apollo 14 astronauts Alan Shepard, Stuart Roosa, and Edgar Mitchell, their parents, their wives, and then their children. Amid the glare of television lights, Chairman Miller pronounced:

We are opening hearings in the most crucial period of our space program. The decisions made based on these hearings will largely determine the direction and emphasis in our national space program in the 1970's.

A month later, when Chairman Teague had his Manned Space Flight Subcommittee visiting TRW Space Systems in Redondo Beach, Calif., he observed:

There is more pressure for the Federal dollar today than there has ever been since I have been in Congress—and it gets down to it's the easiest thing in the world to vote against the space program.

On May 3, 1971, Chairman Miller in an address to the Third International Conference on Space Technology in Rome, Italy, tried very hard to paint the bright side of the space picture, but he had to acknowledge with some realism:

The mood in our country is entirely different than it was in the past decade. The very success of the Apollo program has diminished the sense of urgency in space competition—at least in my country at the present time. The talk in the political arena is about new national priorities, not necessarily including the space program.

When Teague commented on the NASA authorization in his annual floor speech on June 3, 1971, he confessed:

The budget recommended to you today is a minimum budget. It is a budget that delays and defers programs which are in the national interest to move along at a faster pace. Three lunar exploration flights remain. Our ability to conduct lunar exploration is then at an end.

The committee's role in 1971 and 1972 was to insure that as many of the scientific experiments on the canceled Apollo flights as possible should be transferred to the concluding flights. Vigorous oversight was maintained to verify that safety and reliability were not sacrificed, and that the funds, personnel, equipment, and facilities being utilized for the Apollo flights were being transferred as quickly as feasible to future projects like Skylab and Space Shuttle.

On April 20, 1972, true to tradition, the committee brought the NASA authorization bill to the House floor on the same day that Apollo 16 astronauts John W. Young and Thomas K. Mattingly landed on the Moon. In an unusual move, Teague's Manned Space Flight Subcommittee recommended no increases over the NASA budget. The subcommittee's deliberations were thorough and wide-ranging. Teague, in presenting what was to be his last authorization request as a subcommittee chairman, told his colleagues that the subcommittee action "was taken after the most extensive hearings ever undertaken by the Subcommittee on Manned Space Flight." If the hearings were extensive, the markup session, frequently long drawn-out and contentious, set a record for brevity in 1972. The following is the complete transcript of the markup by Teague's subcommittee:

Mr. TEAGUE. Gentlemen, what is your pleasure?

Mr. FUQUA. I will move, Mr. Chairman, unless Mr. Winn wants to move, that we report the budget request submitted, the $1.58 billion, for manned space flight.

Mr. WYDLER. I second the motion, and suggest that we fight like the very devil to hold it, too. Because it is their lowest request; we know that.

Mr. TEAGUE. Shall we put in the report that the committee looked into it, that we think they came in with a minimum, very austere budget, and that we are supporting them?

Mr. FUQUA. And we should point out the budget is $52 million less than last year.

Mr. FREY. And point out that this is the first time we have not come in above it, have not increased it at all. We have always in the past increased it.

Mr. TEAGUE. Mr. Winn, further discussion?

Mr. WINN. No further discussion, Mr. Chairman.

Mr. TEAGUE. The motion has been made that we report the $1,580,652,000. Those in favor say "aye."

[Calls of "aye."]

Mr. TEAGUE. Opposed "no."

[No response.]

Mr. TEAGUE. The vote is unanimous. Let's go home.

*   *   *   *   *   *   *

APOLLO 17 AND CHAIRMAN MILLER'S RETIREMENT

On December 19, 1972, the Apollo 17 astronauts completed their journey to the Moon—the sixth manned landing and the last of the Apollo series. Dr. Harrison H. Schmitt, a trained geologist, became the first astronaut-scientist to make the Moon landing; he later joined another former astronaut, John H. Glenn, in the U.S. Senate.

The ending of the Apollo program almost coincided with the close of Chairman Miller's fruitful career, the last 11 years of which he served at the helm of the Science and Astronautics Committee. In the early years of his chairmanship, Miller was always compared favorably with his predecessor, Representative Overton Brooks, under whose chairmanship subcommittees automatically were held in tight rein. It was Chairman Miller's generous delegation of authority to Manned Space Flight Subcommittee Chairman Teague that enabled the most heavily financed aspects of NASA's space program to succeed so admirably in the leadership and oversight received from Congress.

In his own right, Chairman Miller pioneered in the establishment of of the Science, Research and Development Subcommittee which came to its fullest flower under its first chairman, Congressman Daddario. Miller's deep interest in general science led to the broadening of the charter of the National Science Foundation, the legislation to establish the Office of Technology Assessment, and the movement toward converting the United States to the metric system.

In 1967, Chairman Miller was awarded the Robert H. Goddard memorial trophy for "his sustained leadership in the formulation and execution of national policy contributing immeasurably to the remarkable accomplishments of the U.S. space effort." Miller's adulation for Dr. Goddard, the acknowledged "father of American rocketry," is indicated by the fact that he arranged to have Goddard's portrait displayed in the main committee room, 2318 Rayburn, although the portrait of Representative Overton Brooks, the first chairman of the committee, is conspicuously absent.

Through Chairman Miller's leadership, the Panel on Science and Technology, begun by Chairman Brooks, was expanded and strengthened. An additional scientific panel, the Research Management Advisory Panel, also worked closely with the Subcommittee on Science, Research and Development and helped broaden the dialogue between scientists and Members of Congress.

Representative Wayne N. Aspinall (Democrat of Colorado), who served for two years on the Science and Astronautics Committee and later became chairman of the Interior and Insular Affairs Committee, had this to say about his fellow chairman:

I have always found George Miller to be understanding and friendly, ready for a light remark when it was in order and for a serious one when it was in order.

Chairman Miller ran his committee in a quiet and conversational sort of fashion, never flamboyant, given to occasional flashes of petulance or anger but not by nature combative, always unhurried and full of frequently lengthy anecdotes. In 1971, when the House leadership decided to make the Science and Astronautics Committee a nonmajor committee to accommodate members who wanted the chance to serve on an additional committee, some of the higher ranked members like Congressman Karth were angry with Chairman Miller. Miller's defense was that the move was made to accommodate members and to prevent wholesale defections from the committee. In any event, Chairman Miller's agreement was characteristic of his general approach toward both the House leadership (a team-playing spirit of cooperation) and toward his own committee members (to favor and accommodate those committee members who were team players rather than mavericks).

His colleagues in the Congress respected Chairman Miller. They listened intently, and applauded vigorously as he interrupted debates on any issue to announce the latest successes in manned or unmanned space flight.

Chairman Miller pioneered in building strong relationships with leaders of science in other nations. He sponsored the establishment in 1971 of a new Subcommittee on International Cooperation in Science and Space. He traveled more extensively than any other committee member, and from Stockholm to Sydney, Rome to Romania, he was always eager to board a plane to deliver an address and cement relationships with those interested in scientific development abroad. At the age of 80, Chairman Miller made a special trip to the South Pole at the special invitation of the National Science Foundation. And he swelled with natural pride when the National Commission on Geographical Names in September 1972, designated an extensive range of high plateau in Antarctica as "Miller Bluffs."

Exactly one week after the Apollo 17 astronauts had splashed down to mark the end of the last manned flight to the Moon in the Apollo series, Chairman Miller wrote a farewell letter to NASA Administrator Fletcher. He observed:

> The conclusion of the Apollo program leaves me with very mixed feelings indeed. As I look back over the years to 1959, when the American people committed themselves to the exploration of space, I am struck with a deep sense of quiet, profound pride at what we have accomplished, especially in manned space flight. * * *
>
> The plethora of benefits of our program going directly to people today and to generations to come are, to me, immeasurable, but nonetheless real. And they are rooted in almost every discipline—medicine, geology, geodesy, astronomy, planetary physics—the list is much too long to enumerate. It is results such as these that vindicate and highlight the faith in the promise given to the American people at the very beginning of the space program in 1959. * * *
>
> I depart from the Committee on Science and Astronautics with boundless pride and satisfaction with the signal success of the relationship between NASA and the Congress. I will watch in the years to come for even greater results that will undoubtedly come from that warm, cooperative effort.

TEAGUE SUCCEEDS MILLER

With the beginning of the 93d Congress in 1973, Representative Olin E. "Tiger" Teague moved up to assume the chairmanship of the Science and Astronautics Committee, a position he held for six years until his voluntary retirement from Congress. Following Chairman Miller's defeat in the California Democratic primary in June 1972, Teague slyly attempted to build up a little suspense over whether he would give up the chairmanship of the Veterans' Affairs Committee to move over to the more prestigious Science Committee. When it became apparent that he would do so, he then predicted darkly that the McGovern ticket might cause the Republicans to capture control of Congress and thus switch the chairmanship to Representative Charles A. Mosher (Republican of Ohio). Some NASA officials were apprehensive lest the next Democrat in line, Hechler, might move up, because he had taken somewhat critical positions on NASA's programs. But according to a Washington Post article following the November 1972 elections, Hechler personally urged Teague to give up the Veterans' Affairs chairmanship in order to take the Science chairmanship.

One of Chairman Teague's first decisions was to find a new executive director. As Teague explained it:

> Ducander and I had an understanding before I became Chairman that he wasn't going to stay because I felt like you had to have a technical man for staff director.

Chairman Teague (center) meets with Apollo 13 astronauts, John L. Swigert, Jr. (left) and James A. Lovell, Jr., in his Capitol Hill office.

Teague found his man in John L. "Jack" Swigert, Jr., command module pilot on the April 1970 Apollo 13 flight which had been forced to return to Earth after an oxygen tank had ruptured. A graduate of the University of Colorado, Swigert had obtained a master of science degree from Rensselaer Polytechnic Institute, and a master of business administration degree from the University of Hartford. In addition to serving as an Air Force fighter pilot in Japan and Korea, Swigert had been an engineering test pilot for North American Aviation and also Pratt and Whitney. In making his appointment effective on April 24, 1973, Teague stated:

Jack Swigert brings to the Committee on Science and Astronautics a broadly based experience, skill and enthusiasm that will aid in the expanding effort being made by the Committee to assure that our national space program and federal research and development will receive adequate support in the mid-1970's.

Teague's interest in having a staff director with technical background he explained this way:

When NASA came up here to testify with Jack Swigert here, I'm telling you, they were careful about what they said.

Prior to joining the staff, Swigert was the subject of a feature article in The New York Times, at the time of the Apollo 13 flight, with a four-column headline reading: "Swigert, 38, Had Girl In Every (Air)Port." As the first bachelor to fly in space, Swigert was characterized as "a man who carries several reputations with him wherever he goes—swinger, student, sportsman, and systematizer." The word "systematizer" was applied because he "likes things neat, in their place." The article explained:

"When he cleaned out my freezer one time," his sister recalls, "he had all the juice cans lined up, with the lemonade before the orange juice. He said he did it that way because L comes before O."

Swigert employed the same style of systematic approach to organizing the committee staff. Under his direction, the staff members were grouped into "task teams" to tackle broad problems arising in several categories, so that if a subcommittee staff member completed work on one problem he could move on to work on another subject matter within the task team.

In the four years and four months Swigert was executive director, the committee staff grew steadily in size, taking a quantum jump when the committee jurisdiction expanded, starting in 1975. Upon Swigert's arrival in 1973, there were 22 members of the staff, and when he left in August 31, 1977, to start an unsuccessful campaign for the Republican nomination to the U.S. Senate in Colorado, the staff had grown to 79.

THE FOUR CHAIRMEN OF SPACE SCIENCE AND APPLICATIONS

1. Representative Joseph E. Karth of Minnesota

Karth was the longest reigning and first chairman of the Subcommittee on Space Science and Applications. During the period of his chairmanship from the early 1960's until his departure to join the House Ways and Means Committee in October 1971, Karth championed the applications side of NASA's work. As noted in chapter VII, Karth provided strong leadership in his subcommittee and on the full committee to furnish more support for the Earth resources technology satellite program (later renamed "Landsat"). Within two months of the first manned landing on the Moon, Karth, in September 1969, bluntly told a Princeton University symposium:

I predict that the pressure on Congress to reduce the space budget still further will increase unless the future orientation of NASA is based less on space spectaculars and more on the production of tangible and economic benefits.

In January 1970, Karth challenged the members of the American Astronautical Society:

I am convinced that now we should more aggressively pursue the many potential applications of existing space technology to practical problems of Earth.

Karth deplored the fact that space applications in 1970 constituted a meager 5 percent of NASA's budget, and he continued to attack the recommendations of Vice President Agnew's Space Task Group for huge new manned programs in the 1970's. Karth advised his House colleagues during the NASA authorization bill debate on April 23, 1970:

I think I can speak for all the members of our subcommittee in stating to the House that we feel strongly that the space applications program—the practical, end-result, benefits-on-earth type of space activities—needs greater emphasis and attention than it has had in the past.

Representative Joseph E. Karth (Democrat of Minnesota), center, discusses Apollo flight training at Johnson Space Center, Houston, Tex. At left is Joseph G. Gavin of Grumman Aircraft Corp., and at right Aleck C. Bond of NASA.

In the face of heavy adverse pressure from Chairman Miller and Manned Space Flight Subcommittee Chairman Teague, Karth fearlessly plunged ahead with his crusade on behalf of space applications. He did not cease his efforts once the hearings and authorization bills had been passed. Karth hit the luncheon circuit in a series of sharp public addresses, sounding the alarm on behalf of people programs instead of manned space spectaculars. Before a largely unfriendly audience at a luncheon meeting of the American Institute of Aeronautics and Astronautics, held at Fort McNair in Washington, D.C. on April 1, 1971, Karth was at his best:

> The people of our country, the taxpayers, find it difficult to see the relevance of the space program to the whole sweep of our economic and social problems. And because some refuse to say it at all, I'm persuaded to say it more often. Our citizens are clamoring for a reordering of our priorities, and unfortunately space doesn't seem very high on their list. We may not like the facts, but that is no excuse for being so stupid that we can't recognize them.* * *
>
> I have urged for the past six years that we place equally strong emphasis upon those activities in space that will result in economic payoff for our people. Specifically, I have urged an increased effort in applications satellite systems—communications, meteorology, earth resources survey, navigation and air traffic control.

Karth pointed out that too much of NASA's effort was devoted to applying space benefits to help individual consumers, like the development of teflon frying pans. He observed:

> Unfortunately, it seems to me that our most pressing problems today are *not* those of individuals, so much as they are those of communities and institutions. While the individual's standard of living has improved, the quality of life has deteriorated. The American housewife now has teflon frying pans, but we stand by helplessly while Lake Erie dies (and) the people of Los Angeles suffocate in smog. * * * Somehow we need to address the problems of mass transportation, of pollution of our atmosphere and our fresh water resources, of urban renewal.

Karth's subcommittee strongly supported his personal emphasis on space applications and people-oriented benefits. In 1971, the following members served on the Subcommittee on Space Science and Applications:

Democrats	Republicans
Joseph E. Karth, Minnesota, *Chairman*	Charles A. Mosher, Ohio
Thomas N. Downing, Virginia	Larry Winn, Jr., Kansas
James W. Symington, Missouri	Robert Price, Texas
John F. Seiberling, Jr., Ohio	Barry M. Goldwater, Jr., California
Morgan F. Murphy, Illinois	

In his farewell address on the NASA authorization bill in 1971—the last subcommittee report he handled before going over to the Ways and Means Committee—Karth on June 3, 1971 told the House once again that every year since the mid-1960's, his committee had emphasized the need for greater priority treatment of space applications. He added:

> I would like to take this opportunity to recommend to the new Administrator of NASA, Dr. James Fletcher, that serious consideration be given to reorganizing

the Space Agency to include a new Office of Applications to be headed by an Associate Administrator of Applications. In this way, we may achieve a new direction for our space program with appropriate emphasis on practical applications until it becomes a reality.

Although Karth did not remain on the committee long enough to see the change made, NASA finally did decide on December 3, 1971, to set up the very office which Karth recommended.

Exactly a week after Karth transferred to the Ways and Means Committee, Representative Thomas N. Downing (Democrat of Virginia) sent the following note to Chairman Miller:

> DEAR MR. CHAIRMAN: Now that Joe Karth has transferred to the Ways and Means Committee, I would deeply appreciate your giving me consideration to my being named as Chairman of the Subcommittee on Space Science and Applications.
>
> I would consider this a challenging assignment and I would very much like to try it.
> Sincerely,
>
> /s/TOM.

This was a little ticklish for Chairman Miller, who was a stickler for following the seniority system. Under ordinary circumstances, Hechler, who was next in line by seniority to Karth, would have had the option to move up from his chairmanship of the Subcommittee on Advanced Research and Technology to take over the Karth subcommittee, a move which Chairman Miller did not view with relish. If Karth had been a troublemaker by using the Applications issue, there was no telling what would happen if the subcommittee fell into the hands of a real maverick like Hechler. And as if there weren't enough problems, Fuqua asked Miller whether it would be possible to split the old Karth subcommittee and pave the way for Fuqua to become chairman of a new Applications subcommittee.

Chairman Miller decided to cool it for a few months. After all, no hearings were scheduled until January of 1972, and the situation might work itself out if there were some delay. Informal soundings were made to see whether Hechler intended to insist on his seniority rights, which would have produced a sticky situation. Hechler surprised his colleagues by opting to stay exactly where he was, thus paving the way for Downing to chair the old Karth subcommittee. Chairman Miller then told Fuqua that since there was one line item in the NASA authorization bill for Applications, that it would probably be best to keep Space Science and Applications together in one subcommittee. Delay had served the useful purpose of eliminating the potential conflict among competing aspirations. Miller did not announce his decisions until the new session of Congress convened in January 1972.

Representative Thomas N. Downing (Democrat of Virginia) meets with Skylab astronauts. From left, Alan Bean, Downing, Jack R. Lousma, and Dr. Owen K. Garriott.

2. *Representative Thomas N. Downing of Virginia*

Congressman Downing, a product of the Virginia Tidewater and lifelong resident of Newport News, Va., is the very epitome of a southern gentleman of the old school. One can picture him riding with the hounds, his 6-foot frame dominating the scene near white-columned mansions, as southern belles with long, flowing dresses sip mint juleps on a wide veranda. Amid the conviviality of a story-telling evening, it seemed natural for Downing to launch into an impassioned plea to restore Gen. Robert E. Lee's citizenship and pass his bill to correct a century-old injustice—which is precisely what Congress got around to doing.

Elected to Congress in 1958 with a huge class of newcomers which also included Karth, Hechler, and Daddario, each of whom ascended to subcommittee chairmanships early in the 1960's, Downing did not join the Science Committee until 1962. This meant he had to wait his turn until 1971 before getting his own subcommittee. He was then named head of the Oversight Subcommittee, and moved up to the chairmanship of the Space Science and Applications Subcommittee in 1972. After he had announced in 1976 that he planned to retire from Congress at the end of that year, Speaker Albert persuaded Downing to take the chairmanship of the Select Committee on Assassinations for the last few months of 1976.

A graduate of the University of Virginia Law School, Downing won a Silver Star as a mechanized cavalry troop leader with General Patton in France, where he commanded the first troops to invade Germany in 1944. "Historically, the man who represents my district has to get on the Merchant Marine Committee," Downing relates. The location of the Newport News Shipbuilding and Drydock Co., Virginia's largest employer, was one of the reasons Downing originally applied for and was appointed to membership on the Merchant Marine and Fisheries Committee. There he developed a close friendship and working relationship with Miller and Karth, who also served on that committee.

The presence of Langley Research Center and Wallops Station, NASA installations in Downing's district, caused him to seek and obtain membership on the Science Committee in 1962. There he rose in seniority on Karth's subcommittee, as well as on the Subcommittee on NASA Oversight. A staunch conservative in contrast to Karth's liberalism, Downing worked closely with Karth and supported him on all the major decisions made by the subcommittee.

There was a reshuffling of personnel on the Space Science and Applications Subcommittee after Karth left the chairmanship and Downing took over in the middle of the 92d Congress in January 1972. The following served under Downing's chairmanship during 1972:

Democrats	*Republicans*
Thomas N. Downing, Virginia, *Chairman*	Robert Price, Texas
James W. Symington, Missouri	Larry Winn, Jr., Kansas
John F. Seiberling, Jr., Ohio	Barry M. Goldwater, Jr., California
Morgan F. Murphy, Illinois	R. Lawrence Coughlin, Pennsylvania
Mendel J. Davis, South Carolina	John N. Happy Camp, Oklahoma
Bob Bergland, Minnesota	

As a subcommittee chairman, Downing's style differed from Karth's. He was inclined to be more courteous and tolerant toward witnesses, and perhaps less incisive in the type of combative questioning which Karth pursued. But like all chairmen of the Subcommittee on Space Science and Applications, he placed heavy stress on NASA's need to place a higher priority on programs for the benefit of all mankind. In his first subcommittee report to the full committee, Downing deplored the fact that NASA was stressing "certain expensive scientific projects" instead of putting more money in applications. He added:

For several years the Subcommittee has urged NASA to give greater emphasis to Space Applications. These recommendations have been largely disregarded by NASA. * * * The announcement of the creation of the Office of Applications was greeted by enthusiasm by many in Congress who have long believed that public support for the national space program in the future will depend very heavily upon these practical applications of space technology.

While supporting NASA's efforts in space science, including such projects as the Orbiting Solar Observatory and the Orbiting Astronomical Observatory, Downing voiced strong support for less expensive suborbital programs through the use of balloons and sounding rockets. He reported and supported a big jump in Space Science funding in 1972 to build and equip the two Viking spacecraft which so successfully orbited and soft-landed on Mars in the bicentennial year of 1976. But he told his colleagues during the authorization bill debate in 1972:

> It is my conviction that the current level of funding for space applications is inadequate, and I intend to urge a substantial increase in the budget. * * * NASA's stated goal of increased emphasis on space applications can be achieved only if sufficient financial support for this work is forthcoming.

Downing remained as subcommittee chairman through 1972, after which he was succeeded by Representative Symington.

3. Representative James W. Symington of Missouri

Paraphrasing Kipling, Congressman Symington is the kind of person who walks with kings without losing the common touch. As President Johnson's Chief of Protocol, his experience covered not only kings, but all foreign heads of state and visitors plus a good cross section of American dignitaries visiting the White House. Only Symington would have the common touch to grasp a fellow-Congressman's arm and plunge into a crowd of young protesters, as we prepared, in black-tied splendor to enter the Century Plaza Hotel in Beverly Hills, Calif., for a lavish dinner President Nixon was throwing for the first astronauts to land on the Moon. The protesters demanded that we tell them, as Symington expressed it, "Why the Moon with so much left to do on Earth in housing, pollution, and education?" For a full hour we talked with them and achieved the goal of communication. Writing in his book, *The Stately Game*, Symington relates:

> We went on to discuss space technology and its relevance to earthbound life: the weather satellites, which give advance warning of impending storms; the communications satellites, which can bring education and new knowledge into the remotest parts of the world; the earth resources satellites, whose infrared sensors may soon tell us far more about the subsurface of the Earth than we know today, telling farmers what and where to plant and fertilize and fishermen where the schools are headed.

Representative James W. Symington (Democrat of Missouri), right, with Soviet Cosmonaut Alexei A. Leonov.

The son of a Democratic U.S. Senator, grandson of a Republican Senator, and great-grandson of a Secretary of State, Symington enjoyed a rich variety of opportunities before coming to Congress—Marine, member of the soccer and boxing teams at Yale, Columbia Law School graduate, professional musician (guitar) and nightclub performer, assistant city attorney in St. Louis, special assistant to an ambassador (London), administrative assistant to Attorney General Robert F. Kennedy, deputy director of food-for-peace program, and executive secretary of President's Commission on Juvenile Delinquency and Crime.

Within the broad area of applications, Symington is best remembered for the extremely vigorous support he gave toward faster development and more user interest in Earth resources technology satellites (later termed Landsat). During his second term in Congress, Symington was chosen to be the Moderator of the 13th meeting of the Panel on Science and Technology, January 25–27, 1972, which was devoted to the subject of "Remote Sensing of Earth Resources." This proved to be the last Panel meeting before the Panel was abandoned by Chairman Teague. In both his opening, welcoming remarks and in his summary statement at the end of the three-day meeting, Moderator

Symington had a chance to spotlight the importance of hurrying along with the development and use of Earth resources satellites. At one point in the proceedings, spying Congressman Karth in the rear of the audience, Symington modestly observed:

> I would like to call attention to the fact that we are graced at this time by the presence of the former chairman of the Space Science and Applications Subcommittee, Congressman Joseph Karth of Minnesota, who is sitting benignly in the back when he should be here chairing these proceedings.
>
> He has done a great deal over the past years to stimulate the work of this committee and this panel. Congressman Karth was, somewhat like Elijah, drawn up to the Ways and Means Committee from this terrain, where he now serves with great distinction.

When he assumed the chairmanship in January 1973, Symington was assigned the following members to his subcommittee:

Democrats	Republicans
James W. Symington, Missouri, *Chairman*	Marvin L. Esch, Michigan
Thomas N. Downing, Virginia	Larry Winn, Jr., Kansas
Bob Bergland, Minnesota	Barry M. Goldwater, Jr., California
George E. Brown, Jr., California	John N. Happy Camp, Oklahoma
Dale Milford, Texas	

Symington, a literary master of the bon mot, regaled his listeners and readers with the best prose which was ever developed on the Science Committee. Serving successively as chairman of the Subcommittees on International Cooperation in Science and Space, Space Science and Applications, and Science, Research and Technology, Symington not only provided vigorous leadership but also attracted the support and esteem of scientific experts throughout the world. The dialogue with the scientific community was lifted to a new level through his smooth handling of complex and technical issues espoused by those approaching genius in the areas of science and technology.

In opening the first Space Science and Applications Subcommittee over which he presided, Symington on March 1, 1973 paid tribute to the efforts of his two predecessors, Karth and Downing for their support of a "more aggressive and vigorous applications program", adding:

> I would like to take this opportunity to add my voice to theirs, and to note that there is no NASA activity which is better understood or more widely supported by the American public than the applications program.

Both Symington and Bergland jumped on NASA witnesses for not expressing sufficient enthusiasm for solar energy conversion. Bergland put it this way:

> I don't like to sound like an alarmist, but * * * within ten years' time, qualified experts in the field tell us we will see 75 cents a gallon gasoline in the United States. * * * Mr. Chairman, I submit that we simply can't afford to proceed with such a token effort in the field of solar energy conversion.

Symington also needled NASA witnesses on why so many budgetary reductions were made in NASA's applications program. This prompted the following exchange with Charles W. Mathews, head of the Office of Applications:

Mr. SYMINGTON. Was that decision occasioned by an overall budget review, as distinct from a purely Agency decision?

Mr. MATHEWS. That sort of thing, Mr. Chairman, generally would occur as a somewhat iterative process.

Mr. SYMINGTON. Repeat that word?

Mr. MATHEWS. Iterative. That means a back and forth process.

When a subsequent discussion revealed that the process ended up with NASA always recommending less money in the Applications area, Symington was prompted to observe: "Who put the 'it' in that iterative process?"

Appalled by the failure of NASA to budget for a navigation satellite also desired for use by the Maritime Administration, Symington had this sprightly colloquy with Mathews:

Mr. SYMINGTON. So they want it very badly, and you would like to give it to them, but it's not in either budget?

Mr. MATHEWS. That is correct.

Mr. SYMINGTON. That is confusing to me.

Mr. BERGLAND. Mr. Chairman, I submit it doesn't make any sense.

Mr. SYMINGTON. The gentleman has expanded my thought. * * * Was there a third silent partner in the decision by these two great agencies not to place this item in their respective budgets?

Mr. MATHEWS. Usually that happens when there are two agencies involved.

Mr. SYMINGTON. Divide and conquer, or unite and conquer in this case.

There were a great many aspects of NASA's programed budget which Symington in 1973 labeled as "incredible." First, Dr. Fletcher announced that the establishment of a separate Office of Applications meant that Applications would be given greater emphasis in the future; yet the budget request in 1973 was only $153 million as contrasted with $195 million in 1972. Then there was the little item of phasing out NASA research on communications satellites, when the Communications Satellite Corporation through private industry indicated no desire to pick up that big tab. When NASA decided to delay the launch of another Earth resources technology satellite for budgetary reasons, Symington's subcommittee stepped in and added $8 million to NASA's authorization to speed up this valuable program. His efforts were unanimously supported by both his subcommittee and the full committee, with strong assistance in the debate by Downing, Esch, Goldwater and Camp.

It had been the practice of most subcommittees to accomplish their major assessments of agency programs through the annual authorization hearings, plus oversight through field visits and careful analysis during the year. Symington added a new twist to this process in 1973 by calling a series of four "informal briefings" between September 24 and October 4. The briefings were designed to ascertain progress, technical difficulties, costs, launch schedules, and any program changes. The briefings also enabled Symington and his committee to reiterate their strong support for the entire Applications program, and specifically satellites like ERTS.

Symington and Esch put some more heavy pressure on NASA to speed up the launch of an additional ERTS satellite, which Congress had authorized. The conversation went like this:

Mr. MATHEWS. We are still actively pursuing your desire to accelerate ERTS–B——

Mr. ESCH. It seems ridiculous to me if the Congress mandates a position and proposes it through the Authorization Committee, that NASA cannot respond. * * * It would seem to me that if NASA is doing its job they would be asking for supplemental appropriations, if necessary, or asking how to implement the appropriation so it will be expedited. * * * I would like to know how we can help you convince the Administration and the other NASA officials that Congress means this. Do you think maybe we ought to have a GAO report, for instance?

Mr. MATHEWS. I think that both NASA and others in the Executive Branch of the Government fully understand the Congressional position on this point and the fact that it is a strong position.

Mr. SYMINGTON. It seems anyone with a grain of sense would want to keep the momentum going and operating. I do think you ought to take a strong message back, that I think a lot of people felt this thing was going to be on track. They didn't realize that it was getting off again.

Mr. ESCH. I'm just very much concerned * * * that something as highly successful as this would be caught in the bureaucratic bowels of OMB or NASA.

As was customary in any Symington hearing, everything was not completely sober and serious. At one point the colloquy went like this:

Mr. CAMP. Have you done any work with windmills?

Mr. MATHEWS. My office has not, but the Office of Aeronautical and Space Technology has been working on it.

Mr. SYMINGTON. The Congress is said to contribute in this area.

Mr. MATHEWS [diplomatically]. I think a lot of positive energy is produced by the Congress.

Although more sparks flew during the Applications hearings, Symington spent considerable time in bringing the subcommittee up to date on recent developments in the space science field, including lunar and planetary programs as well as physics and astronomy. One day when Vincent Johnson of the Space Science Office was discussing "comets that we know are coming back that we do want to intercept and rendezvous with," this prompted Symington to relate:

It reminds me of the story of a little boy who is asked, to test his intelligence, what he would do if he saw a train coming down the track from the north at 70 miles an hour and about three or four miles away he could barely see, but certainly hear, a train coming from the south at about 50 miles an hour. The professor asked the boy, "What would you do?" and the boy responded: "I would call to my brother."

"Why would you do that?" the professor wanted to know.

"Because he has never seen a train wreck like that before."

In 1974, the Symington subcommittee continued to stress support of Applications, as well as continued funding of space science projects. The subcommittee was enthusiastic about the results achieved and data obtained from communications and weather satellites, the Earth resources surveys, and programs for monitoring pollution. In addition, the subcommittee under Symington helped fund the highly successful Applications Technology Satellite-6 which communicated education and health services information to millions of people, via television, in India and remote areas of the United States such as Alaska. During its 1974 markup, the Symington subcommittee adopted an amendment by Congressman Winn authorizing NASA to set aside $2 million for research on tornadoes and other short-term weather phenomena.

Reflecting in 1978 on his experiences with the committee, Symington related that one of the most crucial decisions he had to make concerned the Viking program—by far the most expensive and most complex unmanned project authorized by the committee for a 1975 flight to Mars. Two spacecraft, each containing an orbiter and a soft lander, also included a biomedical package designed to perform a number of tests with Martian soil to ascertain the possibility of life on Mars. The total program exceeded $1 billion in cost, and Symington's subcommittee soon discovered that the costs were escalating far beyond the preliminary cost estimate of $346 million. After personal visits to the contractors and NASA's field installations, Symington's subcommittee decided that in view of the heavy cost overruns, oversight hearings should be held on November 21 and 22, 1974 "to determine the nature of these development problems, and why they were unexpected at such a late stage in the Viking project." Symington announced at the opening of the oversight hearings:

We shall also review the financial history of the project, and seek an expert assessment of the probability of mission success.

Symington vividly recalls the nature of the big decision he and his subcommittee had to make. Everything was in readiness to meet the Mars "window" for launch in the summer of 1975, except the biomedical package—a highly complex and super-miniaturized set of delicate instruments which had been packed into a box one cubic foot square, capable of performing what customarily took up the space of several huge laboratory rooms on Earth. Symington relates:

> The biomedical package was sterilized with intense heat to make sure no Earth-bound micro-organisms were carried to Mars, and this was one of the reasons the biomedical package wasn't ready and was behind schedule. They were having trouble with it, and finally we went out to look at it. Here I am an English major and a lawyer, and they showed me this black box with a lot of wiring in it. * * * Well, the question was: should we fly this thing with no assurance that the biomedical package is going to work, which was a package to test if there is life on Mars, and if it did not work, what is the sense of sending it up there?

Symington graphically described his dilemma: if it flies and doesn't work, "you have sent up a package for nothing". But if you don't fly because you are worried about the biomedical package, then you wait for the next Mars window about 26 months hence, which means a huge increase in the costs.

During the oversight hearings, Winn became exasperated with the parade of problems, cost increases, and complex technical difficulties which required costly new research while the development was proceeding:

> I am telling you right now * * * I am going to start saying "no" to a lot of these programs. The American people are fed up with cost over-runs. If you look at some of the people in both parties, they campaigned on this issue. I am saying you guys are not going to be able to stick this stuff down our throats anymore.

Symington immediately added:

> There is another dimension to Mr. Winn's point. From your testimony, it appears that you would have flown a 1975 mission if OMB had not stepped in to stop you, true?

Dr. Edgar Cortright of NASA responded circumspectly:

> I can't remember whether OMB said that. I believe the Administrator made the judgment.

This prompted Symington to make the tongue-in-cheek response:

> We know OMB has nothing to do with budget constraints. [Laughter.]

It was a serious oversight hearing, dealing with extremely complex dialogue on new forms of instruments which were to play a part in discovering and pushing back vast frontiers. Every now and then, Symington's somewhat puckish sense of humor bubbled to the surface. After a dreary recital of a long series of unanticipated delays, at one point Symington mused:

One way to delay the mission would be to require an environmental impact statement of the impact on Mars.

The Viking mission technically was perfect in execution. At 5:12 a.m. on July 20, 1976, the mission controller at the Jet Propulsion Laboratory in Pasadena, Calif. declared simply: "Touchdown. We have touchdown," as the first Viking lander smoothly reached the surface. Less than an hour later, Viking-1 began transmitting the first of an incredible series of photographs of the ridges, sand, bluffs and rocks on the surface of Mars. Later, long arms reached out from the spacecraft, scooped up and analyzed samples of the Martian soil and sent the results back to Earth. Two months later, Viking-2 landed at a different location on Mars. In the areas of landing, no unambiguous evidence of any form of life appeared to exist.

4. Representative Don Fuqua of Florida

Shortly after 10 o'clock one morning early in January 1975, Chairman Teague assembled the top senior Democrats of the committee for a very informal, unrecorded meeting in room 2317 of the Rayburn Building, the small anteroom adjoining the main committee room (2318). Going down the seniority list, Teague asked each member to choose the subcommittee of his preference. To nobody's surprise, when it came time for Fuqua to choose, he opted for the vastly expanded Subcommittee on Space Science and Applications, with jurisdiction over all of NASA's work except aeronautics. The committee had at first been dubbed "Space Flight, Science and Applications," which Fuqua with consent of the committee changed to its more permanent title.

By the start of the 94th Congress, the full committee which back in 1959 had perhaps 90 percent of its work dealing with NASA, in 1975 was devoting only about 20 percent of its effort in that area. Nevertheless, the greatly enlarged jurisdiction meant that Fuqua would preside over all the issues and programs once handled by two subcommittees (Manned Space Flight and Space Science and Applications) plus a portion of the work once handled by a third subcommittee (Aeronautics and Space Technology); the new Fuqua subcommittee took over jurisdiction dealing with tracking and data acquisition, technology utilization, and all forms of basic and advanced research once handled by the Hechler subcommittee. The work in aeronautics went to a new Subcommittee on Aviation and Transportation Research and Development.

At the opening of the 94th Congress in 1975, the following were the members of the Science Committee in order of seniority:

Democrats	Republicans
Olin E. Teague, Texas, *Chairman*	Charles A. Mosher, Ohio
Ken Hechler, West Virginia	Alphonzo Bell, California
Thomas N. Downing, Virginia	John Jarman, Oklahoma
Don Fuqua, Florida	John W. Wydler, New York
James W. Symington, Missouri	Larry Winn, Jr., Kansas
Walter Flowers, Alabama	Louis Frey, Jr., Florida
Robert A. Roe, New Jersey	Barry M. Goldwater, Jr., California
Mike McCormack, Washington	Marvin L. Esch, Michigan
George E. Brown, Jr., California	John B. Conlan, Arizona
Dale Milford, Texas	William M. Ketchum, California [1]
Ray Thornton, Arkansas	Gary A. Myers, Pennsylvania
James H. Scheuer, New York	David F. Emery, Maine
Richard L. Ottinger, New York	
Henry A. Waxman, California	
Philip H. Hayes, Indiana	
Tom Harkin, Iowa	
Jim Lloyd, California	
Jerome Ambro, Jr., New York	
Christopher J. Dodd, Connecticut	
Michael T. Blouin, Iowa	
Tim L. Hall, Illinois	
Robert (Bob) Krueger, Texas	
Marilyn Lloyd, Tennessee	
James J. Blanchard, Michigan	
Timothy E. Wirth, Colorado	

[1] Ketchum was replaced by Larry Pressler of South Dakota.

In 1975, the Subcommittee on Space Science and Applications included the following:

Democrats	Republicans
Don Fuqua, Florida, *Chairman*	Larry Winn, Jr., Kansas
Thomas N. Downing, Virginia	John W. Wydler, New York
James W. Symington, Missouri	Louis Frey, Jr., Florida
Walter Flowers, Alabama	David F. Emery, Maine
Robert A. Roe, New Jersey	
Jim Lloyd, California	
Tim L. Hall, Illinois	
Henry A. Waxman, California	
Michael T. Blouin, Iowa	

Although under tremendous time pressure, Chairman Fuqua held almost nonstop hearings, field trips, and conferences from February 5 until just one month later when he assembled his subcommittee to mark up the NASA authorization bill in 1975. Not only was the subcommittee dealing with a vastly new area of jurisdiction previously handled by other subcommittees, but the members also had to wrestle

Inspection of solar heating panel. From left, Representative Don Fuqua (Democrat of Florida), John L. Swigert, committee executive director, Dr. Rocco Petrone of NASA, Representative William M. Ketchum (Republican of California), and Representative Larry Winn, Jr. (Republican of Kansas).

with funding a "transition quarter" to bridge the gap while the Government was moving its fiscal year from July 1 to October 1. Fuqua said:

We held some 23 different hearings, both here in Washington, and at NASA Centers, and with the key industrial contractors.

The Fuqua subcommittee also managed to take testimony during February from the Air Force, the European Space Organization, and the American Institute of Aeronautics and Astronautics. Winn, commenting on the thorough and speedy work, told his colleagues: "We have done as good a job as I have ever seen done." With Symington serving as a member of the Fuqua subcommittee, the transition was smooth to pick up the work and also carry through the emphasis Symington had placed on applications, indicated by this exchange during the subcommittee markup session:

Mr. WINN. The subcommittee is very aware of the importance of applications as a part of the new assignment in this committee, and the work that Mr. Symington and his subcommittee did in the past few years had not been forgotten. As a matter of fact, I think this subcommittee has benefited much by the work Mr. Symington's subcommittee did and we will follow through in that field of applications.

Mr. Symington. Thank you, gentlemen. I am glad I arrived in time.

In 1975, the Fuqua subcommittee put in a total increase of $6.5 million, including severe storm research and earth resources surveys. In action supported by the Congress later, the Fuqua subcommittee directed NASA to "take a more affirmative approach to the planning of application missions with a view toward the ultimate user." The Fuqua subcommittee also put in a strong plug for additional work to bring down the costs on a large space telescope to accompany the Space Shuttle. The greatest emphasis, of course, throughout the 1970's was placed on speeding the development of the Space Shuttle, discussed in the preceding chapter. When the conference committee met, the House conferees persuaded the Senate to go along with the increase in applications research, and the conference report stated:

> The committee of conference adopts the House position authorizing $181,530,000 (for applications), emphasizing that the additional $6,500,000 authorized is to augment and strengthen research and development programs in the area of severe storm research, earth resources development and Space Shuttle payload studies. The conferees also note the need for timely action to assure continuity of remote sensing of earth resources data from space.

Throughout his subcommittee chairmanship, Fuqua as well as all the members of the Subcommittee on Space Science and Applications continued to exert pressure on NASA to emphasize projects of practical benefit. At the same time, Fuqua and the subcommittee pushed NASA to accomplish more long-range planning of its objectives, and to pursue an aggressive program to share the knowledge acquired with the general public. Chairman Teague and Fuqua both felt that first NASA should plan its future objectives both realistically and with sufficient idealism, and then translate the programs clearly enough to win public support.

Midway in the first year of Fuqua's chairmanship of the subcommittee, he launched a series of productive hearings and reports entitled "Future Space Programs 1975." Between July 22 and July 30, 1975, scientists, industrialists, professors, physicians, economists, environmentalists, editors, management experts, and administrators paraded before the subcommittee in a truly mind-expanding experience. In September 1975, the subcommittee made its report, and Fuqua noted in his letter of submittal:

> It is apparent that the imagination, skill and technology exist to expand the utilization and exploration of space. The positive benefits of a bold space program are compelling.

The report stated that NASA should demonstrate a sense of urgency in its future program planning and development. The subcommittee warned that the key element in future programs should be measured by the following yardstick:

Substantial return on past and current investments in space through clear and immediate benefits to the society on earth in the form of greatly expanded services and direct contributions to solution of earthbound problems.

The subcommittee stressed the need for more space systems for education and medical satellite services—like the highly successful Applications Technology Satellite which had been used for this purpose in India, Alaska, and other areas; and earth resources surveys. The report also made this recommendation:

NASA should develop and implement a comprehensive cost benefit analysis for each major program which will include the relative social and economic benefits as well as the potential for public support and international cooperation.

In 1976, the Fuqua subcommittee took several steps to implement the report. Following 27 hearings, the subcommittee recommended and Congress agreed to set up a new line item in the budget for "Earth resources operational systems"—the new Landsat satellite (which was formerly termed the Earth Resources Technology Satellite), and the user systems like the Departments of Agriculture and Interior. Among the other items recommended by the House subcommittee and approved in conference with the Senate was $3 million to start work on the large space telescope.

Representative Wes Watkins (Democrat of Oklahoma), second from left, converses with three Speakers of the House, all of whom played a role in the formation of the Science Committee. From left, former Speaker Carl Albert of Oklahoma, whose resolution established the Science Committee; Watkins; former Speaker John W. McCormack, who chaired the select committee which created the Science Committee; and Speaker Thomas P. O'Neill, Jr., who authored the report from the Committee on Rules which established the Science Committee.

During the 95th Congress which started in 1977, the following Members served on the full committee:

Democrats	Republicans
Olin E. Teague, Texas, *Chairman*	John W. Wydler, New York
Don Fuqua, Florida	Larry Winn, Jr., Kansas
Walter Flowers, Alabama	Louis Frey, Jr., Florida
Robert A. Roe, New Jersey	Barry M. Goldwater, Jr., California
Mike McCormack, Washington	Gary A. Myers, Pennsylvania
George E. Brown, Jr., California	Hamilton Fish, Jr., New York
Dale Milford, Texas	Manuel Lujan, Jr., New Mexico
Ray Thornton, Arkansas	Carl D. Pursell, Michigan
James H. Scheuer, New York	Harold C. Hollenbeck, New Jersey
Richard L. Ottinger, New York	Eldon Rudd, Arizona
Tom Harkin, Iowa	Robert K. Dornan, California
Jim Lloyd, California	Robert S. Walker, Pennsylvania
Jerome A. Ambro, New York	Edwin B. Forsythe, New Jersey
Robert (Bob) Krueger, Texas	
Marilyn Lloyd, Tennessee	
James J. Blanchard, Michigan	
Timothy E. Wirth, Colorado	
Stephen L. Neal, North Carolina	
Thomas J. Downey, New York	
Doug Walgren, Pennsylvania	
Ronnie G. Flippo, Alabama	
Dan Glickman, Kansas	
Bob Gammage, Texas	
Anthony C. Beilenson, California	
Albert Gore, Jr., Tennessee	
Wes Watkins, Oklahoma	
Richard A. Tonry, Louisiana	
Robert A. Young, Missouri	

The Subcommittee on Space Science and Applications included the following:

Democrats	Republicans
Don Fuqua, Florida, *Chairman*	Larry Winn, Jr., Kansas
Robert A. Roe, New Jersey	Louis Frey, Jr., Florida
Jim Lloyd, California	Harold C. Hollenbeck, New Jersey
Thomas J. Downey, New York	Eldon Rudd, Arizona
Ronnie G. Flippo, Alabama	
Bob Gammage, Texas	
Albert Gore, Jr., Tennessee	
Wes Watkins, Oklahoma	
Timothy E. Wirth, Colorado	

EARTH RESOURCES INFORMATION SYSTEM

Opening three days of hearings in June 1977, on the Earth Resources Information System, Chairman Fuqua stated:

> The number of users of Landsat data and the market for Earth resources information have increased dramatically. Many users have expressed great concern over the lack of commitment to insuring a continuing source of these data. In the minds of many persons the time has arrived to prepare for transition of Landsat from experimental status to an operational status. A policy needs to be established which outlines the respective roles of Government and industry.

The committee contracted with Charles W. Mathews, formerly NASA Associate Administrator for Applications, to synthesize the views of industrial firms, Government agencies and other individuals on the definition and structure for an Earth resources information system. Mathews produced two reports, one of the definition and scope of the system, and the second on the institutional arrangements required for a transition from an experimental to an operational system. The subcommittee then prepared a report recommending early and positive action to make Landsat a truly operational worldwide system.

Following the hearings, Fuqua suggested to Dr. Frank Press, Director of the Office of Science and Technology Policy, that the Federal Government should take a more active leadership role in "organization and communication" of the data developed. In a July 15, 1977 letter, Fuqua stressed that this was especially important for the private sector and state and local governments, which had expressed some dissatisfaction with the confusing lack of a central coordinating authority. Dr. Press replied on July 22 that he would establish a task group under the leadership of NASA Deputy Administrator Alan Lovelace for this purpose.

Once again, on November 29, 1977, Fuqua wrote Dr. Press, citing some of the significant issues in this area needing attention:

> Clarification of agency roles, clarification of government/private sector roles, a commitment to provide continuity of data, a mechanism for user input to the Federal planning process, and enhanced technology transfer activities are needed in the near term. Furthermore, the Federal Government should commit to a 5-year Earth resources information system validation program with a stated goal of an orderly transition to an operational system.

Dr. Press responded on December 6, 1977:

> The administration shares the sense of the Congress that remote sensing technologies can be of ever increasing benefit to the nation and the world, and is committed to a positive program that will advance these promising applications of space science and technology.

The subcommittee continued to place a high priority on development of a more active Earth Resources Information System.

On May 2 and 3, 1979, the subcommittee held hearings to review the progress being made by the executive branch in planning for an operational Earth Resources Data and Information Service, and to consider legislation for such a Service. Fuqua stated in opening the hearings:

There is good reason and evidence to be confident about the health and prospects of the technology of remote sensing and its use. The potential has been demonstrated in geology and oceanography, meteorology, land management, crop prediction, and a host of other disciplines. However, remote sensing of Earth resources involves a number of major policy issues, many of which are of an institutional nature.

On May 3, Fuqua observed:

The subcommittee is encouraged by a recent statement by Dr. Frank Press, head of the Office of Science and Technology Policy, that the administration is committed to an operational Earth resources system. However, no timeframe has been announced and there appears to be little progress in establishing a lead agency and assignment of roles and responsibilities to Federal agencies and little progress in defining the role of the private sector. No mechanism has been established to provide on a continuing basis input by users other than Federal users.

Brown was equally sharp in his reaction:

I think there may be a mismatch between NASA and at least some Members of Congress in the perception of the urgency with which we should move ahead in this area. * * * I think the executive branch needs to be pushed. That is my personal opinion.

FUTURE SPACE PROGRAMS

In 1977 and 1978, the Fuqua subcommittee demonstrated more sharply the difference in initiative between the legislative and executive branches. Despite declining budgets, the Ford administration and the NASA Administrator, Dr. James C. Fletcher, shared a clear understanding with the committee on the objectives and rationale of the space program. President Carter and his NASA Administrator, Dr. Robert A. Frosch, did not have the same rapport with the committee. The Fuqua subcommittee in particular was disappointed with the failure of high officials in the Carter administration to give inspiring leadership to the space effort.

When Dr. Frosch made his first appearance before the full committee on January 26, 1978, to discuss the future programs of NASA, he turned off some members with this comment:

I have been cast in some of the testimony as a conservative bureaucrat. I would like to submit that this is one of the roles I should be playing.

Fuqua reacted quickly:

One of the things that leaves me somewhat troubled is the lack of long-range planning and what seems to be a lack of more specificity in what may be the plans for the future.

Wydler declared:

These hearings have not given me the feeling of confidence that anybody is really trying to push the space programs for the future.

When Dr. Frank Press, head of the Office of Science and Technology Policy, generally echoed Dr. Frosch's theme, Congressman Winn, with some exasperation declared:

I don't quite know how to say this without sounding rude, and I don't mean it that way, but most of us on this committee are really excited about the space program and about our accomplishments. After listening to the testimony of you two gentlemen, you leave us very bored.

Fuqua jumped into the debate and asked the administration witnesses:

What are you talking about for the future? The administration was able to get along for a while saying "We are new and we're trying to formulate our policy." We are into the fourth quarter now and almost at the 2-minute warning, and we need to be getting on with the program of what we are going to be doing down the road.

In his floor statement urging adoption of the NASA authorization bill on April 25, 1978, Fuqua added:

The committee views with increasing concern the apparent lack of interest of the executive branch in consistent and continuous future planning for a strong national space program. * * *

When President Carter announced his new American Civil Space Policy on October 11, 1978, it also met a cool reception from Fuqua, who wrote to Dr. Press on October 20:

I am writing you this letter to express concern about the tone and content of the release. * * * It would be easy for the public to draw the conclusion that the American civil space policy will be maintained at its present dollar level or reduced.

Fuqua also raised questions about the speed with which crucial decisions were being made in a number of areas.

Despite the generally lackadaisical attitude displayed by the administration, Chairman Teague and the Fuqua subcommittee continued to press for better support for expanding applications, for more imaginative program planning, for development of a more thoroughgoing agenda for space industrialization, and a more inspired effort to give leadership to all the space programs across the board. The committee initiative may not have resulted in a wholesale beefing up of all the NASA programs, but at least there was a positive response among Members of Congress. On April 25, 1978, the House of Representatives passed the NASA authorization bill by 345–54.

THE COMMITTEE IN 1979

In 1979, the full committee included the following:

Democrats	Republicans
Don Fuqua, Florida, *Chairman*	John W. Wydler, New York
Robert A. Roe, New Jersey	Larry Winn, Jr., Kansas
Mike McCormack, Washington	Barry M. Goldwater, Jr., California
George E. Brown, Jr., California	Hamilton Fish, Jr., New York
James H. Scheuer, New York	Manuel Lujan, Jr., New Mexico
Richard L. Ottinger, New York	Harold C. Hollenbeck, New Jersey
Tom Harkin, Iowa	Robert K. Dornan, California
Jim Lloyd, California	Robert S. Walker, Pennsylvania
Jerome A. Ambro, New York	Edwin B. Forsythe, New Jersey
Marilyn Lloyd Bouquard, Tennessee	Ken Kramer, Colorado
James J. Blanchard, Michigan	William Carney, New York
Doug Walgren, Pennsylvania	Robert W. Davis, Michigan
Ronnie G. Flippo, Alabama	Toby Roth, Wisconsin
Dan Glickman, Kansas	Donald Lawrence Ritter, Pennsylvania
Albert Gore, Jr., Tennessee	Bill Royer, California
Wes Watkins, Oklahoma	
Robert A. Young, Missouri	
Richard C. White, Texas	
Harold L. Volkmer, Missouri	
Donald J. Pease, Ohio	
Howard Wolpe, Michigan	
Nicholas Mavroules, Massachusetts	
Bill Nelson, Florida	
Beryl Anthony, Jr., Arkansas	
Stanley N. Lundine, New York	
Allen E. Ertel, Pennsylvania	
Kent Hance, Texas	

The Subcommittee on Space Science and Applications in 1979 included the following:

Democrats	Republicans
Don Fuqua, Florida, *Chairman*	Larry Winn, Jr., Kansas
Ronnie G. Flippo, Alabama	Robert K. Dornan, California
Wes Watkins, Oklahoma	Ken Kramer, Colorado
Marilyn Lloyd Bouquard, Tennessee	
Bill Nelson, Florida	
George E. Brown, Jr., California	

On June 15, 1979, Darrell R. Branscome was named acting subcommittee staff director to replace James E. Wilson, who took a position with McDonnell Douglas Corp. With B.S. and M.S. degrees in mechanical engineering from Virginia Polytechnic Institute, Branscome had served at NASA's Langley Research Center and the NASA Headquarters. He started working for the committee in 1974, rising to

Among new members of Space Science and Applications Subcommittee in 1979 was Representative Bill Nelson (Democrat of Florida), who represents the district in which Kennedy Space Center is located.

become deputy staff director of the Space Science and Applications Subcommittee prior to being chosen as acting staff director with Wilson's departure. On October 24, 1979, he was named staff director.

Following its customary procedure, the subcommittee held three advance hearings in September 1978 on the NASA authorization, capped by 15 sessions during the first three months of 1979. NASA, the Air Force, the European Space Agency, and members of the industrial and scientific community testified on NASA-related programs.

In addition, the subcommittee on February 14, 1979, reviewed the President's civil space policy. At this hearing, Fuqua expressed his apprehension that NASA and OMB were projecting no new starts for several years:

> I am very concerned that we are not looking down the road at new programs and where we are going, and we are dying on the vine. It is very alarming.

Wydler echoed this view, indicating:

> What we are worried about here today is what is the program going to look like 10 years from now, with the decisions you are making today.

Referring to the gung-ho attitude which had prevailed in the 1960's, Wydler asked:

Would we be leading the world in space if we had followed this kind of approach that you are really now recommending for our country for the next decade?

In presenting the NASA authorization bill to the House on March 28, 1979, Fuqua reviewed the accomplishments of the Earth resources satellites (Landsats). He also called attention to the achievements, past, present and future, of the Voyager and Pioneer interplanetary spacecraft with relation to Jupiter and Saturn. But he pointed out that the value of NASA's 1979 effort in constant 1968 dollars had sagged to $1,653 million—less than 50 percent of the 1968 buying power.

Aside from a brief skirmish over NASA funding of supersonic research, there was little opposition to the NASA authorization in 1979. The house rejected by 246–137 the perennial effort of Representative Ted Weiss (Democrat of New York) to cut the NASA effort by $23 million for supersonic research. Then the House passed the bill on March 28 by a 323–57 vote. The conference report, adopted by the House on July 27, 1979, also supported two initiatives by the House:
  —The addition of $2 million, which had been cut out by OMB, to initiate development of a "Multi-Spectral Resources Sample"— an advanced remote sensor instrument for improved resolution and higher reliability.
  —The addition of $4 million to start development of a National Oceanic Satellite System.

During May and June 1979, the subcommittee held a series of hearings on Fuqua's bill to establish a Space Industrialization Corporation. In opening the hearings, Fuqua defined the objective of his bill:

To provide a means for financing the development of new products, processes, and industries using the properties of the space environment.

In the hearings, the subcommittee examined a number of issues associated with the prospects for commercial ventures in space, the role of the Federal Government, and the appropriate mechanism for fostering cooperation with the private sector.

TRACKING AND DATA ACQUISITION

From 1970 to 1974, jurisdiction over the tracking network was handled by the Subcommittee on Advanced Research and Technology (later renamed "Aeronautics and Space Technology"), and after 1975 jurisdiction passed to the Subcommittee on Space Science and Applications.

In reviewing the tracking network budget, which amounted to approximately $300 million annually, the subcommittees through oversight and field visits attempted to ascertain how a greater degree of efficiency could be instilled into the NASA operation. Most of the top officials administering the tracking network were veterans of long service with the National Advisory Committee for Aeronautics, they were a proud group, and it was difficult to penetrate the veil of self-protection which occasionally surrounded the group. To suggest in the abstract that money could be saved through automation and computerization was frequently met with the NASA argument that new equipment was needed. In general, the efficiency of the network operations was high, morale was excellent, and mission failures were never caused by shortcomings in the tracking system.

In 1970, the Hechler subcommittee recommended a reduction of $4.2 million from the $298 million asked by NASA, on the grounds that some of the requested new equipment could be deferred for another year. In battling the issue out with the Senate in the conference committee, it was finally decided to make the cut a compromise $2.8 million.

In 1971, with the NASA budget request at its lowest point since 1962, only $264 million was programed for the tracking network. The Hechler subcommittee recommended the full amount, after ascertaining that the NASA request "was carefully examined and found to be austerely based." The members of the Subcommittee on Advanced Research and Technology in 1971 were the following:

Democrats	*Republicans*
Ken Hechler, West Virginia, *Chairman*	Thomas M. Pelly, Washington
John W. Davis, Georgia	John W. Wydler, New York
William R. Cotter, Connecticut	Barry M. Goldwater, Jr., California
Charles B. Rangel, New York	Marvin L. Esch, Michigan
Mike McCormack, Washington	

In the Hechler subcommittee hearings in 1972, Congressman Rangel drew the admission from NASA that the tracking stations in the vicinity of Johannesburg, South Africa, employed blacks at a top salary of $1,428, whereas the lowest salary paid to a white South African at the station was $1,680, and housing and other facilities were segregated. To a question by Chairman Hechler, as to whether there had "ever been an attempt to try and (increase) the number of blacks employed, or the salaries, or the facilities," the NASA witness responded that discussions were going on.

During the consideration of the NASA authorization bill on April 20, 1972, Congressman Rangel offered an amendment to delete funds for any tracking station located in South Africa because of the effect of its apartheid policies on practices at the NASA tracking stations around Johannesburg. Rangel argued:

The question before us today is a moral one: Will the Congress permit our tax dollars to continue to be used to pay for racism, or will we continue to allow the world to believe that we still think about people as well as progress in our scientific programs?

Subcommittee Chairman Hechler advised Chairman Miller that he would not speak against the Rangel amendment, and in fact planned to vote for it. Representative Alphonzo Bell (Republican of California) undertook to oppose the Rangel amendment, with this argument:

This is not a situation we can influence, something we can directly control. No one can say there is any racism in my soul. I have voted for every civil rights measure brought to the floor of the House. I simply think it is ridiculous for us to attempt to interfere in the internal affairs of the South African Government.

Pelly told the House that "the author of this amendment in the subcommittee, on which I have the honor to be a member, did some very effective work, and directed some very penetrating questions on this whole subject." Pelly added:

I would hope, though, that if we are going to eliminate Johannesburg as a tracking station that we would also eliminate Chile, which has a Communist government, and Ecuador, which seizes our fishing boats, and I wish the gentleman would bulk them in all together.

Rangel responded succinctly:

I think that this Congress is sophisticated enough to take on one moral problem at a time, and I think I would be susceptible to any suggestion concerning any nation that is violating the rights of men.

Teague and Miller both spoke against the Rangel amendment, Miller stating:

These stations are located geographically irrespective of the government in the area where they are located.

Representative Ronald V. Dellums (Democrat of California) put it this way:

The only justification I have heard in the past 20 minutes for maintaining this station in Johannesburg is to protect safety of astronauts. I certainly do not want anything to happen to astronauts, but I raise this rhetorical question: Why are we so committed to a program that would allow the astronauts to walk in the tranquility of the Moon when we have not found the ability to come together on the floor of this Congress to adopt policies that would enhance our ability to walk as brothers and sisters on the face of this Earth?

Miller concluded:

I cannot see how you are going to do the things you have got to do and decide that you cannot do it because geographically the part of the country in which you want to place a tracking station happens to be under a government we do not like.

The Rangel amendment was defeated in 1972 in a nonrecorded vote, but Rangel brought the issue up again in 1973. By 1973, the composition of the subcommittee had changed, and the following were members of the Subcommittee on Aeronautics and Space Technology:

Democrats	*Republicans*
Ken Hechler, West Virginia, *Chairman*	John W. Wydler, New York
John W. Davis, Georgia	Barry M. Goldwater, Jr., California
William R. Cotter, Connecticut	John B. Conlan, Arizona
J. J. Pickle, Texas	Stanford E. Parris, Virginia
Ray Thornton, Arkansas	

Even though he had left the Science Committee, Rangel's efforts to eliminate NASA funding of the Johannesburg-area tracking facilities stirred a great deal more support in 1973. Rangel argued that no black NASA official had ever visited the South African tracking facility, that an alternative to the station should be found, and that continued financing of the station violated our national policy on civil rights. Chairman Teague took on the defense of funding the South African facility, and in opposition to the Rangel motion:

> That station in South Africa is one of the most important tracking stations we have. * * * I do not feel this is an item that should be a factor in our authorization bill.

To Rangel's contention that blacks were not being trained for the higher paying positions exclusively held by whites at the NASA facility in South Africa, Teague offered:

> I will go with the gentleman to the State Department or any place he wants to go and try to see if we can confer with that country and see if they will not train some of the black people in technical areas where they can get some of the better salaried positions.

At this point, Hechler broke with his chairman and decided to speak out on the floor for the Rangel amendment, declaring:

> Because of the efforts of the gentleman from New York (Mr. Rangel) a number of improvements in housing, educational facilities, and medical care have been made at the Johannesburg installation. But the relative salaries of black and white personnel are shockingly unequal and inequitable.

On a rollcall vote on May 23, 1973, Rangel's amendment was defeated by 294 to 104. Members of the Science Committee who went against their chairman and voted for the Rangel amendment were Bergland, Hechler, and Mosher; the other members of the Science

Committee remained loyal to their chairman. Of the ten members of the Hechler subcommittee who were responsible for overseeing tracking and data acquisition, all except Hechler voted with Chairman Teague.

Hechler also clashed with Teague on a recommended cut in the tracking network. Wydler, the ranking Republican on the Hechler subcommittee, successfully amended the authorization bill in subcommittee, cutting the tracking authorization from $250 million down to $240 million. This action so upset top-side officials at NASA that Deputy Administrator George Low paid a secret visit to Chairman Teague, bearing a strongly worded argument in a letter from Administrator Fletcher which contended that U.S. dollar devaluations had already robbed the tracking program of $8 million. Dr. Fletcher added that a further cut of $10 million would endanger both the applications program and manned space flights. It was a bold move which NASA made to catch the subcommittee completely by surprise. The secret was tightly guarded. Chairman Teague completely disarmed Hechler at the opening of the full committee's markup session by saying:

> The Chair would like to state to the subcommittees that he's had nothing but compliments on their work on this bill.

After Hechler had completed his usual plug for increasing the aeronautics authorization, Teague dropped his bombshell. He read a few sentences from the NASA Administrator's letter, blasting the $10 million reduction as endangering manned space flights and the applications program. Flabbergasted, the gentleman from West Virginia just opened his dry mouth and no sound came forth; about all he could think of was to ask that the full text of the letter, which the subcommittee had not seen, be placed on the record. He counterattacked:

> I must express some surprise that a communication like this should not reach any members of the subcommittee. I think it is a rather unusual procedure.

Wydler defended his amendment to cut $10 million on the grounds that $10 million had been reprogramed out of the tracking appropriation the prior year, that the testimony indicated a 4-percent cut would not hurt that much, and that this helped offset increases in aeronautics recommended by the subcommittee. Then the tide began to turn against the subcommittee.

"I always thought that tracking and data acquisition was extremely important," Downing led off. He pointed out that his subcommittee was concerned that their science and applications satellites were sending out more data than the tracking and data network could accommodate. "You are cutting a very essential program," charged Downing.

"This particular cut distresses me," echoed Milford. He labeled the cut as "bad economy."

Winn was even sharper in his criticism. He challenged the way the Hechler subcommittee had cut the tracking authorization to counterbalance increases in aeronautics:

I just hope that this wouldn't be the way our committee would try to dipsy-doodle funds back and forth to make room for their increases. But I am more concerned, and I want to stress again, I think we ought to pay more attention—and some of the members of the full committee were not here when the Chairman read the letter from Dr. Fletcher that involved this—and I, for one, just cannot support a $10 million decrease in Tracking and Data Acquisition.

This colloquy then ensued:

Mr. HECHLER. I think it's very, very unfortunate that the subcommittee, which is charged with the responsibility of making recommendations, had absolutely no knowledge of this communication from Dr. Fletcher until the very minute the chairman of the full committee read it, and I don't stand on ceremony or protocol. It just seems to me that if our subcommittee is going to make a judgment or recommendation that we ought to have the benefit of at least some small caucus to discuss the details in this issue.

Mr. WINN. I would be glad to move to the Chairman that we adjourn so that you can have a small caucus if that would help you.

Mr. TEAGUE [the chairman]. The Chairman would state that he has no idea when I received this communication. Yesterday morning they contacted me. They didn't get in touch with me until afternoon, and I understand that Mr. Mosher got a copy of the same communication. * * * This item can't be put off, and I would like you to have a chance, and you should have had a chance.

Hechler was convinced, after backpedaling in the face of the criticism from both Democrats and Republicans, that a vote at the time would certainly have reversed the subcommittee's action. Spotting an opening in the last comment of Chairman Teague, Hechler observed:

I think the inference of the gentleman is entirely correct, that perhaps we should have an opportunity in our subcommittee to review this letter that Dr. Fletcher sent up to the full committee, because I think that's the only orderly manner on which we can proceed.

By unanimous consent, the vote on the tracking cut was deferred until the next meeting of the full committee on May 1, 1973. When the Hechler subcommittee convened, its chairman did not have to say a word; there was a voluble unanimity among both Democratic and Republican members who strongly urged that the new information from Dr. Fletcher did not justify changing the recommendation.

By the time the full committee met on May 1, the entire atmosphere had radically altered. In Chairman Teague's absence, Hechler was asked to chair the full committee, and he reported the unanimous action taken by his subcommittee. Some active advance lobbying by its members produced an amazingly different reaction: Without any comment whatsoever, the motion to cut $10 million from the tracking authorization was unanimously adopted by the full committee.

The Subcommittee on Aeronautics and Space Technology had been reviewing NASA tracking and data activities for a dozen years, and most members had personally visited tracking stations to enable them to become fully briefed on most aspects of the program. But the twin shocks of the Johannesburg floor fight and the brouhaha during the authorization hearings in 1973 convinced the subcommittee that more drastic action was necessary. In a letter to NASA Administrator Fletcher on June 18, Hechler outlined plans for an intensive oversight hearing on the tracking network. He reminded Dr. Fletcher that he expected monthly progress reports from NASA in improvements in the Johannesburg area stations, adding:

It is our intention to examine closely the status and progress of the Johannesburg Tracking Station. I urge you to do everything you can in working with the South Africans to accelerate improvements related to working conditions and opportunities for Black South Africans at the station. Action in the past has been very slow.

Dr. Fletcher responded affirmatively on June 29, promising to deliver the first report on Johannesburg "in early July."

That report, when hand delivered on July 10, proved to be a real shocker. After persuading the committee leadership how vital the South African tracking facilities were to the entire space program, NASA announced they had decided to close down their operations in South Africa. It did little good to protest after the fact that NASA should have maintained better lines of communication concerning its plans for the South African facilities. Hechler told his colleagues: "If they can communicate with deep space, they ought to be able to communicate better with us." Obviously, the strong opposition expressed during the 1973 authorization debate influenced NASA's decision. The action had no visible adverse effect on the reliability of the tracking network. The tracking station in Spain took up the slack.

Proceeding with the general oversight investigation, Hechler obtained permission from Chairman Teague to borrow personnel from the General Accounting Office and Department of Defense. To help cement better understanding in the other subcommittees, a July 31 memorandum to Teague indicated:

I would welcome the participation of Don Fuqua and Jim Symington in the conduct of the Tracking and Data Acquisition review because of the close relationship of this program to their areas of interest.

During extensive hearings and field trips throughout the summer of 1973, including well attended public hearings in October 1973 and January 1974, the Hechler subcommittee examined each tracking station, its role, its manpower, and activities of supporting contractors, as well as relationships with host nations. The committee also examined the proposed new Tracking and Data Relay Satellite System, a leased system which enabled better transmission and reception of data through what were essentially two network satellite stations in synchronous orbit.

In general, the thorough oversight investigation gave the NASA tracking network high marks for efficiency and economy of operation. It also strengthened the subcommittee's ability to field any and all questions which had been popping out of the blue concerning the program. But in 1974, NASA revealed the details of the new financing proposal for the planned new satellite relay tracking network. NASA asked that the basic 1958 Space Act be amended to give them authority to enter long-term leasing arrangements. NASA attempted to prove that leasing would be cheaper, but the subcommittee produced its own figures to show that it was far cheaper to buy the services outright. New hearings, investigations, and conferences wrestled with this issue during the early months of 1974. Finally, the subcommittee came up with a compromise which amended the authorization bill in 1974, and required NASA to come back for committee approval after the RFP's (requests for proposals) had been circulated to the bidding contractors. This brief summary covers many weeks of sweating out an extremely complex series of decisions, clearly demonstrating the impact of the committee on public policy—albeit not as spectacularly dramatic as being present at the creation of the Apollo program.

When the subcommittee presented its proposal in 1974, there was a universal rush by both Republicans and Democrats to praise the care and attention given in protecting the interest of the taxpayers. At Hechler's request, NASA also produced a written promise to speed up its cost-benefit analyses in such a way as to give the authorizing committee in the spring of 1975 a clearer roadmap for the future. The subcommittee was still slightly troubled that the central NASA argument for leasing instead of purchasing was that it would stretch out the total expenditures over many years instead of lumping them in one year. But the subcommittee agreed that the compromise procedure would give Congress one more crack at it in 1975.

Contrary to the buffeting which the subcommittee took in 1973 when they tried to save $10 million, in 1974 the complex compromise went through unanimously. The extensive hearings and staff work on the tracking and data acquisition program certainly confirmed the fact that knowledge is power.

In 1975, jurisdiction over the tracking and data acquisition program passed to the Fuqua Subcommittee on Space Science and Applications. In presenting the tracking authorization to the House on April 9, 1975, Fuqua reported the successful data obtained from

Pioneer and Mariner spacecraft photographing and obtaining other valuable data from Saturn, Jupiter, Venus, and Mercury. The full committee and the House supported the subcommittee's recommendation to cut the tracking authorization in 1975 by $2.2 million. Unlike earlier protests against reductions in this area, the committee ratified the cuts without a murmur. Wydler successfully put across his amendment to insure that NASA would have the option to purchase the expensive new Relay satellite system at the conclusion of the leasing period. Once again in 1976, the Fuqua subcommittee engineered a reduction—this time of $4 million—in the NASA tracking request, and succeeded in having $3 million of that reduction stick by the time the bill emerged from the conference committee. Despite inflation which affected costs at the Madrid tracking station, the committee felt that the $255 million actually authorized would fully support the 40 individual programs being tracked. In 1977 and 1978, the authorization for the tracking network crept up to exceed $300 million annually, primarily due to inflation. In addition, larger requirements were imposed on the tracking network as interplanetary flights like Voyager proceeded to Jupiter, Saturn, and the sizable moons near those planets. Late in the 1970's, NASA also began building a new ground terminal at White Sands, N. Mex. to supplement the two geosynchronous satellites to be used in the new Tracking and Data Relay Satellite System scheduled for operational use in 1980.

The committee kept a watchful oversight eye on the rapidly changing nature of the tracking network, with special emphasis on how the new tracking facilities would cope with the specialized demands of the Space Shuttle missions. The Fuqua subcommittee followed very closely the terms of the lease contract with Western Union Space Communications, Inc., a subsidiary of Western Union Telegraph Co., to insure that the public and taxpayer interest was fully protected. From 1977 through 1979, Dan Cassidy of the committee staff completed several detailed reports on program cost, performance, and schedule, which further enabled the committee to ride herd on the extremely complex procedures in the tracking program.

As the 1970's drew to a close, the Science Committee through the Fuqua subcommittee was buttressed with more than enough specialized data to enable sound decisions to be made on funding, general oversight, and keeping tabs on program developments. The committee had clearly come of age since the days when hundreds of millions of dollars were authorized pretty much "on faith." A mutual respect had developed between the committee and NASA, as the complex tracking operation moved into the transition period toward the installation of a new system of high-speed transmission and fuller coverage through the Tracking and Data Relay Satellite System of the 1980's.

CHAPTER X

A New Name and Expanded Authority for the Committee

The Science and Astronautics Committee was initially established as a major committee, as part of the plan hatched to insure that Representative Overton Brooks would transfer from the Armed Services Committee to become the first chairman of the Science Committee (see page 15). With a broader jurisdiction than the comparable Senate committee, the House committee nevertheless experienced some difficulty in attracting Members with interests outside of space and science. The turnover in committee membership became unusually large, as many Members sought to be on those committees which helped their own districts to a greater extent. This was especially true after the Moon landing in 1969, as it became apparent that the decline in the space program might mean a decline in the significance of the Science Committee.

As chairman, Brooks did a remarkable job in preserving and even extending the jurisdictional frontiers of the committee. He successfully fended off numerous attempts by other committees, notably Armed Services and Interstate and Foreign Commerce, to hem in the scope of Science Committee activities. Miller, who had greater prestige among his colleagues, did not go out of his way to expand jurisdiction and thereby create conflict. Both Brooks and Miller were well protected at the highest levels of the House by the membership of two successive Speakers, McCormack and Albert, on the committee at the time each served as House majority leader. In 1963, Miller moved positively to strengthen the jurisdiction of the committee through the establishment of the Daddario Subcommittee on Science, Research and Development. By the end of the decade the committee had authorization power over the National Science Foundation, but the power to authorize the funding of the National Bureau of Standards did not get asserted until the 1970's.

Aside from his brilliant initiative in establishing the new Science Subcommittee, and his imaginative utilization of panels of distinguished scientists, Miller's greatest contribution toward expanding the power and influence of the committee came in the international area. This was done primarily through international visits and his somewhat reluctant consent to adopt Fulton's recommendation to set up a special Subcommittee on International Cooperation in Science and Space.

MONRONEY-MADDEN JOINT COMMITTEE

In the mid-1960's, there was a good chance to broaden the committee's jurisdiction which lacked only the leadership of the committee to capitalize on it. Speaker McCormack appointed Hechler as one of the three House Democrats on the Monroney-Madden Joint Committee on the Organization of Congress in 1965. Working from the inside, Hechler had frequent opportunities to enhance the prestige, defend the good reputation, and even broaden the committee's jurisdiction. For example, when Senator Proxmire suggested that space was simply a matter of communication and transportation and should be merged with the Commerce Committee, Hechler responded:

> Being a member of the House Committee on Science and Astronautics, our committee handles a good deal more than "communication and transportation." We deal with the entire spectrum of scientific research and development, and building of the scientific strength of the Nation.

Occasionally, coaxing questions to witnesses produced good results, as when Hechler asked the Director of the Budget, Charles L. Schultze how he would characterize the relationships between Congress and the scientific community. Schultze responded:

> The recent hearings of Chairman Daddario's subcommittee of the House Committee on Science and Astronautics on the National Science Foundation evidenced mutual respect between the legislator and the scientist, no doubt based on some mutual education in recent years.

Despite discreet proddings, Miller did not recognize the advantage of this great forum for building the strength of the committee. Only two committee members testified before the joint committee, Wydler and Rumsfeld. Wydler advocated application to Congress of many of the computerized techniques developed in NASA. Beyond that, he and Rumsfeld also zeroed in on the desperate need for more staff on the Science Committee and particularly staff assigned to the minority. It is unfortunate that the type of organized effort utilized in 1973 in connection with the Bolling committee was not also put forward in 1965.

THE LOST OPPORTUNITY

The final report of the joint committee recommended that the jurisdiction of the Science and Astronautics Committee be broadened to include "jurisdiction over environmental sciences." Hechler was able to persuade the joint committee to include this language in its final report in 1966:

> Science and engineering have acquired in recent decades a crucial importance in governmental affairs. They influence and help shape not only our national security policies but a broad range of domestic and international public policies as well.

Congress now authorizes and appropriates over $15 billion annually for scientific research and development. It is also faced with the task of monitoring a complex array of 42 technical programs that cross agency lines.

At present, no single committee in either House has comprehensive and coordinating jurisdiction over these activities. * * * We therefore recommend that the committee in each House that now most nearly approaches such concentration have its present jurisdiction expanded to encompass the necessary coordination.

At this point, once again the Science Committee lost its opportunity. There was no rallying of the troops, and in fact there was sheer apathy toward the recommendation of the joint committee on the part of the Science Committee. Soon other committees began to object to any action to change committee jurisdictions. The report gathered dust. The Committee on Rules did not act because there was simply not enough pressure to act. Finally, the Rules Committee decided to hold hearings in 1970 on a stripped-down version of the 1966 recommendations. Here again, the Science Committee leadership neither testified nor seemed to express any interest. Perhaps the Rules Committee would have smothered such an effort, yet there were those who felt at the time that the effort was at least worth a try.

EXODUS FROM THE COMMITTEE

In 1970 and 1971, the first effects of the impending decline in space funding and space interests began to be felt. The younger members of the committee began to look for greener pastures—committees which could produce more direct benefits for their districts. At the close of the 91st Congress in 1970, the Democratic side of the committee was hit with a wave of resignations. Daddario left to run for Governor of Connecticut, and Brown went off to try for the Senate in California; seven other members voluntarily left to join other committees. This meant that out of the 18 Democrats on the committee in 1970, only nine opted to be assigned to the committee in 1971.

For Miller and Teague, this represented a crisis in the life of the committee. As chairman of the Veterans' Affairs Committee, Teague had frequently discussed with Speaker Rayburn the problem of keeping good Members on that committee, which Teague realized could only be done if they were allowed to serve at the same time on other committees. Since the caucus rules permitted service on only one major committee, the status of the Science Committee as a "major committee" was effectively preventing some Members from serving on any other committee. This issue, of course, cut both ways: Changing the Science Committee to a nonmajor committee would enable more good Members to bid for assignment, but at the same time it seemed to reduce the prestige of the committee.

The entire issue came to a head at the start of the 92d Congress in 1971. Symington, going into his second term in the House, had made his mark as an active participant in committee affairs. Genial, co-operative, imaginative, with a sharp sense of humor, one of the two father-and-son teams in the Congress (Symington and Goldwater, whose fathers were both in the Senate), Symington had a high standing in the scientific community as well. Symington mentioned to Dr. William D. McElroy, Director of the National Science Foundation, that he had been offered a vacancy on the House Interstate and Foreign Commerce Committee. Symington told Dr. McElroy that he probably would have to leave the Science Committee because of the Democratic caucus rule which prohibited him from serving on more than one major committee. In an unusual gesture of support for a Congressman who sat on the other side of the witness table during NSF hearings by the Subcommittee on Science, Research and Development of which Symington was a member, Dr. McElroy mentioned to Miller and Davis how unfortunate it would be if the rules prevented Symington from continuing on the Science Committee. Since Interstate and Foreign Commerce gave Symington the leverage he needed in his home district in Missouri, he was prepared to leave the Science Committee.

HOW THE COMMITTEE BECAME "NONMAJOR"

What subsequently transpired is very clear: The Science Committee in 1971 was redesignated as a nonmajor committee and Symington was able to serve on both committees of his choice. Exactly how and precisely when this deed was accomplished has been lost in the fading memory of the participants, and the lack of precise documentation. Speaker McCormack, who relinquished his office at the end of 1970, and Speaker Albert, who took office in 1971, both served on the Science Committee and did not in 1978 recollect the move. Neither the chairman of the Ways and Means Committee (Representative Wilbur D. Mills—Democrat of Arkansas) nor his staff director, John M. Martin, Jr., recall the circumstances, although Chairman Mills remembers it was done to accommodate one of the Members who wished to serve on two committees. Neither the Parliamentarian's Office, the Democratic Steering and Policy Committee (which inherited from the Ways and Means Committee the power to recommend committee designations and assignments) nor the Democratic caucus have a record of how it happened. Nor do Miller or Teague recall the precise chain of events which caused the redesignation of the committee.

One senior subcommittee chairman, Karth, has a very vivid recollection of his reaction to the move. Karth was furious that Miller, as chairman, did not put up a fight against making the committee "non-major." This one of the reasons that Karth gave for leaving the committee to join the Ways and Means Committee in October 1971. Karth reflected:

> I don't think that a chairman accepts those things without first going back to his committee and saying: "This is what the leadership is talking about, and I want to discuss it with you because you're affected as much as I am and probably more." But he didn't do that. * * * I didn't think that the Chairman should just accept it without saying anything.

BOLLING COMMITTEE HEARINGS

On May 2, 1973, the House Select Committee on Committees, popularly known as the Bolling committee, started its six weeks of public hearings prior to recommending major jurisdictional and other reforms in the House. Even before the hearings started, Teague and his staff director, Jack Swigert, had huddled on the strategy to use in preparing for a major presentation to the Bolling committee. Swigert introduced the subject at several of his weekly staff meetings, stressing that he wanted ideas, suggestions and input for several different approaches, ranging from a single appearance by Teague to separate presentations by the subcommittee chairmen. Dr. Holmfeld was assigned to monitor the hearings, and he made periodic reports on the nature of the presentations, the types of questions being raised by the committee as witnesses appeared, and the particularly effective techniques being used by witnesses. For example, Dr. Holmfeld reported that the testimony of Representative Albert H. Quie (Republican of Minnesota) had been well received. Swigert forthwith forwarded copies of Quie's statement to all his task team leaders with this note:

> A good example in the use of appendices for historical information and material for the record. We are going to need devices like this, or other innovations, to cover the spectrum of the committee's areas of interest with the depth needed.

Initially, it was planned to divide up the 60 minutes of testimony time allocated to the committee with several minutes for each subcommittee. In a memorandum to the staff, Swigert indicated:

> The objective of the staff will be to prepare this testimony so that it is the most concise, factual and with the most depth of any testimony presented yet to the committee.

In this fashion, a whole sheaf of valuable material was assembled, specifying the work and future capabilities of the committee. Jack Kratchman, detailed to the committee from the National Science Foundation, prepared a voluminous report analyzing current and possible future energy jurisdiction options for the Congress. All of this material was then used with telling effect to bolster the complete and persuasive testimony which was subsequently delivered by Teague and Mike McCormack on June 8. Originally scheduled for May 11, Teague and McCormack decided to wait until the last day of the first phase of the hearings so their combined testimony could be fully assembled and have a greater impact. From the standpoint of committee influence, this developed into better strategy because it enabled Davis and Mosher to present their case on May 11, followed by Cronin, Pickle, and Ketchum who appeared later in May.

DAVIS AND MOSHER TESTIMONY

Both Davis and Mosher called attention to the wide range of activities and accomplishments of the Science Committee. They also quoted House Minority Leader Ford, who a few days before had mentioned he was in on the creation of the Science Committee, and had stated:

> With our space program now more or less stabilized, it seems to me that this committee could justifiably be given additional responsibility.

Representative Dave Martin (Republican of Nebraska), vice chairman of the Bolling committee, asked Mosher: "Do you have any specific suggestions, Charlie, as to additional jurisdiction that you do not now have?" Mosher responded:

> John Davis had the temerity to suggest that NOAA might come within the purview of the Science Committee. It happens that I am the ranking Republican on the Subcommittee on Oceanography of the Merchant Marine Committee and sort of grandparent of NOAA in many ways and so that puts me in the middle. However, it is true that NOAA is an agency which, I think, has great potential as precisely the type of agency that could well be assigned to the Science Committee.

Both Davis and Mosher elaborated on the new work which McCormack's task force and Subcommittee on Energy had done, and Mosher advised: "We do have the capacity, the interest, and the willingness to accept added responsibilities."

TEAGUE AND McCORMACK TEAM UP

Because of the rising importance of energy in the work of the committee, Teague and McCormack decided to testify in tandem on June 8, 1973. The stack of supporting documents which they presented to the Bolling committee was so impressive that Bolling observed at the outset:

> Mr. Teague, I have had the chance to read your statement. I know how much effort has gone into it and many other documents that were submitted. We are very

grateful to you and the staff for taking it seriously. * * * These are very extensive documents and are very helpful to us, as you know, in terms of specific approaches of your committee and other committees related to it. But in terms of the general problem, we truly are grateful for the very creative and constructive efforts that you have made.

As Teague got ready to begin his statement, Bolling also gave him credit as one of the original backers of the reorganization idea:

When the Speaker talked to me first about the idea of this, he mentioned you and a few others who felt that we needed to do something very badly about reorganizing, looking at the problem of committee structure. So, in a sense, you are one of the parents of this committee.

Teague mentioned that he wanted McCormack to testify along with him because of the importance of energy. This was the main thrust of his testimony, aside from detailing the principal achievements of the committee. He spoke broadly, not parochially, focusing on issues of national concern. Teague advocated distributing committee workloads more evenly, providing for joint referrals to minimize future jurisdictional conflicts, and clarifying responsibilities. He sketched in the rigorous oversight which his committee had accomplished in high technology areas. Making a telling point concerning the 2½-year study of civil aviation research and development, Teague added:

Our committee has been the only congressional unit to hold hearings on this study's conclusions and recommendations even though the problem areas identified extended across multiple committee jurisdictions: Armed Services, Banking and Currency, Interstate and Foreign Commerce, Joint Economic Committee, Judiciary, and Ways and Means. Civil aeronautical research and development should be concentrated in a single committee, if we are to legislate effectively in the aeronautical field.

Teague wove a very evenly meshed pattern of the relationships between health research and development and other forms of scientific R. & D., the importance of patent policy decisions in translating research results into useful technology, and the growing importance of technology assessment. He raised new questions: The relations of computers to privacy, the ethical and moral implications of genetic engineering, the proper balance between energy and environmental research, finally leading into the qualifications of McCormack in the energy area:

Two years ago, when Mike was a freshman, Chairman George Miller of this committee appointed Mike chairman of a task force on energy. The task force did a splendid job, and this year we upgraded it to a full Subcommittee on Energy, and made Mike its chairman. I think we are fortunate to have scientists of Mike McCormack's caliber and dedication available to chair such an important subcommittee at this time.

McCORMACK AND ENERGY

McCormack led into his testimony by describing the energy crisis, and commenting that this had produced a crisis within Congress in organizing to meet it. He pointed out how fragmented authority, dispersed among many committees, had produced confusion and impotence in the legislative branch. He stated that there was no integrated team of top flight scientists, engineers, economists and other specialists working as a unit to tackle the energy crisis. McCormack characterized the response of Congress "insipid," and blamed the nature of the response on the diffused committee structure. He was not bashful in his prescription:

> I believe that these responsibilities logically fall to the Committee on Science and Astronautics. This committee has established a tradition of dealing with technological problems, and of doing so in a scientific manner.

To bolster his argument, McCormack even drew on the example of the defunct Panel on Science and Technology, which had not met for over a year and which Teague apparently had no intention of reviving. He then presented 52 tightly drawn printed pages of analysis which supported his arguments. The analysis went into deep detail on the current House committee system for energy matters, assessments of that system from the standpoint of efficiency, output, and other criteria of operation, and a complete evaluation of existing committee jurisdictions pertaining to energy. There was included an identification and evaluation of alternative jurisdictional systems, along with carefully presented interpretations and conclusions. From the hearings and reports of the various House committees, charts were prepared indicating their interests and output in relation to various sources of energy, and why a more centralized jurisdiction made sense. He described the work of his Energy Subcommittee as "constructive, deliberate, sincere, not excessive publicity, and nonpartisan." Much of McCormack's material was drawn from the study which Kratchman had produced while detailed to the committee from the National Science Foundation.

Teague then wound up the presentation with a challenge that "the House of Representatives must be ready to respond with timely and effective legislation and vigilant oversight."

The questions were sympathetic, from Bolling, Martin, and other Members. The main thrust of the questions was on oversight, and Teague pointed out the "absence of scandals and overruns" in the space program as an illustration of the value of oversight. Bolling offered a nostalgic reminder of the committee chaired by Teague, on which Bolling has served, which had reviewed the operation of the G.I. Bill of Rights: "That certainly was an oversight operation."

DR. SEAMANS SUPPORTS SCIENCE COMMITTEE

Following the completion of the first phase of the hearings, the Bolling committee turned to panels of experts, political scientists, and outside analysts and observers for advice and assessments. Dr. Robert C. Seamans, Jr., President of the National Academy of Engineering and former Deputy Administrator of NASA, told the Bolling committee:

> From my own experience in the fields of research and development, I think we have been very fortunate in the House to have the Science and Astronautics Committee which has looked not just at astronautics as somewhat of a special area within research and development, but has looked at research and development on a broad scale.

Representative John C. Culver (Democrat of Iowa) questioned Dr. Seamans about the need for more integrated planning in the congressional committee system. In response to one of Culver's questions, Dr. Seamans responded:

> Maybe it is time for the Committee on Science and Astronautics to be looking at broader issues because the NASA program is obviously less now than it was 5 or 6 years ago. Maybe this committee shouldn't be considered primarily to oversee one agency, but rather should review the aggregate of all Federal R. & D.

There were so many Members of Congress eager to testify before the Bolling committee that the time was extended for Members of Congress. On September 13, Brown presented his customarily broad-gauge approach to congressional reform, including a recommendation for a Select Committee on Energy. Brown's suggestion eventually came to pass in 1977, and resembled the 1971 resolution of Tennessee's Congressman Richard H. Fulton. At that time, however, the Bolling committee was not inclined to proliferate more committees, but rather to make sense out of the existing structure. Faced with a negative reaction, Brown then stated:

> If a new source requires new technologies which are not now available, it might be a part of Science and Astronautics, which deals with basic science, research and development.

DR. SHELDON'S INFLUENCE

Once the hearings were completed, the Bolling committee at the end of the year went into the crucial premarkup period when the tentative draft of a bill was prepared. At this point, the importance of the staff of the Bolling committee became of critical importance. Bolling's Chief of Staff was Dr. Charles S. Sheldon II, who had been assistant director of the Select Committee on Astronautics and Space Exploration and also technical director of the House Committee on Science and Astronautics. When Teague testified on June 8, he remarked:

> I think the members should know that Dr. Sheldon was a senior staff member on the Science and Astronautics Committee about ten years ago.

Also on the Bolling staff was Spencer C. Beresford, another alumnus of the Science Committee staff, and Robert C. Ketcham, who later became a counsel for the Science Committee. Dr. Sheldon played a key role in seeing that the first draft bill and committee print included a substantially increased jurisdiction for the Science Committee. Although this fact cannot be documented with any precision, Dr. Sheldon's constant presence and advice certainly did nothing to hurt the Science Committee during this critical period.

THE DECEMBER 7 WORKING DRAFT

The Bolling committee produced a 119-page committee print— a "Working Draft"—on December 7, 1973. One of the objectives stated in the preface, entitled "Basic Organizing Principles," paralleled the testimony of Teague and McCormack:

> House committees should be organized to give coherent consideration to a number of pressing policy problems whose handling has been fragmented, e.g., * * * energy research and development.

Under the draft resolution, the Science Committee was to lose jurisdiction over science scholarships (which went to the Education committee) and biomedical R. & D. (which went to the Commerce and Health committee). But there were vast gains: Overview of military R. & D., to be shared with Armed Services; oceanic and atmospheric sciences, from Merchant Marine and Fisheries; energy R. & D. (including nuclear) from Interior, Commerce, Joint Committee on Atomic Energy, and Merchant Marine; environmental R. & D. from Interior, Public Works, Commerce, and Merchant Marine; and weather, from Commerce. The working draft also spelled out the extension of the principles of oversight to include "legislative preview"—or "foresight" and advocated a strengthening and expansion of the oversight role of each committee. This turned out to affect the Science Committee to a major extent, because of the special oversight role assigned to R. & D.

SOURCE OF THE NEW NAME: "SCIENCE AND TECHNOLOGY"

In the December 7, 1973 draft, the new name "Science and Technology" first surfaced. The only recorded source of inspiration for this name change is in the May 11 testimony by Mosher who included the following in his prepared remarks:

> I would like to enter a vigorous protest and disavowal here against the careless habit of many Members who refer to us as "the Space Committee." That label obviously derives from the concentrated emphasis, now ten years or so ago, that our committee once devoted to the space program, and with proud success. But that distorted emphasis, I now assure you, is a thing of the past. As John Davis has em-

phasized in his testimony here, and other committee members will also agree, an ever-increasing amount of our committee time and effort is necessarily devoted to a variety of science and technology problems and proposals other than those involved in the space program. We maintain a strong allegiance to NASA, and in fact NASA's needs and goals are of increasing variety and interest; but it is essential to consider astronautics and aeronautics in the perspective of all our other R. & D. activities, and to coordinate them all.

Therefore, I personally would welcome—although I consider it a relatively minor matter—a recommendation from your Select Committee on Committees that our S and A Committee might best be renamed, to indicate the true breadth of its mission. Perhaps we should be the Committee on Science and Technology, or on Science and Engineering. (I do emphasize this is only my personal recommendation.)

When he actually delivered his prepared remarks, Mosher said to the Bolling Committee:

I think I have emphasized why it is not correct to call it the Space Committee. I think it much more accurate to call it the Science Committee. It might well be that you might want to recommend—this is sufficiently superficial and not an important item—a new name for the committee such as the Science and Technology Committee or some such terminology as that.

Once the new name was included in the December 7, 1973 working draft, it stuck, and was included in the revised legislation passed in 1974 to become permanent in 1975.

WORKING BEHIND THE SCENES

Neither the Bolling committee nor the Science Committee were idle following the printing of the December 7 working draft. The Bolling committee was carrying on intensive discussions with individual Members, committee chairmen and staff, and outside groups who could help build a consensus through making additional suggestions for improvements in the highly tentative draft. There were many memoranda floating back and forth within the Science Committee as various staff members analyzed the implications of the December 7 draft. Teague was on the phone frequently, and engaged in numerous conversations with Bolling to encourage the Bolling committee to stick to its initial inclination to enhance the power and jurisdiction of the Science Committee. Sheldon was working quietly behind the scenes toward the same objective. As Chairman of the Democratic caucus, Teague was in a strategic position to exert his prestige in an influential way. Completely sold on the importance of coordinating energy research and development, encouraged by the successful efforts of McCormack as leader of the task force and Subcommittee on Energy, and convinced that the Science Committee must expand or descend toward oblivion, Teague found a very sympathetic audience in Bolling and other members of Bolling's committee.

The decisions of the select committee were not very difficult to make on the issue of expanding the jurisdiction of the Science Committee. It was either a case of adding to the responsibilities, or phasing out the committee completely, and nobody suggested the latter course. Instead, additional duties were transferred in the interests of logic, sound substantive reasons, and the successful interchanges between the principals and the staffs involved. Dr. Sheldon's constant presence then helped seal in and protect the decisions.

SWIGERT MOBILIZES THE TROOPS

Swigert did not relax his efforts. He bombarded his Task Team Leaders with memoranda early in January 1974, offering a number of alternative subcommittee alignments assuming the December 7 recommendations were put into effect. Of course, there was still a long way to go before the Bolling committee resolution eventually went to the floor of the House, where, as we shall see, it was defeated in favor of a less thorough structural reorganization of Congress. Nevertheless, Swigert lined up his troops and made the telling point that the months ahead carried the potential of vast benefits for the future of the committee. Swigert insisted that carefully-laid plans were necessary to take full advantage of the opportunities.

Meanwhile, the Bolling committee began its public markup sessions on February 4, 1974. Officially, these markups were termed "Open Business Meetings." Right out in the open, the five Democrats and five Republicans went through the text of the December 7 draft, ratifying certain sections, making some changes, freezing the language as they went along, and occasionally taking votes when issues stirred differing opinions. A clue as to how various Bolling committee members felt toward expanding the Science Committee jurisdiction was contained in some of the comments made by Representative C. W. Bill Young (Republican of Florida), a member of the Armed Services committee, who fought against giving the Science Committee oversight over military R. & D. The colloquy went like this, with Bolling and Dr. Sheldon defending the jurisdiction:

Chairman BOLLING. The intent there is to give them a look at and a hope that they will be able in a nonlegislative way to help coordinate the kinds of things that go in a variety of different places in R. & D. Do I have that correct?

Dr. SHELDON. That is correct. * * *

Mr. YOUNG. I want to make sure that we are not making this a super committee over and above the Armed Services Committee or the other committees we are talking about.

HOW NOT TO ELIMINATE RUMORS

This process of discussion had only been under way a few days when Swigert felt compelled to dispatch a rather peremptory memorandum to "All Professional Staff," dated February 8, 1974, which read in its entirety:

> As you can well imagine, there is much maneuvering going on with regard to the jurisdictional evaluations of the Select Committee on Committees.
>
> Since we in the staff are not privy to all the positions taken by the Chairman and Members of our Committee, there should be no opinions expressed or discussions held by staff members relative to the Select Committee's recommendations.

Perhaps Swigert intended for this memorandum to mean that staff members should not speculate with the news media, although this is not clear. In any event, it was akin to an officer assembling his unit on the parade ground and handing down an edict that henceforth all latrine rumors must cease immediately. It was difficult to carry on intensive planning, when the committee was at the crossroads and could contemplate a bright and challenging future, without some weighing of possible alternatives.

SCIENCE COMMITTEE RECOMMENDED AS MAJOR COMMITTEE

In March 1974, the Bolling committee made its recommendations. They brought joy to the Science Committee, to be raised once again to the status of a major, exclusive committee—one of the 15 nominated for that prized category. Beyond existing jurisdiction, the March bill extended authority over all research and development in energy (including nonmilitary nuclear and thermonuclear energy), environmental, civil aviation, and scientific prototypes; that is, working models. Authority was added over oceanic and atmospheric sciences and sea grant programs. Although the committee was already exercising jurisdiction over science policy and technology assessment, these functions were specifically spelled out in the bill accompanying the report. The Bolling committee also set up a new function called "special oversight," which authorized general investigations without legislation, and specifically singled out the following R. & D. oversight areas for assignment to the Science Committee: biomedical, agricultural, military and water research and development.

The rationale offered for these decisions was indeed heartening. The March report noted:

> As far as practicable, the related components of science and technology are united in a single committee that can provide the necessary expertise and develop comprehensive and coherent policies.

After listing the various new categories of work being assigned to the Science Committee, the Bolling committee added:

> The select committee proposes to place all other elements of scientific research and development under the jurisdiction of this committee. It further proposes to give the committee a responsibility not previously assigned to any committee of Congress—over-all review of Federal research and development. * * *

The significant effects of the Teague-McCormack testimony and the behind-the-scenes work of Dr. Sheldon emerge in this comment by the Bolling committee:

> In energy research and development particularly, the committee has taken strong initiative and undertaken significant preparatory work.

The March report of the Bolling committee was duly circulated and publicized. Intensive work began to line up support. Symington wrote Bolling:

> I think it is a landmark effort where both fools and angels fear to tread, thus opening the way for a natural man who wants to see his country governed more rationally. For this you are to be congratulated and thanked.

Wydler, fascinated by the prospects of an interesting floor fight ahead, wrote Bolling:

> I look forward to a consideration of the House committee organization on the floor. It will certainly be exciting and I'm sure it will be a moment of high drama. Nothing seems to generate emotion as much as committee jurisdiction.

DEMOCRATIC CAUCUS TORPEDOES PLAN

On May 9, 1974, the Democratic caucus assembled to consider the Bolling committee recommendations. Teague chaired the raucous caucus debate. Unlike formal proceedings in the House of Representatives, he had no experienced parliamentarian by his side to advise him concerning rules, precedents and points of order. The opposition to the Bolling report was a strange coalition of every committee member who felt his jurisdiction was being reduced, plus ambitious Members who saw the recommendations as disturbing their power bases. Labor fought the plan because it divided the Education and Labor Committee, and numerous other special interests joined the fight to defeat a plan they felt might upset their influence. Three experienced infighters teamed up to gouge the Bolling plan: Representatives Phillip Burton (Democrat of California), Wayne L. Hays (Democrat of Ohio), and John D. Dingell (Democrat of Michigan.)

Defeat of the Bolling plan in the Democratic caucus turned on an unfortunate ruling by Caucus Chairman Teague. After some heavy artillery had been fired at Bolling, Burton moved to refer the entire matter to the Democratic Committee on Organization, Study, and Review, headed by Representative Julia B. Hansen (Democrat of Washington.) Mrs. Hansen chaired a committee which had made several reports on the seniority system and committee operation in the past. Burton asked for a secret vote on the referral motion, and Bolling immediately asked for a rollcall on whether the vote should be secret. This was a crucial point, because there were some Democrats who secretly opposed the Bolling recommendations, but were publicly committed to reform, hence did not want to be caught voting against reform in a public rollcall vote. Teague ruled that Burton's motion did not require a recorded rollcall, but could be conducted by secret ballot. Enraged at an obviously incorrect ruling, Bolling charged up to the front of the caucus and persuaded Teague that he ought to at least go out and consult the House Parliamentarian, which Teague agreed to do. When Teague returned, he announced that the Parliamentarian agreed with his ruling. Bolling simply threw up his hands; he did not learn until later that Teague had presented the whole issue to the Parliamentarian in a somewhat confusing fashion, and the Parliamentarian, who had not been present, gave an equally confusing answer.

The applecart was upset. Teague, who should have been strongly on the side of the Bolling reforms, was perhaps leaning over backward to be a scrupulously impartial chairman. By secret ballot, the caucus voted 95–81 to take a secret ballot vote on referring the Bolling reforms to the Hansen committee. A glum Bolling saw the handwriting on the wall. By the further secret ballot, the caucus voted 111–95 to sidetrack the carefully devised Bolling recommendations and send them on for study and report by July 17 by the Hansen committee.

HANSEN COMMITTEE RECOMMENDATIONS

"That was a very stormy committee," recalls Mrs. Hansen as she reflected on the complex negotiations during the early summer of 1974. "They would each go out and talk to their constituents," Mrs. Hansen said, referring to the crush of lobbyists outside the secret meeting room. During the negotiations, Teague kept in frequent touch with Mrs. Hansen and his other good friends on her committee. "Teague was very, very nice to work with," Mrs. Hansen recalled. "He was interested in protecting his turf, but he was never adamant on it." As head of the Interior Appropriations Subcommittee, Mrs. Hansen had more than a passing acquaintance with what was being done in the energy field, and she related: "I was convinced the United States was doing very little energy research," hence she was equally interested in concentrating energy research in the Science Committee. She termed the issue of what to give the Science Committee "not a battle at all."

For Ralph Nader, a long-time opponent of nuclear power, and friend of Burton, there was a somewhat different issue involved. He knew McCormack's strong pronuclear attitude and appreciated the fact that although Holifield on the Joint Committee on Atomic Energy felt the same as McCormack, Holifield was ready to retire and McCormack would probably last many years. Nader feared placing jurisdiction over nuclear research in the Science Committee where McCormack held an energy subcommittee chairmanship. Eventually, the Hansen committee decided to keep the Joint Committee on Atomic Energy with the feeling it might be phased out two years later. This meant taking back the jurisdiction over nonmilitary nuclear research originally conferred on the Science Committee under the Bolling proposal.

Power politics operated behind the scenes in the Hansen Committee. On at least two occasions, galley proofs of suggested jurisdictional arrangements revealed that the Science Committee had been stripped of its future power in the energy R. & D. area. On both occasions, Teague personally got on the telephone to Speaker Albert, Mrs. Hansen, and key members of the Hansen Committee. Miraculously, the jurisdiction was each time restored—a tribute to Teague's personal prestige in the House.

The Hansen committee scuttled the Bolling design of a "one-track" system on committee service. This meant that the Hansen recommendations allowed Members to serve on more than one committee (with the exception of Rules, Appropriations, and Ways and Means). This also meant that the Science Committee would not be a major, exclusive committee as under the Bolling recommendation.

The elaborate plans of the Bolling committee to have an Energy and Environment Committee to concentrate work in these fields beyond the research and development phase were also ditched by the Hansen committee. These functions were retained by the Commerce, Interior, and Joint Atomic Energy Committees, among others. Bolling put this way:

> Thirteen different committees and subcommittees of the House have jurisdiction over aspects of the energy squeeze, and there is, thus, little wonder that a forward-looking national energy policy is still beyond our grasp. All the chairmen of those 13 committees and subcommittees fumbling over energy policy want to keep their chairmanships.

DIFFERENCES BETWEEN HANSEN AND BOLLING PLANS

There were a few other differences between the Hansen committee report and the Bolling plan as they affected the Science Committee. The Hansen committee specifically omitted the following references to legislative jurisdiction for the Science Committee which had been added by the Bolling committee: Science policy, scientific prototypes, sea grant programs, and oceanic and atmospheric sciences (however, the Hansen Committee specifically gave jurisdiction over the National Weather Service to the Science Committee). There was an interesting difference in the wording of the "special oversight" clauses. The Hansen committee language read:

> The Committee on Science and Technology shall have the function of reviewing and studying, on a continuing basis, all laws, programs, and Government activities dealing with or involving nonmilitary research and development.

The major difference, of course, was that the Hansen committee removed any special oversight over military R. & D. from the Science Committee.

SIMILARITIES IN HANSEN RECOMMENDATIONS

The big similarities between the Hansen and Bolling recommendations, to the benefit of the Science Committee, were the addition of civil aviation research and development, and environmental research and development. All energy research and development except nuclear (the latter being dropped from the Bolling recommendations by the Hansen committee) was pegged for the Science Committee. This gave the Science Committee pieces of jurisdiction formerly held by Commerce, Interior, Merchant Marine, and Public Works Committees.

So far as the nuclear area was concerned, the Hansen report stated:

> Nuclear research and development is specifically excluded, but it seems clear that the jurisdiction of this committee is meant to include those matters relating to nonnuclear research and development presently handled by the AEC labs, for example.

The Hansen proposals were presented to the Democratic Caucus on July 23, 1974. Unlike the earlier tug-of-war within the caucus, this time by voice vote it was agreed to send both the Hansen and Bolling proposals to the House floor for debate and disposition. An open rule from the Rules Committee also allowed amendments to be presented freely on the floor. But before the Rules Committee voted, there was a vast amount of filibustering as numerous Members who opposed any change at all asked to appear and use up time before the Rules Committee, hoping to prevent any action at all.

The Science Committee staff was not idle. On September 26, Swigert distributed to all members a huge summary, with charts, of the alternative proposals. On the day the House debate opened, the full committee met for a briefing on the proposals. Finally, on September 30, the House commenced its extensive debate and amendment of the plans.

TEAGUE SPEAKS AGAINST MINORITY STAFF

As the debate opened, Teague became the first Science Committee member to lob a shell at the Bolling committee. He objected to the requirement that the minority be entitled under the rules to one-third of the staff, and one-third of the committee funds. Teague thundered:

> Mr. Chairman, I think that provision is wrong. I say that it does the committee a great disservice. There has never been any politics in this committee, and there never have been in the Committee on Veterans' Affairs. To make this provision mandatory is wrong.

Bolling and others hastened to add that the Bolling recommendations did not mandate for the minority to have its own staff if it did not want to do so. Bolling said:

> There is no compulsion on the minority to demand a staff. Obviously that is the situation that prevails in the gentleman's committee.

While this colloquy was taking place, it was apparent that some of the Republican members of the Science Committee did not fully catch the meaning of Bolling's last phrase, or certainly they would have arisen to object. As introduced on September 30, the Hansen committee resolution authorized each subcommittee chairman and each ranking minority member (up to six) to hire and compensate one staff person—rather than the authority under the Bolling provisions for the minority to have a total of one-third in case they asked for it.

During the general debate, Representative Paul S. Sarbanes (Democrat of Maryland), who along with Representative William A. Steiger (Republican of Wisconsin), headed up the drafting team for the Bolling committee, stated:

> Research and development has been brought together in the Science and Technology Committee in recognition of the importance of that area for the future of this country. I am very frank to say that I believe many greatly underestimate the significance of that jurisdiction and what a properly strengthened Science and Technology Committee could do in anticipating the problems that are facing the Nation down the road.

SCIENCE COMMITTEE WINS FIGHT FOR AVIATION R. & D.

The first effort to chip away at some of the additional jurisdiction given the Science Committee was attempted by Representative Dan H. Kuykendall (Republican of Tennessee) who offered an amendment which, among other things, would keep jurisdiction of civil aviation research and development in the Commerce Committee. Milford and Wydler both spoke against the amendment. Wydler indicated that "there is totally fractured jurisdiction at the present time," with FAA R. & D. under the Commerce Committee and NASA R. & D. under the Science Committee. The amendment lost, with Wydler noting:

> This has caused serious problems. As a matter of fact, there is a very serious one in the jet noise retrofit program where both agencies were working on and have worked on a different type of retrofit program. So I think this step being offered in this amendment would be a great step back and should be rejected out of hand.

Several Science Committee members inexplicably supported Kuykendall's efforts to take jurisdiction away from the Science Committee: Brown, Davis, Downing, Frey, Symington, and Winn.

OCEANIC AND ATMOSPHERIC SCIENCES

On the final day of debate, Hechler and Mrs. Hansen engaged in a carefully planned colloquy designed to firm up the jurisdiction of the Science Committee in oceanic and atmospheric sciences. The Hansen committee had not, like the Bolling committee, given full jurisdiction over oceanic and atmospheric sciences to the Science Committee, and had narrowed the jurisdiction only to the "National Weather Service." Therefore, Hechler and Mrs. Hansen worked out the parameters of a mutually agreeable colloquy. The colloquy developed along these lines:

Mr. HECHLER. For several decades we have been moving in the direction of integrating oceanic and atmospheric research because of their complex interactions * * *. The National Weather Service has forecasting responsibility for oceanography as well as weather, and includes such items as sea state, swell, ocean temperature, and storm effects. A major finding from the space program is to reinforce the first point and to add to it the understanding that the Sun, the atmosphere, and the oceans are closely interrelated. In summary, I would hope that it is the intent of House Resolution 1248 to encourage integration of oceanic and atmospheric research rather than to divide the research effort.

Mrs. HANSEN. If the gentleman from West Virginia will yield, it is indeed our intent to encourage such integration of research work in the oceanic and atmospheric research area.

Mr. HECHLER. Based on this intent, would it appear to be a reasonable interpretation that the Committee on Merchant Marine and Fisheries and Committee on Science and Technology should cooperate closely in legislative and oversight matters affecting the National Oceanic and Atmospheric Agency, and that, in the case of subjects having a high content of research and development, joint referrals and oversight would be appropriate?

Mrs. HANSEN. That is a reasonable interpretation, if the gentleman from West Virginia will yield further. I am sure our atmospheric and oceanic research program could only benefit from the expertise available on both the Committee on Merchant Marine and the Committee on Science and Technology.

THE VOTE ON THE HANSEN PROPOSALS

At the close of all the amendments, debate, and clarifications, the House finally had an opportunity to make its choice between the Bolling proposal and the Hansen substitute as amended. The House voted 203 to 165 in favor of the less drastic Hansen proposal, which did not disturb the status quo as much and generally preserved committee prerogatives. There was a wide divergence of opinion among Science Committee members. Those voting for each of the approaches are listed below:

Bolling	Hansen	Not voting
Hechler	Downing	Teague
Davis	Fuqua	Hanna
Symington	Flowers	
Roe	Cotter	
McCormack	Milford	
Bergland	Thornton	
Pickle	Frey	
Brown	Goldwater	
Gunter	Camp	
Mosher		
Bell		
Wydler		
Winn		
Esch		
Conlan		
Parris		
Cronin		
Martin		
Ketchum		

SUMARY OF REFORMS

To summarize the new jurisdiction which the committee received effective at the start of the 94th Congress in January 1975, the following was added:

Civil aviation research and development.

Environmental research and development.

All energy research and development except nuclear research and development (but including nonnuclear R. & D. handled by the Atomic Energy Commission laboratories.)

National Weather Service (a floor colloquy indicated the intent is to "encourage integration of oceanic and atmospheric research.")

Special oversight over laws, programs, and Government activities dealing with nonmilitary research and development.

THE BOAT TRIP

Just as soon as the House of Representatives had voted the committee reforms in October 1974, Swigert got busy preparing plans and charts in order to weigh the options for staff expansion and subcommittee jurisdictions in the next Congress. Teague faced a major decision in the fall of 1974: How best to capitalize on the expanded jurisdiction of the committee and at the same time maintain the leadership and control necessary to carry out a unified policy. He decided to use a thoroughly democratic procedure to ascertain

what a majority of the committee wanted on subcommittee jurisdictions, and at the same time he decided to exercise his chairman's
prerogative to make staff decisions himself. Neither of these decisions
were announced, but they became apparent as events unfolded.

For the first time, the 1974 reform amendments required the
Democrats and Republicans to have their presession caucuses in
December of each election year. Committee assignments were made in
December, but subcommittee chairmen were selected in January.
To resolve a very heated argument over jurisdictions, Teague decided
he would invite those majority members who would be returning
to the committee in 1975 to take a boat trip on the Potomac River.

The night before the Democratic caucus assembled in December
1974, was rainy and squally, so the boat did not move from the dock
for awhile. It was an ideal craft for discussions; nobody could escape.
Teague borrowed the boat from LTV Corp., announcing that he
wanted to get the Members away from the Capitol so they wouldn't
be interrupted by telephone calls. Before the boat had left the dock,
the phone rang and Speaker Albert was on the line for Teague.

For a long period into the evening, very little was done except to
drink and socialize. By the time the Members lined up at the galley to
enjoy a tasty plate of food, almost everbody was in a high mood.
Still no shop-talk. Swigert and Wilson, the only non-Members who
went along, were enigmatic. Swigert had done his job by circulating
in advance a detailed memorandum on the various options, with this
conclusion: "The staff makes no recommendations." Teague was
equally noncommittal.

Finally, after a leisurely dinner and more drinks, Teague assembled
the group in the cabin. It was not necessary to deliver a pep talk on
the challenging new responsibilities facing the committee in 1975:
Everybody appreciated this already. But Members were eager to detect
if there were some signal from Teague as to how he preferred to organize
the committee, so they could act or react. Swigert gave a recap of the
memorandum he had already circulated on the options. Then Teague
went around the circle, asking each Member (by seniority) to give his
views on how the jurisdictions of the subcommittees should be
arranged.

McCORMACK ADVOCATES ONE ENERGY SUBCOMMITTEE

Unquestionably, the most articulate and best-structured case was
presented by McCormack. He strongly urged that energy had been
split up too long, and now that the committee had a rare opportunity
to pull it together, the chance should not be muffed. To separate energy
research and development into two subcommittees would merely

contribute to the confusion which had prompted the reform itself when all energy R. & D. was concentrated in the Science Committee, he contended. It was a hard sell, and an impressive one.

There were arguments on the other side, some of which Swigert had summarized in the options he presented to Teague. In terms of work load distribution among the subcommittees, it was fairly obvious that there would be a disproportionately large concentration in a single energy subcommittee. Also, if every committee member got a chance to serve on his first choice subcommittee, a single energy committee would have been overloaded with applicants.

When it came time for Flowers to speak, he came as close as he could without discourtesy in denouncing his host, Teague, for the brutal fashion in which the task force on energy and especially the Subcommittee on Energy had been created by passing over Flowers in 1971 and 1973. Everybody spoke very freely that night. Nobody minded at all what anybody else said. It was all done with the high good humor of old friends and drinking buddies.

McCormack had set a target to shoot at. Pretty soon, a lot of the discussion seemed to center around the issue McCormack had presented. Teague bluntly interrupted and inquired: "All right, do you want two energy subcommittees or one energy subcommittee?" There was a babble of voices. Members started to line up in two camps: It was them or us.

WE'LL JUST TAKE A VOTE

Finally, Teague surprised everybody by suddenly announcing: "If you guys can't reach a decision, we'll just take a vote on it." The babble of conversational argument stopped. He was serious. This wasn't all cut and dried. He really wanted to decide this in a democratic fashion. There were a couple of jibes about a "secret ballot." There were other remarks about "open meetings" and "sunshine laws." It was finally decided, by nobody in particular, to resolve this burning issue by a show of hands. By a very slim margin—nobody remembers exactly how much—the Members voted for two energy subcommittees. McCormack had all the strength of logic on his side, but he had one handicap: Nobody was very sure who might opt for the second energy subcommittee, and it is likely that several of the aspirants lined up in favor of two subcommittees. Because of West Virginia's role among the leading coal producers in the Nation, it was assumed that Hechler as second-ranked member of the committee would have a strong interest in heading up the Fossil Subcommittee. He soon let it be known that he would bid for that subcommittee. Flowers bided his time until 1977, because in 1975 he preferred to retain his Judiciary Committee subcommittee chairmanship.

The next day at the Democratic caucus, Hechler and McCormack sat down together to chat about the future. McCormack indicated once again that the energy crisis demanded a unified approach among the interrelated sources of energy, and it simply did not make sense to proceed in a divided fashion as two subcommittees would perforce do. McCormack suggested that a fully independent sphere of action could logically be carved out through Hechler being named chairman of a task force on coal, oil and natural gas which would be coordinated through an overall Energy Subcommittee such as McCormack had chaired in 1973 and 1974. Hechler responded: "No. But to insure that our subcommittees work together, I suggest that you serve on my subcommittee and I serve on yours." This was done.

NAMING THE ENERGY SUBCOMMITTEES

The preliminary agreements reached on the boat trip were quickly and painlessly ratified when the Democratic caucus met at the opening of the 94th Congress in January 1975, followed by an organization meeting of the full committee. McCormack named his energy sub-committee first, calling it "Energy Research, Development and Demonstration," with jurisdiction over solar, geothermal and other advanced energy systems, energy conservation and utilization, and special oversight over nuclear energy R. & D. It was an attractive and challenging field, and 23 Members bid for McCormack's subcommittee as against only 15 for Hechler's. The naming of the subcommittee set off the first of a series of polite arguments between the two sub-committee chairmen. Hechler contended that the title of McCormack's subcommittee inferred that it covered the entire energy field, and that the title should be more explicit and not all-encompassing. McCormack countered that each Member had the right to name his own sub-committee title, and Hechler had perfect freedom to do what he pleased. Although he personally preferred a shorter title, Hechler then named his subcommittee "Energy Research, Development and Demonstration (Fossil Fuels)." The other subcommittees were:

Space science and applications (Fuqua).
Science, research and technology (Symington).
Domestic and international scientific planning and analysis (Roe, succeeded by Thornton).
Environment and the atmosphere (Brown).
Aviation and transportation research and development (Milford).
Ad hoc Subcommittee on Special Studies, Investigations, and Oversight (Teague)—established in 1976.

NUCLEAR R. & D. JURISDICTION

Additional jurisdiction over nuclear research and development was added to the Science Committee in 1977 with the abolition of the Joint Committee on Atomic Energy. Representative Jonathan B. Bingham (Democrat of New York) submitted the amendments in the Democratic caucus December 8, 1976, which split the joint committee jurisdiction among the Interior, Commerce and Science Committees. Bingham stated in the Caucus:

> Generally, it is the intention of this resolution that the jurisdiction of the Interior Committee would parallel that of the Nuclear Regulatory Commission, and the jurisdiction of the Science and Technology Committee would parallel that of ERDA.

When the House convened on January 4, 1977, to adopt the new rule for the 95th Congress, Bingham placed in the Record a "memorandum of understanding" which clarified the nuclear regulatory functions which would go to the Interior Committee, and the public health and environmental protection against radiation functions conferred on the Commerce Committee. It was strictly a Commerce-Interior deal which made no mention of the Science Committee. This impelled McCormack to follow up with a colloquy prearranged with Majority Leader Jim Wright (Democrat of Texas) to insure that the Science Committee jurisdiction was protected and clarified as the full intent of the new rule. The colloquy went in part as follows:

> Mr. McCormack. Would the majority leader state that the intention of the new rules involving this jurisdiction is as indicated by Mr. Bingham, that is, that jurisdiction for all activities of the Energy Research and Development Administration, except for weapons research and fabrication, falls within the jurisdiction of the Science and Technology Committee? * * *
> Mr. Wright. The gentleman from Washington is completely correct. * * * This includes fusion energy research.
> Mr. McCormack. It is my understanding that the intent of the amendment is to transfer the jurisdiction for energy research and development activities performed under contract for the Energy Research and Development Administration at the National Laboratories to the Committee on Science and Technology.
> Mr. Wright. Yes; that is correct. * * *

Having established that point, McCormack went on to congratulate the majority leader for his announced intention to establish a Select Committee on Energy, "a goal I have sought for many years." McCormack then lashed out at the "chaotic situation" produced by the Bingham amendments. He denounced the "splintering of energy jurisdiction in the House" as a result of the Bingham amendments, which redistributed among three committees what was once within the Joint Committee on Atomic Energy. He said he hoped the new Select Committee on Energy would be able to pull things together through its "wiser deliberations."

FLOWERS BIDS FOR FOSSIL AND NUCLEAR IN 1977

A few weeks later, Flowers surprised McCormack by bidding to take over the chairmanship of a new Fossil and Nuclear Energy Subcommittee, as McCormack retained a subcommittee handling advanced energy technologies.

These, then, were the major jurisdictional expansions in the Science Committee authority. As always, there were minor forays which other committees made to establish footholds—as when Rep. John M. Murphy (Democrat of New York) persuaded Speaker Albert in 1976 to appoint him to chair an Ad Hoc Committee on the Outer Continental Shelf. Murphy proceeded to consider legislation which included R. & D. authority, which he only dropped from his bill after Science Committee protests. On the other side of the coin, Scheuer raised some eyebrows in several other committees through his aggressive use of "special oversight" powers, when he served as chairman of the Subcommittee on Domestic and International Scientific Planning, Analysis and Cooperation commencing in 1977.

RELATION WITH ARMED SERVICES COMMITTEE

The abolition of the Atomic Energy Commission and the Joint Committee on Atomic Energy presented some new jurisdictional problems. The Joint Committee interface with the Armed Services Committee had been frequent, but not many Science-Armed Services negotiations had been necessary since the old Brooks-Vinson days. In 1977, the ERDA authorization bill included funding over which the Science and Armed Services staffs each claimed jurisdiction, in areas relating to Laser Fusion and Naval Reactor Development. When an impasse was reached, Representative Mel Price (Democrat of Illinois), Chairman of the Armed Services Committee and Teague negotiated an agreement. Teague related:

> I went to Mel Price and Mel and I worked it out. Our staffs couldn't agree. We just got together and worked it out.

On September 13, 1977, in a floor colloquy, Teague and Price had this exchange:

> Mr. Teague. We should all recognize that the fiscal year 1978 budget submission was the first opportunity for the Committee on Armed Services and the Committee on Science and Technology to exercise their new responsibilities for nuclear energy legislation which was given to the two committees by rules adopted in the 95th Congress. ERDA was not prepared to submit legislation to the two committees in a form that would coincide with their jurisdictions. Is that not true?
>
> Mr. Price. That is correct. * * * The Science and Technology Committee which has legitimate concern for the continuation of research that could eventually lead to dramatic civilian applications for laser fusion, has added $9.2 million which we agree is a modest yet appropriate addition to the authorization bill. * * *
>
> Mr. Teague. I thank the gentleman for his cooperation, his recognition of the interest to both of our committees in laser fusion development, and feel that we have reached a reasonable compromise at this point.

In the same bill, the Science Committee recognized the traditional jurisdiction of the Armed Services Committee over Naval Reactor Development, but agreed that certain new R. & D. activities were civilian in nature and would be handled by the Science Committee in future year budget requests. In 1978, Teague wrote Price on May 9:

> I am pleased that this year our two committees have not had the type of jurisdictional controversies that we faced last year.

Teague raised some questions prompted by language in the Armed Services Committee report, resulting in an exchange of letters with Price.

JURISDICTIONAL PROBLEMS WITH ENERGY

The House establishment of an Ad Hoc Committee on Energy did not by any means resolve the complex issues raised when the Department of Energy in 1978 started sending up its authorization requests which involved the jurisdiction of four other committees—Interior, Commerce, Armed Services, and International Relations. The joint or sequential referral process established by the 1974 reforms could not solve the problem of overlapping jurisdictions. Nor could the separate distribution of certain titles of the bills eliminate conflict, because there were heated arguments over substantive areas within titles. Since the Department of Energy was formed not only from ERDA, but from the Federal Energy Administration and portions of the Department of the Interior, the organizational structure almost invited jurisdictional fights on Capitol Hill.

The most contentious jurisdictional squabbles were had with Representative John D. Dingell (Democrat of Michigan). Dingell, a member of the Commerce Committee, carried on a running fight over numerous issues which were being handled by the Science Committee. Teague made this observation about the Chairman of the Commerce Committee, Representative Harley O. Staggers (Democrat of West Virginia) and one of his subcommittee chairmen:

> Harley Staggers was easy to work with. John Dingell—now that's another story.

Matters came to a head with the Commerce Committee in the spring of 1978, when an attempt was made in that committee to assert jurisdiction over a number of items in the R. & D. areas of solar, conservation, nuclear and fossil energy in the Department of Energy authorization bill. The Commerce Committee contended, for example, that items such as the R. & D. program for the gas-cooled thermal reactor were clearly on their way to commercialization and therefore within the purview of the Commerce Committee. This was only one of the many differences of opinion which led to extensive negotiations among Teague, Staggers, and Udall, who finally arrived at an agreement on the jurisdictional issues.

DALE MYERS SUPPLIES DEFINITIONS

In order to help clarify the meaning of R. & D. and other terminology utilized by the Department of Energy, Teague hit on the idea of writing to Dale D. Myers, Under Secretary of the Department of Energy, asking him to provide a set of clear definitions of various processes on the long road toward commercialization. Myers responded on May 10, 1978, with a copy of the definitions to Dingell. When the DOE authorization bill was being considered in the House on July 17, Fuqua offered an amendment incorporating definitions of the following terms into the bill: Basic and applied research; exploratory development; technology development; concept and demonstration development and operational systems development. Dingell, having a copy of the definitions, strenuously objected to their inclusion into the legislation. After some sharp words exchanged with Fuqua, Dingell attempted unsuccessfully to get the amendment knocked out on a point of order as not germane to the bill.

Perhaps the most objective and statesmanlike commentary on the entire issue was the brief conclusion written by Brown, printed as "Additional Views" appended to the DOE authorization bill:

> Generally, committees fight out their jurisdictional struggles until everyone tires of the process, with the result that legislation begins to be drafted more and more narrowly to avoid future jurisdictional conflicts. * * * I believe it is impossible to organize our committees, or the agencies of the Executive Branch, in a manner which would eliminate overlaps and conflicts. We continuously reform or reorganize our structure to minimize conflicts, but it would be foolish to pretend that we can ever achieve unambiguous organizational lines in an ever-changing society. Instead, we must focus on procedures for resolving the inevitable ambiguities, and concentrate on setting precedents for more effective techniques of resolution than those represented by the handling of this first Department of Energy authorization bill.

Brown went on to recommend that the problems Congress encountered in dealing with the energy bill be looked at as a general organizational problem and not as a special case. He cited similar jurisdictional conflicts in the areas of welfare reform, urban policy, water policy, and health care. He concluded:

> We can continue to deal with these questions in a fragmented, ad hoc fashion, such as the establishment of ad hoc select committees, or simply letting the disputes be settled on the House floor, or we can attempt an approach of negotiation and arbitration, perhaps with the assistance of the Rules Committee. * * * I simply believe that there must be a better way than our recent actions indicate, and everyone's time can be put to more productive use if we find it.

MINORITY STAFF

The successive committee chairmen fought consistently against a separate staff for the minority. Brooks, Miller and Teague shared the same philosophy—that the committee staff was there to serve all committee members without reference to party. The issue did not arise while Brooks was chairman, but Fulton as ranking minority member brought it up frequently during the 1960's. As pointed out on pages 183–184, Fulton was finally successful in appointing the first minority staff member, Richard E. Beeman, in 1968.

From that point on, the struggles of the minority to obtain fairer staff representation increased each year. Since the Republicans were in a minority during the entire period and had not controlled the Congress since 1953–55, it is unfortunate that the debate took on strictly partisan overtones. This made it more difficult for the minority to argue, which they did very effectively, that the caliber of legislation was raised through better data compiled by staff working directly for minority members. It was undeniable that no matter how fair the chairman, and no matter how nonpartisan the staff, there were occasions when minority members took a position on legislation which needed staff aid on issues like drafting amendments, researching arguments, and presenting minority testimony. In addition, of course, it was obvious that the regular staff responded with higher priority to requests from the chairman, subcommittee chairmen and executive director. The minority wanted a staff which would be more responsive.

During the 1960's when Fulton was the ranking minority member of the full committee, there was a growingly favorable sentiment in the Congress and among political scientists and journalists toward better minority staffing. While the Democrats controlled the White House up until 1969, this became more marked. But even with the period of Republican administrations from 1969 until 1977, the support for minority staffing grew. There were a few maverick Democrats who dared to buck party lines to lend support for minority staffing. For example, when the Monroney-Madden Joint Committee on the Organization of Congress made its 1966 report, of the six House Members, three Democrats and three Republicans had an early standoff on the issue. Hechler broke the tie by voting with the three Republicans to support staffing for the minority. When the House Rules Committee finally came out with a congressional reform bill in 1970, it also contained a provision for minority staffing. Very quickly after the 1970 election, Fulton sent word to Miller in California that the minority was ready to exercise its rights under the 1970 legislation. There followed the precipitous action of the 1971 Democratic caucus, ratified by the House, wiping out the short-time gift bestowed on the minority.

MOSHER AS RANKING MINORITY MEMBER

With the death of Fulton in October 1971, Mosher was next in line to be ranking minority member. He had been elected in 1960, and joined the committee in 1961. A smalltown newspaper editor in several communities, graduate of Oberlin College, and former Ohio State senator, the 6-foot 4-inch Mosher stood out as one of the most liberal members of his party—so much so that there was a question within Republican ranks whether he should be allowed to move up to be ranking minority member of the committee in 1971. He was one of the first opponents of the war in Vietnam, a strongly positive supporter of education and scientific advancement, coauthor of the legislation establishing the Office of Technology Assessment and Vice Chairman of the Technology Assessment Board, and coauthor of the bill restoring the Science Adviser and scientific machinery in the White House. "Supportive" is one of Mosher's favorite words, and in practice he was intelligently supportive of the successive committee chairmen and also the programs generally sponsored by a majority of the committee. Scholarly in manner, tolerant of differing opinions, even tempered, a good negotiator, Mosher approached issues with the equable grandfatherliness of a senior academician.

Wydler, Rumsfeld, Winn, Goldwater, and younger committee Republicans pushed Mosher to put up more of a scrap for minority staff. In 1973, there was a Republican confrontation of sorts with Teague and his newly appointed staff director, Jack Swigert. With Teague's blessing, Swigert had set up a system of task team leaders in all areas of the committee's jurisdiction. He indicated that as the committee's responsibilities grew (in areas like energy) it was necessary to shape the staff structure and operation toward goals of greater productivity and efficiency. Swigert stressed the importance of clear lines of command and authority reaching up to the executive director.

On July 18, 1973, Swigert outlined his plans for a reorganization of the committee staff to implement these ideas. That afternoon, following his presentation, 11 Republican committee members met to discuss the implications of Swigert's staff plans. They unanimously signed approval of a three-page memorandum which Mosher drafted to Teague and Swigert, reiterating their strong feeling that the integrity of the hard-won minority staff must be preserved at all costs. At that time the minority staff consisted of Carl Swartz, Joseph Del Riego and Theresa Gallo (secretary). Mosher's memorandum stated in part:

> We consider it of the utmost importance that the identity and reality of the minority staff (all three members) shall be maintained as a working team, with a high degree of autonomy, responsible basically to the ranking minority member, but cooperating and participating as fully and productively as possible in the work of the full staff of the committee * * * and that certainly means working closely with Swigert and his "team leaders" and amenable as far as possible (without losing minority identity) to their planning, procedures and programming.

We believe it essential, in order to maintain the minority staff identity, that the three minority staffers continue to have office space which allows them to be grouped together, working closely together as a team, as they now are; but we recognize, of course, that it may be necessary to shift them to a different location.

SWIGERT AND MINORITY STAFF

On the way over to the Capitol on the House subway late that same day, Mosher and Teague talked about the problem. Mosher could get no immediate resolution of the issue, hence the memo, which suggested that Teague and Swigert meet with the minority members "for whatever further discussion may be necessary to arrive at a complete understanding." Swigert penned on the memo when he gave it to Teague: "Can't Mr. Mosher decide for minority?" The upshot was a lengthy meeting between Mosher and Swigert, at which Swigert outlined "areas of agreement" and "areas of disagreement." One central bone of contention was whether the minority staff could or should be moved around by Swigert to even out the workload, and whether or not the minority staff should, for command purposes, report directly to Swigert rather than to the minority members led by Mosher. Teague made clear his personal feelings, reiterating that although Swigert had told him he was a Republican when he had been hired, Teague warned him he would be fired if he ever mentioned again he was a Republican on a staff which Teague insisted must be nonpartisan.

The issue of relationships remained basically in disagreement. Mosher's attitude was best expressed in the final paragraph of his cheerfully worded covering letter to Teague:

> I hope that both you and Jack will recognize that the positions we have asserted in our memo represent a completely friendly and genuine desire on our part to coordinate and cooperate with you in establishing a very effective, efficient staff operation for the committee, even though we are very firm in our conviction that the identity of the minority staff and its prime responsibility to the minority members must always be recognized and very real.

In point of time, these negotiations were proceeding while the Bolling committee was holding hearings on congressional reform during the summer of 1974. The major effort of the committee was pointed toward putting its energy foot forward and bidding for expanded jurisdiction on that attractive base. When Teague and McCormack testified before the Bolling committee in June, their far-ranging arguments effectively demonstrated the value of centralizing expanded energy jurisdiction in the Science Committee, they did not stoop to arguing against minority staffing, nor did they even mention it. For his part, Mosher brought up the issue in his May 11 testimony:

I have a very strong devotion to our former colleague, George Miller. He vigorously resisted appointing or allowing any minority staff because he felt it would be a divisive influence in the committee and in the staffing. I believe that now, after only very brief experience, that has not proved to be so and I am glad that it has not.

Mosher also added:

Based on my own personal experience and observations, I can testify that even the minimun of minority staffing we now have in the Science and Astronautics and Merchant Marine Committees has produced very positive, valuable results. Those two committees are excellent, practical examples of how a minority staff can operate with considerable independent autonomy—responsible immediately to the needs of the minority members, yet of genuine service to the whole committee, and without creating any partisan disruption.

BOLLING RECOMMENDATIONS ON MINORITY STAFF

The Bolling committee recommended that professional staffs of committees be expanded from 6 to 18, and clerical staffs from 4 to 12. The minority members of a committee were allowed the opportunity to select one-third of the staff of 30, including one-third of the funds available through the annual expense resolutions voted by the House Administration Committee. Although the Hansen committee report did not include a provision for minority staff, Representative Frank Thompson, Jr. (Democrat of New Jersey), a member of the Hansen committee, recognized the need to include this provision to insure enactment of the Hansen package, so he offered the amendment on the House floor and it was adopted.

At the opening of the 94th Congress in January 1975, the House adopted a resolution which spelled out more specifically the legislative foundation for minority staff. The resolution authorized the ranking minority member of up to six subcommittees "to appoint one staff person who shall serve at the pleasure of the ranking minority party member." The House rules also contained the authorization for a total of six professional and four clerical personnel to be assigned to the minority when so requested by a majority of the minority members.

1975 marked the beginning of the expansion of the minority staff in conformity with the House rules. At the organization meeting of the committee on January 23, 1975, Teague declared:

Ladies and gentlemen, as far as the staff is concerned, there is no question that we are going to add a number of staff members to this committee. We can not do it at this time because we don't have any money and because we don't have space. I am going to ask the committee to approve the present staff as is and then as soon as we get some money to hire further staff members the Chair will come back to the membership of the committee for other staff members who may be hired. Is there discussion?

The following interchange then occurred between Mosher and Teague:

Mr. Mosher. Mr. Chairman, as I understand it, the approval of your motion does not condition in any way or prejudice in any way further discussion of the minority staff and that sort of thing?

The Chairman. It does not. The Chair will do everything in his power to completely comply with the new rules concerning staffing as soon as we know what they are. As of this moment, I don't think we do.

Mr. Mosher. We appreciate that policy on your part.

THE STRUGGLE OVER APPOINTMENTS

Between 1975 and 1978, the minority gradually filled their statutory quota of up to 16 professional and clerical staff members. This was not achieved without a struggle. Both Teague and Swigert fought a rearguard action along several fronts. The minority staff continued to be integrated through the task team leaders, and only gradually did they begin to assume a separate identity. There were a few debates about qualifications of individual staff being recommended. Also, there was always an issue of how many minority staff could be allocated to the statutory (standing) committee staff as against the investigative staff. The advantage of being assigned to the standing staff was that there seemed to be a greater degree of permanency (the standing staff was hired by the committee and paid for by the House without the need for a special expense resolution from the House Administration Committee) and the salary levels could be higher on the standing staff. The investigative staff, funded by annual resolutions through the House Administration Committee, was not as desirable an assignment because there was a salary ceiling, hiring was determined by the chairman, and therefore status and permanency were not as great. Teague resisted what he considered a too-rapid expansion of the minority staff and their understandable desire to obtain more standing committee slots rather than investigative positions.

Prior to 1978, the various minority staff members were scattered throughout the subcommittees and other administrative areas of the committee staff. From the start, there was an effort by the minority to consolidate its efforts at a central location, in order to achieve coordinated direction. Not until June of 1978, however, was this goal achieved with the assignment of Room 2320 of the Rayburn Building to the minority. Minority staff was still assigned to work directly with subcommittees, but after June of 1978 the minority members had a central office to call their own.

Senior Democrat Don Fuqua helps ranking minority member John W. Wydler (Republican of New York) cut the ribbon on the new minority headquarters in Room 2320 Rayburn Building. From left, Fuqua, committee executive director Mosher, Representatives Hamilton Fish, Jr. (Republican of New York), Harold C. Hollenbeck (Republican of New Jersey), and Wydler.

WYDLER BECOMES RANKING MINORITY MEMBER

At the end of 1976, the minority staff was up to nine members. With the retirement of Mosher from the House of Representatives, Wydler moved up to become the committee's ranking minority member. Elected in 1962, Wydler represents the southern part of Nassau County, Long Island, a district he labels the "Fabulous

Fifth." A lawyer and former prosecuting attorney, Wydler's interest in space first drew him to the Science Committee because of the location of the Grumman Corp. at Bethpage, N.Y. The dish-rattling decibels of noise from aircraft at Kennedy Airport soon propelled Wydler into the most outspoken leader in the Congress on aircraft noise. Mosher describes this personality difference:

> I never pushed as vigorously as Wydler on minority staffing. I was more in the role of a mediator.

In September 1977, Mosher was summoned out of retirement to become executive director of the committee for the final 16 months of Teague's service as chairman. Once again, he served more as a mediator, with the basic staff work on the minority problem being performed by Colonel Gould in presenting the facts to Teague. But the point man on minority staff was clearly Wydler, who went to bat and refused to accept delay or opposition. In commenting on the minority staff during his testimony before the House Administration Committee on March 1, 1978, Wydler stated:

> During the past year, we have expanded the minority staff. These additions have been highly qualified, competent people who have contributed significantly to enhancing the professional capabilities of the committee. * * * (They) have a very substantive role in the achievement of the committee's mission. I feel that an autonomous, capable minority staff is extremely important in helping the minority meet its responsibilities.

One of the notable contributions of the minority staff has been the minority briefing book, including objective, pro and con views on some of the major issues confronted by all the subcommittees. Updated quarterly, this briefing book has been found to be a useful tool in interpreting the issues which surface in the committee.

The briefing book was developed by Paul A. Vander Myde, who became minority staff director on August 15, 1977. A tall, personable man with a smooth and easy manner, Vander Myde uses quiet persuasion rather than bombast to get his points across. Following his undergraduate and graduate work at the Universities of Iowa and Minnesota, Vander Myde served at the National Security Agency, as Legislative Assistant to U.S. Senator Bob Packwood (Republican of Oregon), and Executive Assistant to the Vice President from 1971 to 1973. After 6 months as a staff member of the Domestic Council in the White House, he was appointed Deputy Assistant Secretary of Agriculture in 1973, his last position before being selected as minority staff director.

WYDLER CLASHES WITH TEAGUE

In the spring of 1978, Wydler clashed with Teague on their respective interpretations of the size, assignments and qualifications of minority staff. Wydler told the House Administration Committee that "we continue to experience difficulties in bringing new staff members on board." He contended:

> Last year when I appeared before this committee, I testified that a ratable portion of the new investigative personnel—four professionals and a proportionate number of clerical personnel—would be coming to the minority. This has not happened to date. In fact, we are led to understand that the minority will be allotted only the minimum number of personnel under the House Rules, and even those people have been hard to come by. * * * The minority intends to continue to press for an equitable share of the committee staff allocations.

Both Teague and Wydler refused to budge. Wydler felt that the statutory minimum of 12 professional and 4 clerical employees was all right for a starter, but that as the total committee staff expanded, so should the minority staff. When the traditional one-third allocated by the congressional reforms of 1974 had slipped proportionately down to less than one-fifth, the minority felt it was time to stand up and fight. Aided by several minority members of the House Administration Committee, Wydler maneuvered to have the Science Committee's funding resolution tabled until the minority staffing issue was resolved.

Teague's philosophy was expressed in an April 19 letter to Chairman Thompson of the House Administration Committee:

> The size of our staff is determined by an analysis of requirements and skills as well as the availability of funds to defray related expense. Is it your Committee's position that we should hire staff regardless of need? Mr. Wydler seems to believe that he is entitled to sixteen minority staff members whether or not warranted by the Committee's needs. * * * The staff of the Committee on Science and Technology, in my mind, has never been selected on a partisan basis. Except for one professional staff member, Charlie Mosher, I do not know the party affiliation of any of our 83 staff members; nor am I concerned. Our recruiting system, in my opinion, is second to none on Capitol Hill and is based upon *qualifications*. Selection of personnel is carried out on a competitive basis. We are a research and development oriented Committee and the subject matter under our jurisdiction does not lend itself to partisan politics.

As happens in most confrontations, each side gave a little and Wydler released the funding resolution he had been holding hostage. Despite the tone of the letters, and the occasional angry verbal outbursts, Teague and Wydler remained good friends who were able to work out the problem so the business of the committee could move forward.

SUBCOMMITTEE CHAIRMAN'S STAFF

Although the issue of minority staffing received more attention, there was an analogous situation with respect to designation of one staff member by subcommittee chairmen. In the eyes of Brooks, Miller, and Teague, such a choice would weaken the control of the chairman over coordinated policies within the committee. For a subcommittee chairman, this meant an opportunity to get a competent and qualified person who would make the work of his subcommittee more meaningful and effective.

Throughout the Nation and in the Congress, the spirit of reform was abroad in the late 1960's and early 1970's. To make Congress more responsible and responsive, a large contingent of younger and middle-level House Members were raising questions about the seniority system and how to open up new initiatives for the overwhelming majority of Members who had not been around long enough to rise to become chairmen. In 1970, the Democratic caucus set up the Democratic Committee on Organization, Study and Review which was headed by Mrs. Hansen. The January 1971 Democratic caucus passed by a substantial majority a recommendation of the Hansen Committee, stipulating:

> A subcommittee chairman shall be entitled to select and designate at least one staff member for said subcommittee, subject to the approval of a majority of the Democratic members of said full committee.

MILLER OPPOSES CAUCUS RULE

When the organization meeting of the full committee assembled on February 23, 1971, it occurred to one Member that this might be a good provision to implement in the rules of the committee. Miller made it clear that he did not like any effort to incorporate the caucus rule as a committee rule. At the start of the meeting, he engineered a quick maneuver, as follows:

> The CHAIRMAN. Gentlemen, this meeting will come to order. This is the organizational meeting of this committee, and it has always been a closed meeting, and under the new rules if we have a closed meeting (it) will require a majority vote of the committee. Therefore, I will now entertain a motion that the organization meeting be closed.
> Mr. FULTON. I so move.
> Mr. KARTH. I second.
> The CHAIRMAN. It has been moved and seconded that the organization meeting be a closed meeting. All those in favor signify by saying "aye." Contrary minded? The "ayes" have it. The meeting is a closed meeting.

It all happened so fast that very few Members read any significance into the adroit move to insure that no searchlight of publicity could pick up what was to occur.

Chairman Miller moved ahead smoothly. Members at the opening of the new Congress were in a glowing mood. Nobody wanted a fight, that was obvious. Suddenly, Miller said:

> Adoption of the rules.* * * It is my thought that the committee approve the rules as adopted in the previous Congress. But if there is no objection——

Suddenly, Hechler heard himself shouting: "Mr. Chairman." He quickly submitted two amendments which had been adopted by the Democratic Caucus, one to allow each subcommittee chairman to designate one staff member, and the other to prevent the committee chairman from heading more than one subcommittee and authorizing subcommittee chairmen to handle legislation in the House which emerged from their subcommittees.

REPUBLICANS ATTACK SUBCOMMITTEE CHAIRMAN'S POWER

Fulton immediately attacked the resolutions. He contended that this would make the subcommittee chairman so powerful that it would destroy staff unity and coordination. He said that this was totally different from the concept of a minority staff which would be working for the entire committee, whereas the proposal would enable a staff member to work for only one person.

Frey and Wydler joined in to oppose the amendments. With some sarcasm, Frey wanted to know if this was the same type of reform which "you helped us out on our staffing"—referring to the action of the House in 1970 which had been reversed by the Democratic Caucus in 1971. "I accept the needle with grace," Hechler responded. Wydler joined the battle with zest. Turning to Hechler he said:

> You talk about the caucus as if we are a part of it. This is a foreign group as far as I am concerned, and I don't feel bound in any way.

Now Miller brought out his biggest artillery. He rapped the gavel sharply and stated:

> The Chairman of the Democratic caucus wants to make a comment, who happens to be the ranking member of this committee. Mr. Teague.

Teague put it straight:

> I was also on the Hansen Committee and I was outvoted by 6 to 1 by subcommittee chairmen on this particular (amendment) that Ken has here.

Like a lawyer who knew precisely what this witness would answer, Fulton then asked:

> Does the Chairman of the Democratic Caucus, might I ask, favor the amendment of the gentleman from West Virginia?

Teague answered: "The Chairman of the Democratic Caucus does not." The meeting then got a little wild, and went along like this:

> Mr. WYDLER. I raise the point of order whether it is proper to adopt these rules of the committee, a committee of the House of Representatives, rules that refer to the powers of the Democratic caucus.
> Mr. HECHLER. May I be heard on that point of——
> The CHAIRMAN. You may not. Rollcall has started——
> Mr. HECHLER. May I be heard for five minutes?
> The CHAIRMAN. You may not be heard. You have been heard twice on that matter. * * * All right. Proceed with the rollcall. I vote no.

The amendments were crushed by a vote of 22 to 3. The author of the amendments realized that it was impossible to form a coalition with the Republicans to help them with minority staff, which he had long advocated, if they would help on this issue. But the issue did not die, and although it took a long and at times bitter struggle the committee finally swung around and accepted the idea as though it had been right from the start.

Fulton received his reward for supporting his chairman in the uneven fight to defeat the Hechler amendments. A revealing aftermath was that the events of February 23 contributed to the establishment of a Subcommittee on International Cooperation in Science and Space, for which Fulton had been agitating unsuccessfully for years. On February 24, 1971, Executive Director Ducander wrote a persuasive note to Chairman Miller, urging him to comply with Fulton's repeated requests. One of Ducander's arguments was: "Fulton was strong behind you in the organizational meeting yesterday, let's don't forget that."

With remarkable speed, Chairman Miller on the same day dispatched a memorandum to all committee members, announcing the formation of the new international subcommittee. Fulton had effectively scored his brownie points .

SWIGERT AND SUBCOMMITTEE STAFFING

The Bolling-Hansen reforms which took effect in 1975 resulted in the inclusion in the House rules of the principle that subcommittee chairmen could designate one of their own staff members. This principle was totally unacceptable to Swigert, who argued that it would undermine efficient coordination of the staff and the power of the chairman. Several subcommittee chairmen attempted to make staff recommendations, and the word was circulated that those recommended lacked qualifications for the job. "We don't want political hacks invading our highly competent staff," was the warcry. As the minority staff started to grow, some members ruefully observed that the rules were helping give the minority their staff while handcuffing some of the subcommittee chairmen.

These were not easy issues to resolve. But the trend was clear. The breadth and depth of the subject matter, the wide-ranging nature of the oversight required, and the sheer complexity of the substantive matters being handled all added up to a need for two staff qualities which were very much in demand—high competence and mutual understanding. The subcommittee chairman who hired an incompetent was obviously going to hurt himself, and this fact was never acknowledged by those fighting to preserve one of the last bastions of the status quo. Gradually, the walls came tumbling down, and as time went on the trend was very clearly in the direction of greater freedom for staff selections by the subcommittee chairmen.

OVERSIGHT SUBCOMMITTEE

In the early 1970's, separate oversight subcommittees operated primarily for NASA oversight (see chapter IX.) Teague chaired the Subcommittee on NASA Oversight in 1970, Downing in 1971 and Fuqua in 1972. When Teague became chairman of the full committee in 1973, he stressed that each subcommittee should conduct vigorous oversight functions within its jurisdictional areas, and that general oversight would be conducted through the full committee and the staff. In 1973 and 1974, the oversight work by the committee and its subcommittees proceeded aggressively. Even when bogged down by annual authorization hearings, the subcommittees managed to get out on field trips to review and assess not only research and development but also construction of facilities. The committee's No. 1 specialist in construction oversight was Colonel Gould, who took the lead in oversight up to the time he moved up to become Deputy Director in mid-1975, and Ron E. Williams took over as construction specialist in early 1976.

The Bolling committee report, and the reform legislation enacted in 1974, put a great deal of stress on the need for beefing up the oversight function in all committees. The Bolling report required the establishment of an oversight subcommittee on every standing committee, setting up a network of oversight reports supervised by the House Committee on Government Operations. The legislation as finally enacted in 1974 softened these requirements somewhat, while preserving the central oversight authority of the Government Operations Committee and also requiring the reporting of oversight plans and progress. But instead of requiring a specific oversight subcommittee, the final version of the reforms gave every committee with 15 or more members the alternative of either establishing an oversight subcommittee or conducting oversight through subcommittees. The 1974 reform law stipulated:

> The establishment of oversight subcommittees shall in no way limit the responsibility of the subcommittees with legislative jurisdiction from carrying out their oversight responsibilities.

In addition, as noted above, the Science Committee was given specia oversight over all nonmilitary research and development.

TEAGUE'S PHILOSOPHY OF OVERSIGHT

As 1975 began, Teague reflected on the new challenges of oversight presented to the committee:

> Throughout all of its years, the committee spent much of its time on oversight—the intensive review of agencies under its legislative jurisdiction to determine how well they are doing their job and how they are spending the taxpayers' dollars. Members and staff have spent long periods on comprehensive investigations which involved hearings in Washington, field hearings at government research centers and contractor plants, weekend visits while Congress was in session, and a lot of plain hard work and study. I know—I have done it for years. But, this is how you get to know how well a program really is working, how effective management is, how the dollars are being spent. * * * A House oversight agenda coupled with more emphasis on each committee can result in a more effective Congress. Congress should do more than pass laws and approve budgets; it should see how those laws are carried out and what is done with the money it approves in the budgets.

Teague's philosophy on oversight was that Congress was obligated to check on whether the money authorized was being spent in accord-ance with the intent of Congress. He did not feel that "oversight" should entail actually going down into any agency to tell them how to run their internal operations, unless they were clearly violating the intent of Congress.

During 1975, Swigert analyzed the oversight responsibilities of the committee and came up with a detailed proposal for a "Special Investigations and Oversight Task Team," which he submitted to Teague on October 9, 1975. Although Teague originally had expressed his opposition to a separate oversight subcommittee, Teague eventually approved the hiring of a new "task team leader" for oversight, with this twofold purpose, as outlined by Swigert:

> (1) Provide a special investigative force to be employed by Chairman Teague on matters requiring selective investigating effort, and
> (2) In coordination with subcommittee chairmen provide a special mechanism for independent management review to assist in carrying out assigned oversight responsibilities.
> It is envisioned that the new task team will be headed by a specially selected person with an extensive background in management, engineering, and an intimate knowledge of the programming, budgeting, and legislative processes.

Swigert and his Deputy, Colonel Gould, interviewed a number of applicants and finally agreed that the man who obviously had the best qualifications for the job was Dr. Robert B. Dillaway. "Too good to be true," wrote Colonel Gould on Dr. Dillaway's written application, although he was one of the first to recognize that there was a problem in Dr. Dillaway's performance. With 15 years of impressive experience at North American Aviation, and even more responsible tours of duty with the Secretary of Navy and Army Materiel Command, Dr. Dillaway had served on the faculties of the Universities of Illinois, California, and Stanford, teaching such subjects as engineering systems, nuclear reactor engineering, fluid mechanics, rockets, and controls. On paper, he looked so good that a senior minority staff member, Michael A. Superata asked: "Why does he want to come here?"

After some negotiations, Dr. Dillaway was hired early in January 1976.

On January 20, Teague met with Dr. Dillaway, Colonel Gould, and Swigert. Teague laid down these rules:

> Don't go around the subcommittee chairmen.
> Coordinate with all concerned.
> First thing is to develop a plan.
> Letter would be prepared to introduce Dr. Dillaway.
> Dr. Dillaway to talk with subcommittee chairmen.

ESTABLISHMENT OF SSIO SUBCOMMITTEE

Early in February, Teague established the Ad Hoc Subcommittee on Special Studies, Investigations, and Oversight, which included the following members:

Democrats	*Republicans*
Olin E. Teague, Texas, *Chairman*	Charles A. Mosher, Ohio
Ken Hechler, West Virginia	John W. Wydler, New York
Don Fuqua, Florida	
James W. Symington, Missouri	
Mike McCormack, Washington	

The subcommittee held no hearings, did not meet to discuss the appointment of Dr. Dillaway, and in fact held only one meeting in its entire existence—on August 24, 1976, after the "SSIO" operation got into trouble.

In announcing the establishment of the subcommittee, Teague sent a notice to all committee members on February 3, 1976, reviewing the oversight work and plans under way by the standing subcommittees, adding:

> However, I believe a more intensified program should be undertaken if we are to be fully responsive to the House rule concerning this matter. * * * It is not my intent that this ad hoc subcommittee infringe upon or erode the jurisdictional responsibilities delegated to the present subcommittees. Rather, I visualize this new organizational element will serve to augment the efforts of existing subcommittees and provide independent management reviews as appropriate. * * * Dr. Dillaway has been charged with developing a special studies, investigations, and oversight plan in coordination with the subcommittees on matters pertaining to their areas of jurisdiction. He will be seeking inputs from the subcommittees on areas that they feel should be subjected to additional review or areas beyond their present capabilities, time-wise and staff-wise. Following development of plans for each subcommittee area, the ad hoc subcommittee will prioritize the overall plan, assure there is no duplication, and implement the plan about mid-April 1976.

MIXED REACTION TO DR. DILLAWAY

Dr. Dillaway visited the offices of the subcommittee chairmen and staff. There was a mixed reaction to his mode of operation. Some subcommittee chairmen felt he could perform a useful function in oversight pertaining to activities of the agencies under the Science Committee jurisdiction; others felt that he was simply overlapping or duplicating useful oversight work already in progress at the subcommittee level.

Colonel Gould, who had endorsed Dr. Dillaway's qualifications as looking extremely good on paper, began to have reservations about his methods of operation. In a May 25 memorandum to Swigert, Gould noted:

> As I have indicated before, the Dillaway oversight plan, which apparently has been endorsed, is an overly ambitious undertaking and would probably take 3 or 4 years to complete. Further, some of the issues outlined in the plan are present-day viewpoints, which may not prevail even during the next session of Congress.

Meanwhile, Dr. Dillaway was assembling a rather sizable staff which included personnel borrowed from the Congressional Research Service, General Accounting Office, and other sources. At times, news would filter back concerning strange telephone calls emanating from Dr. Dillaway. NASA Administrator Fletcher was ordered to appear in his office within one hour, and Dr. Fletcher called around to try to find out who Dr. Dillaway was and why Dr. Fletcher's presence was so peremptorily needed. (P.S., he did not come.) There were also strange meetings and private business relationships which appeared to be commingled with committee business. On August 10, a meeting of the full committee was held, at which time Swigert presented the SSIO request to hire two consultants. McCormack, Goldwater, and Hechler raised a number of questions about the nature of the investigations and qualifications of the consultants, who were being hired for oversight over ERDA.

Mosher asked:

> Is it contemplated that these consultants would be hired before the subcommittee has met? As far as I am aware, the subcommittee has never met to consider its role, its jurisdiction.

Temporarily chairing the meeting, Hechler suggested that action on hiring the consultants be deferred until such time as the SSIO Subcommittee could meet. Goldwater in supporting the recommendation, added that the subcommittee should at the same time work out "proper coordination" with the subcommittees. So far as the two energy subcommittees were concerned, there was a strong feeling that Dr. Dillaway was clearly getting into areas which the subcommittee already had in hand, in process, or contemplated in the future.

When the SSIO subcommittee met on August 20, it was obvious that Dr. Dillaway's philosophy of oversight did not coincide with Teague's. He not only visualized a large empire of personnel, but also talked confidently of straightening out several agencies under the committee's jurisdiction. Numerous complaints were raised at the first and last meeting of the SSIO Subcommittee. Once at a later time when he was asked how he controlled Dr. Dillaway, Teague responded: "I didn't control him; I decontrolled him."

Most of the reports drafted by the SSIO Subcommittee were never printed. One, however, proved useful: A report jointly prepared by the SSIO Subcommittee and the Brown Subcommittee on the Environment and the Atmosphere on the Environmental Protection Agency's Research Program, with primary emphasis on the Community Health and Environmental Surveillance System (known as CHESS). This investigative report was prepared largely at the direction of Brown and his staff by the Science Policy Research Division of CRS and a group of consultants from various health agencies. The report grew out of allegations which were first published in the Los Angeles Times at the end of February 1976, charging EPA with falsification of data on the adverse health effects of air pollution. Joint hearings were held, in which SSIO did not participate, but which were conducted by the Commerce Subcommittee on Health and the Brown Subcommittee. SSIO did take part in the investigative report. Although the report upheld the honesty and integrity of the EPA project leader, it did raise a number of questions about proper evaluation of data assembled in the future.

With the assistance of Dr. John V. Dugan of the minority staff, two unpublished but useful studies were completed on NASA's aeronautical R. & T. base effort as related to the Department of Defense, and a review of NASA's energy R. & D. role.

Early in 1977, Dr. Dillaway left the staff and the subcommittee was not heard from again. It was not revived in the 95th Congress.

A brief obituary on the SSIO Subcommittee was relayed to the Government Operations Committee on February 18, 1977, referring to the fact that the subcommittee had been created on an "experimental basis":

The experiment resulted in findings that, although beneficial to the oversight function, some duplication of effort occurred because of jurisdictional overlap, despite controls invoked to preclude same. Investigations and Oversight has not been reconstituted for the 95th Congress.

SUBCOMMITTEE ON INVESTIGATIONS AND OVERSIGHT

One of the early decisions facing the committee when the Democratic members assembled for their caucus on February 1, 1979, was the issue of whether to establish a seventh Subcommittee on Investigations and Oversight. The chief proponent of this concept was Ottinger, who had unsuccessfully attempted to put the idea across at the opening of the prior Congress in 1977. In 1977, the new subcommittee had been voted down for two reasons: Some members were simply voting against giving Ottinger a subcommittee, which he could have claimed in 1977 had a new one been established; and also in 1977 the memory of the failure of Dillaway's operation was fresh in everyone's mind.

By 1979, the climate had changed. By seniority, Ottinger was slated for a different and probably more important subcommittee in any case. In a memorandum distributed to all members, he had argued that the other subcommittees did not have the time or staff to investigate fully the manner in which the committee authorized billions of dollars, the implementation of procurement policies, and compliance with the intent of the Congress. Ottinger also attached to his motion a proviso drawn from the House Public Works Committee rules, that no investigation could be undertaken without consultation with the subcommittee chairman whose jurisdiction was involved, and also requiring the approval of the chairman of the full committee. Scheuer strongly endorsed the Ottinger motion, citing the success of the Investigations Subcommittee in the Commerce Committee, and underlining the opportunities which such a subcommittee offered to junior committee members to make their mark.

While noting the unfortunate past experience with an oversight subcommittee, Fuqua stressed the importance of effective oversight. He concluded: "It may work. It may not." Lloyd and Ambro, either of whom seemed to have an opportunity to become chairman of the new subcommittee if it were established, both spoke in favor of its creation. On a rollcall, Ottinger's motion prevailed, 17 to 2. Lloyd then faced a dilemma.

When it was Lloyd's turn to single out which subcommittee he preferred to chair, the environment and investigations subcommittees were the only two left. Naturally disappointed that he did not get a chance to head up the subcommittee handling his first love—aeronautics and aviation—Lloyd made the decision that the investigations subcommittee might afford him an opportunity to launch some broader inquiries of interest. He ended the uncertainty quickly by opting to take over the Subcommittee on Investigations and Oversight.

Representative Jim Lloyd (Democrat of California), chairman of the Subcommittee on Investigations and Oversight.

JIM LLOYD AS SUBCOMMITTEE CHAIRMAN

Tall, flaxen haired and self-assured, Jim Lloyd had been a Navy pilot for 21 years—including combat as a fighter pilot in the South Pacific during World War II. He had a longstanding practical interest in aeronautics. A native of Helena, Mont., Lloyd had received a B.A. from Stanford and a M.A. from the University of Southern California. Following two years as public relations director for the Aerojet General Corp., he set up his own public relations firm. He served as a member of the city council and also as mayor of West Covina, in eastern Los Angeles County, while he was teaching political science at Mount San Antonio College in Walnut, Calif. Lloyd won his congressional seat as an aftermath of the Watergate upheaval in 1974 in a newly created congressional district, but nevertheless had to run against incumbent Republican Congressman Victor Veysey whose own district had been carved up in the redistricting. He eked out a 705-vote victory in 1974, advancing to wins of 10,500 and 12,000 in 1976 and 1978. Lloyd's voting patterns reflected a somewhat conservative suburban district.

After considerable study and many interviews, with the approval of Chairman Fuqua, Lloyd chose as his staff director 35-year-old Jerry Staub, who had been counsel of the Transportation and Aviation Subcommittee on which Lloyd had served since 1975. With degrees in history from Gettysburg College and law from the University of Florida, Staub's major interest has been in international law and astrophysics. Like Lloyd, he had been a fighter pilot in the Navy. Staub also had two years of experience with the Senate Committee on Aeronautical and Space Sciences.

MEMBERSHIP AND JURISDICTION

The following members were assigned to the Subcommittee on Investigations and Oversight in 1979:

Democrats	Republicans
Jim Lloyd, California, *Chairman*	Manuel Lujan, Jr., New Mexico
Ronnie G. Flippo, Alabama	William Carney, New York
Albert Gore, Jr., Tennessee	Toby Roth, Wisconsin
Bill Nelson, Florida	

The jurisdiction was set forth in the committee rules as follows:

Review and study, on a continuing basis, of the application, administration, execution, and effectiveness of those laws, or parts of laws, the subject matter of which is within the jurisdiction of the committee and the organization and operation of the Federal and private agencies and entities having responsibilities in or for the administration and execution thereof, in order to determine whether such laws and the programs thereunder are being implemented and carried out in accordance with the intent of the Congress. In addition, the Subcommittee on Investigations and Oversight and the appropriate subcommittee with legislative authority may cooperatively review and study any conditions or circumstances which indicate the necessity or desirability of enacting new or additional legislation within the jurisdiction of the committee, and may undertake future research and forecasting on matters within the jurisdiction of the committee. The Subcommittee on Investigations and Oversight shall in no way limit the responsibility of other subcommittees from carrying out their oversight responsibilities, nor shall any investigation be undertaken by the Subcommittee on Investigations and Oversight without (a) consultation with the chairman of the appropriate subcommittee with legislative authority and (b) approval of the chairman of the committee.

THE TRIP TO MEXICO

As the junior of all subcommittees, one of the first handicaps the Lloyd subcommittee discovered was lack of space. With his congressional office in the southeast corner of the Cannon Office Building, and his staff housed in cramped quarters in the old FBI Building, Chairman Lloyd soon found that he was one Metro stop away from his own committee staff. Lloyd pursued a wise course in the first few months of his subcommittee's existence, laying very careful plans for the future instead of rushing in to establish public visibility at the expense of committee good will. The subcommittee's first major activity was a field trip to Mexico, May 3–7, 1979. The subcommittee and staff were joined by Brown, whose Science, Research and Technology Subcommittee also held later hearings on scientific and technological cooperation between the United States and Mexico, with Lloyd chairing one of the hearings. The Mexican trip included a meeting with President Lopez Portillo, as well as members of CONACYT, the Science Council of Mexico. The group examined the potential for the transfer of technology and energy resources between the United States and Mexico, also assessing the role that science and technology might play in recent agreements to expand Mexican-American cooperation resulting from President Carter's Mexican trip.

The group was personally escorted to the Mexican oilfields by officials of Pennex, the Mexican National Oil Co., and visited the International Maize and Wheat Improvement Center, birthplace of the Green Revolution. In the report on the trip, recommendations were made for closer joint cooperation and agreements to speed the development of better trade relations between the United States and Mexico, with particular emphasis on petroleum, natural gas, and technology transfer.

During July 1979, the Lloyd subcommittee held hearings in Washington, D.C., and Los Angeles, Calif., on the aeronautical design of the DC–10, as an aftermath of the worst tragedy in U.S. aviation history when an American Airlines DC–10 lost an engine on takeoff and crashed on May 25 near O'Hare International Airport in Chicago, Ill. The subcommittee investigated the technical aspects and design in order to recommend future modifications.

RESEARCH PROGRAMS TO AID THE HANDICAPPED

For many years, Teague had been interested in and concerned with programs which involve handicapped people. As a disabled combat veteran who had spent many years working with veterans' program in his capacity as chairman of the Veterans' Affairs Committee, Teague had become intimately acquainted with the work being carried on in veterans' hospitals, rehabilitation centers, and the limited amount of research going forward in other agencies like the Department of Health, Education, and Welfare. Teague's personal interest in the handicapped was further enhanced when he suffered a stroke and his left leg was amputated in 1977.

The Bolling-Hansen reforms in 1974 had transferred biomedical research from the jurisdiction jointly held by the Science and Commerce committees to concentrate it exclusively in the Commerce Committee. Nevertheless, the Science Committee was given special oversight over biomedical research as one portion of the generous grant of authority which covered all nonmilitary R. & D.

In 1975, when NASA testified before the committee in their authorization hearings, NASA Administrator Fletcher led off with a series of demonstrations of recent spinoffs from technology developed for the space program. These included a voice-controlled wheelchair for quadraplegics, and a rechargeable cardiac pacemaker for heart attack victims. Teague resolved that it was time to build on what NASA was doing through committee initiatives in research for the handicapped. Brown talked with Teague after the hearing, and they agreed that it would be very useful for the committee to undertake some work in aiding the handicapped. It was decided to assign the work administratively under the umbrella of the Science, Research and Technology Subcommittee.

During his service on the Veterans' Affairs Committee, Teague was impressed with the testimony delivered every year for the Disabled American Veterans by a young Californian named Sherman Roodzant, who had been recognized as California's Outstanding Veteran of the Year in 1971. In 1974–75, Roodzant was elected State commander of the quarter of a million disabled veterans in California. Early in 1976, following the annual DAV testimony before the Veterans' Affairs Committee, Teague asked Roodzant to stop by his office, where they talked about what the committee could do to stimulate more interest by Federal agencies in research to aid the handicapped. Teague offered Roodzant the job of coordinating for the committee a new effort in this area. Then he called Brown over to his office and they continued their conversation on how the handicapped operation would fit into the committee structure administratively. Brown's interest in disability and problems of the handicapped made this a natural for him to generally take under his wing, in addition to the leadership provided by Teague.

PANEL ON HANDICAPPED RESEARCH

Roodzant's first job was to draw up plans for a panel of experts to study the problems of the handicapped, and identify those areas where a more concentrated and intensified effort should be put forward in programs to aid the handicapped. Teague obtained $26,000 in the committee-enabling resolution passed in March to cover the cost of the

panel. On August 10, Roodzant presented to the committee the out-
lines of the study to be undertaken by a 9–member panel (later expanded to 11), headed by Dr. William A. Spencer, director of the Texas
Institute for Rehabilitation and Research in Houston, Tex., and chairman of the department of rehabilitation, Baylor College of Medicine.
The panel included other experts in medicine and rehabilitation, as
well as several handicapped persons, and others who had been active in
the field. Roodzant testified before the committee:

> In an attempt to present to you a representative group of experts on the problem,
> we have solicited some 50 organizations interested in these problems and come up
> with a panel that is proposed before you this morning.

Brown and Krueger spoke in support of the panel, and the committee
unanimously approved the plan presented by Roodzant.

To provide guidance and direction for new research and application of technology to aid the handicapped, Teague announced hearings
of the Subcommittee on Science, Research and Technology on September 22 and 23, chaired by Brown. In his opening statement, Brown
noted that "we have made a commitment to do something about it."
He mentioned Teague's special concern, "which extends for many,
many years," leading the full committee to make a similar commitment. In addition to members of the panel who testified, Edward Z.
Gray, NASA's Assistant Administrator for Industry Affairs and
Technology Utilization presented an updated account of NASA's
application of space technology to aid the handicapped.

THE BROWN HEARINGS

A wide range of witnesses testified, including representatives of the
medical and engineering professions, Federal agencies involved with
handicapped persons, and members of handicapped consumer groups.
Throughout the hearings, witnesses stressed the need to involve
handicapped consumers in the R. & D. phases of technology to benefit the disabled. Acceptance and use by those directly concerned was a
point which some researchers did not fully grasp. For example, the
whole area of architectural barriers was one in which there were
differences of opinion between HUD experts and handicapped people
themselves. Fortunately, the committee had already retained W. R.
"Dede" Matthews, a Texas architect, who was tackling this problem
as a special consultant to the committee. In 1977, John G. Clements
joined the staff to work in this and other areas of research programs
aiding the handicapped. With the departure of Roodzant in 1979,
Clements took over his responsibilities.

In its report in 1977, the panel concluded that a ridiculously low amount—$31 million annually—was allocated for Federal R. & D. for the handicapped. This amounted to $2.92 per disabled person in 1976. Whereas all Federal health R. & D. amounted to 3.7 percent of the total public and private health expenditures, Federal rehabilitation R. & D. was only 0.026 percent of such health expenditures.

The panel also concluded that there was a serious lack of coordination and communication among Federal agencies, private organizations, and handicapped consumers concerning R. & D. for the handicapped. As a result, there was recommended a National Council for Research and the Handicapped, including two bodies under one Director—a Government organization and a non-Government group.

Chairman Teague plugs International Disabled Expo. At an April 7, 1977 news conference, Chairman Teague (standing, center rear) helps publicize new technologies to aid the handicapped. Seated at podium are Max Cleland, Veterans Administrator, Senator Harrison H. Schmitt (Republican of New Mexico) and Representative George E. Brown, Jr. (Democrat of California).

Teague called a news conference on April 7, 1977 not only to release the panel report, but also to focus attention on the upcoming International Disabled Expo, to be held in Chicago in August. Max Cleland, Administrator of the Veterans' Administration and a triple amputee, joined Teague in the news conference. Teague stated:

> I trust that this committee activity will spark a greater national commitment to effectively utilize the scientific and technological resources at our disposal in attacking the problems of the handicapped, thus allowing all handicapped individuals, both in this great Nation and around the world, to enjoy fuller, more complete lives.

THE SECOND PANEL FOR THE HANDICAPPED

A new and larger panel, including the assistance of several Federal agency representatives, was established in 1977. Teague informed committee members in June:

> One-third of the proposed panelists served with distinction on the previous panel, over one-half are handicapped individuals, and over one-half represent the professional community serving the handicapped.

When the panel made its report in March 1978, it pointed out the piecemeal approach in existing research programs, the low level of priority in Federal agencies, and the need for a lead agency to direct the programs. The panel recommended that NASA be designated as the lead agency to coordinate the use of science and technology to aid the handicapped. The panel also recommended a Science and Technology Board for Handicapped Persons. Teague commented in releasing the report:

> The Panel's work has pinpointed the issues and provided a framework for a national program. It is now up to the Congress, the Administration and the American people.

The administration bucked the centralization of authority in NASA. This did not fit in with the traditional concept of organization which dictated that such programs should be located in HEW. With the help of Roodzant and other staff assistance, the charter of NASA was amended in 1978 and $3 million was added to the NASA bill for bioengineering research for the handicapped. In 1978, Congress passed legislation to establish a National Council on the Handicapped, which was recommended by the committee's panel. A vastly increased research program was also placed within HEW. The committee was also instrumental in adding $2 million to the authorization for the National Science Foundation to set up a handicapped-related research program.

On October 18, 1977, Teague announced a joint hearing on the use of computers in aiding education for the handicapped. The hearing involved the Science Committee Subcommittee on Domestic and International Scientific Planning, Analysis and Cooperation (chaired by Scheuer) and the House Education and Labor Subcommittee on Select Education (chaired by Representative John Brademas—Democrat of Indiana). In announcing the hearings, Teague noted:

> The handicapped child faces great difficulties in pursuing his educational goals. Recent advances in technology offer this child the chance to attain an appropriate educational level. It is my hope that the development of computer technology in the education of the handicapped will continue to expand educational experiences for the handicapped child.

Scheuer added:

Communication, perhaps, presents the major obstacle in the education of the handicapped child. Various disabilities such as blindness, deafness and cerebral palsy restrict a good mind from interacting with the world around him, and thus stunt his educational growth. However, the use of computers with their adaptive mechanism give the handicapped child increased learning skills and opportunities unknown until recent years.

ADDITIONAL TEAGUE INITIATIVES

During 1978, Teague launched a whole series of personal and committee initiatives designed to spur both public and private agencies to focus on doing more in a practical and realistic way to aid the handicapped. By insisting that NASA be the lead agency, both the panel and Teague himself shocked the existing agencies into realizing they had better not sit back and relax or their jobs and authority would be preempted by a more aggressive, newer agency. So the agencies started doing more themselves. Working with Brademas, Teague helped push through legislation to coordinate handicapped research, plus an interagency committee which represented all Federal agencies carrying on rehabilitation work. At Teague's suggestion, the Armed Forces, Federal agencies, and leading private employers were polled to determine their plans for hiring handicapped persons. In many other areas, Teague, Roodzant, and his staff worked on amending the social security law to allow recipients to qualify to use motorized Amigo wheelchairs, helped break down resistance to implementation of the new Transbus, thus enabling handicapped people to board intracity buses more easily; and continued his running assault on architectural barriers which hampered the handicapped in Federal buildings. On February 24, 1978, for example, Teague accused the General Services Administration, HUD, and the Department of Defense of violating a 1968 law mandating that public buildings be made more accessible to handicapped persons. Teague got quick action by stating:

The standards specified by public law have not been developed, the inadequate standards that were instead adopted have not been complied with and handicapped individuals are still denied free access to buildings that belong to them as well as every American citizen. We have talked too long. We have asked handicapped people to wait too long. We have not lived up to either the law or our moral responsibility.

From the top to the bottom, Teague made sure that every responsible Federal official was made aware that he and the committee both meant business. On March 22, 1978, Teague wrote to the President:

We have vast scientific and technological resources at our disposal; therefore, a program should be focused which blends our resources with the needs of the country's disabled. The time has come for us as elected representatives, and as a nation, to direct our energies in a manner reflecting our commitment to handicapped persons.

When the Washington Star published an article in April 1978, pointing out the obstacles a young disabled visitor encountered in visiting the Kennedy Center, Museum of History and Technology, and the Library of Congress, Teague fired off letters to the officials concerned to find out why action was not being taken to correct the situation. In remarks for the Congressional Record, Teague colorfully stated:

> Maybe we could understand and appreciate this young man's plight, the dilemma which faces millions of elderly and handicapped citizens who visit or try to conduct their business in Washington, D.C., if we were to lose our parking spaces, restrict ourselves to wheelchairs for a week, and had to come crawling to the floor of the House of Representatives every time we had a rollcall vote.

Wherever Teague went, he looked at the effect of existing facilities on problems which handicapped people encounter, and then went to bat to correct them. When he encountered difficulties in airports for handicapped people to make connecting flights, he wrote to Frank Borman, president of Eastern Airlines and Representative Harold T. Johnson (Democrat of California), chairman of the House Public Works and Transportation Committee. When he heard about a Houston condominium designed especially for disabled people, he sent Roodzant and consultant Matthews down to make some tape-recorded interviews, and helped spread the gospel on the value of enabling the handicapped to enjoy "independent living" in good surroundings outside of institutions. If somebody had trouble boarding a train, he was after Amtrak to get them to live up to their literature advertising they offered assistance to disabled passengers. From all over the country, people wrote Teague about job problems, architectural barrier problems, or simple lack of understanding by people in authority, and all these letters were carefully answered and the situations usually straightened out by Teague or the committee staff.

SPACE-AGE TECHNOLOGY TO AID ELDERLY AND HANDICAPPED

Five Science Committee members also serving on the House Select Committee on Aging—Watkins, Lloyd, Mrs. Bouquard, Hollenbeck, and Dornan joined with Representative Claude Pepper, chairman of the House Aging Committee in a February 13, 1979, letter to Fuqua urging greater application of developing technology to aid the elderly and the handicapped. The letter urged joint action and joint hearings between the two committees, with emphasis on the work of NASA and the Department of HEW.

In a speedy and warm response, Fuqua on February 21 heartily endorsed the idea, suggested joint hearings in mid-July, and also proposed interim measures to encourage "aggressive cooperative efforts" by HUD and the VA as well as NASA and HEW. Following up with requests to these agencies, Fuqua joined the six signers of the February 13 letter in written requests to urge them to get started in working out the necessary cooperative relationships. The letter also stated:

It is important that a senior staff representative from each agency be responsible for developing, implementing and administering such plans and agreements.

The committee was pleased with the agency response, and in particular the enthusiasm which NASA displayed for the idea. In a followup letter on May 15, 1979, Fuqua told NASA Administrator Frosch:

Your appointment of Floyd I. Roberson to be NASA's representative will insure decisive action in this area over the coming months.

Fuqua noted that the joint hearings were scheduled for the week of the 10th anniversary of the first manned landing on the Moon, carrying great symbolism in the new drive to apply space technology to help alleviate the problems of the elderly and handicapped on Earth.

The crowded hearings on July 19–20, 1979, featured noted futurist and inventor R. Buckminster Fuller, NASA Administrator Frosch, National Space Institute President Hugh Downs and author Robert Heinlein, as well as representatives of other Federal agencies, private manufacturers and universities. In announcing the hearings, Fuqua stated:

The demands which are placed on our scientists and engineers in meeting the challenges of NASA's space missions will continue to keep this Nation on the leading edge of technology in many fields. We must, likewise, continue to insure that this technology is made available and not "log jammed" within the agency.

The committee leadership, and the effective efforts of staff members Roodzant and Clements and the committee consultants, resulted in great strides in research and technology to aid the handicapped. It was an area of clearcut accomplishment for the committee. It represented an expansion of jurisdiction under the heading of "special oversight" which proved to be significant.

Epilogue

FUQUA AND THE FUTURE

On January 23, 1979, the caucus of all Democratic Members of the House of Representatives decided by a secret ballot vote of 235 to 10 that Florida's Representative Don Fuqua would become the fourth chairman of the Science Committee. There were only three committee chairmen given a higher number of votes (the most being 238) at the start of the 96th Congress. The House made it official January 24.

At age 45, Fuqua was the first of the four Science Committee chairmen to be born after World War I, as well as the first to become chairman before the age of 60. Teague had already finished at Texas A. & M. before Fuqua was born. The following table gives a comparison of the relative ages of the four committee chairmen:

Chairman	Year of birth	Age became chairman	Congressional term when elected chairman
Brooks	1897	61	12th
Miller	1891	70	9th
Teague	1910	62	15th
Fuqua	1933	45	9th

AN ACTIVE APPRENTICESHIP

Fuqua had served an active apprenticeship on the committee, chairing the following subcommittees since his fifth term starting in 1971: International Cooperation in Science and Space, Manned Space Flight, Oversight, and Space Science and Applications. Although both Brooks and Miller had taken brief flings at handling minor subcommittees at the same time they chaired the full committee, Fuqua was the first chairman to choose to head up a major subcommittee at the same time he piloted the full committee. He decided in 1979 to retain his chairmanship of the Subcommittee on Space Science and Applications (handling all of NASA's activities except aeronautics), a post he had held since 1975.

The circumstances of Teague's severe illness in 1977, when he was confined to the Naval Hospital in Bethesda for major surgery, gave Fuqua the experience of presiding at full committee sessions during Teague's absence for several months.

379

The days of ironfisted committee chairmen who imposed their personal will over the grumbling protests of their committee members are probably gone forever. Fuqua fits well into the modern mold of a chairman whose power and influence spring from creating a democratic consensus within the committee. Observers at the organization meeting of the committee in February 1979 were struck by the fact that Fuqua made his personal preference clearly visible in advance and argued for jurisdictional responsibilities for the energy subcommittees which committee members then proceeded to vote down. The votes were close, there was no rancor, no "hit list" was compiled for later retributions, and the business of the committee went along smoothly without the issue being raised again.

EMPHASIS ON SUBSTANTIVE ISSUES

At the same time, Fuqua had an interest in a number of substantive issues which he encouraged the committee to emphasize. He has exerted leadership in expanding the committee's activity in materials policy, space industrialization, solar-powered satellites, and commercialization of Earth resources information. Fuqua also has helped sponsor overall reviews of all Federal R. & D. programs, as well as the first major review of the National Science Foundation in over a decade. He has taken a particular interest in more rigorous oversight in the construction field, somewhat pushed to the background by the more glamorous fields of operational R. & D. which attract more headlines. In 1978, Fuqua carried the ball for freedom of research in connection with recombinant DNA.

Fuqua's interest in international relations manifested itself not only in sponsoring an oversight trip to Europe, and encouraging members to visit Mexico in the spring of 1979, but a personal role in helping push the Institute for Scientific and Technological Cooperation. He testified before the House Foreign Affairs Committee in support of the latter program, involving an emphasis on transferring "appropriate technology" to underdeveloped countries along grassroots lines originally formulated by the late Vice President Hubert H. Humphrey. In August 1979 Chairman Fuqua led a delegation of seven committee members attending the United Nations Conference on Science and Technology in Vienna.

INTEREST IN SYNTHETIC FUELS

Working closely with House Majority Leader Jim Wright, Fuqua has stressed the development of synthetic fuels and has pressed for greater congressional initiative in this area. In June 1979, he expressed his strong support for President Carter's plan to accelerate the development of solar energy and other renewable energy sources.

At the start of 1979, when the House leadership was anxious to schedule legislation for floor consideration during the early months—when most committees had not yet gotten underway—the Science Committee in March brought five bills to the floor: the NASA, EPA, NSF, and FAA authorizations and a $185 million supplemental authorization for the Space Shuttle. Unlike Presidents, new committee chairmen are not customarily accorded a "honeymoon" during which the opposition stays muted. But Fuqua put his Science Committee legislation through the House with very little trouble, only a couple of months after he had become the chairman of the committee.

SPEEDY ACTION ON SCIENCE COMMITTEE BILLS

It took only a few minutes for the $75 million FAA R. & D. authorization to go through under suspension of the rules by voice vote on March 26. On March 27, the House passed by voice vote the $381.3 million EPA authorization. The billion dollar NSF authorization passed the same day by a voice vote, after sustaining a $14 million cut on the House floor.

Fuqua was concerned about both the $4.76 billion NASA authorization, and the $185 million supplemental funding for the Space Shuttle. Yet both passed on March 28 by votes of 323 to 57 and 354 to 39.

On jurisdictional issues, Fuqua differed markedly from the philosophy and mode of operation of any of the three preceding chairmen. On relations with other committees, Brooks and Teague were inclined to fight and maneuver to expand the committee's jurisdiction, while Miller retreated somewhat passively. Fuqua's approach has resulted in much fewer high decibel confrontations than in past years. In describing his relationship with the Armed Services, Interstate and Foreign Commerce, and Interior and Insular Affairs Committees, Fuqua reflected:

> We have no problem with Armed Services Committee. Dingell and I sat down first, and then we sat down with Udall and agreed that we were going to work together. We got the staff together and repeated that "we're going to work together." And I think it has been very, very good. Last year it was just everybody at each other's throats—trying to assume original jurisdiction.

When the House debate opened on the Department of Energy authorization act on July 26, 1979, Fuqua was able to report to the House:

> The work has been broader than our committee because the Department of Energy has other functions to perform which are handled by the Committee on Interstate and Foreign Commerce and the Committee on Interior and Insular Affairs.

Mr. Dingell and Mr. Udall and their staffs have been most cooperative in working with us and the fruits of our labors are reflected in the fact that there are no disagreements between the committees in the bill before us today.

Chairman Fuqua inspects see-through nuclear reactor with its inventor, Professor Glen J. Schoessow of the University of Florida (right). At the chairman's invitation, Professor Schoessow demonstrated the reactor to the committee during a special hearing.

Mrs. Don Fuqua, Dr. W. R. Lucas, Representative Fuqua and Gerald E. Jenks (minority staff), at the Marshall Space Flight Center.

Chairman Don Fuqua addresses remarks during a committee field trip to Marshall Space Flight Center. From left, Representative Larry Winn, Jr. (Republican of Kansas), Representative Ronnie G. Flippo (Democrat of Alabama) and Dr. William R. Lucas of NASA's MSFC. "HEAO" refers to High Energy Astronomy Observatory.

RELATIONS WITH APPROPRIATIONS COMMITTEE

Fuqua developed the feeling over the years that all too frequently the Science Committee would "pass legislation and then let it proceed on its own merits." He added:

With the Appropriations Committee, we are now actively working with Eddie Boland (chairman of the House Appropriations Subcommittee which handles NASA, NSF and many programs in the Science Committee's jurisdiction) to try and be sure the programs are not eliminated—that they understand what we are trying to do. We've had, I think, reasonably good cooperation.

Within the committee, Fuqua says:

I think one of the differences between myself and Teague is that I have given subcommittee chairmen more autonomy than they have had before, without turning the committee over to them.

This policy has relieved a lot of the personal antagonisms formerly developed when subcommittee chairmen frequently had to fight for months, sometimes unsuccessfully, to obtain the kind of staff assistance they deemed essential to get their jobs done. Under Fuqua, new staff assistants have been brought aboard expeditiously, without the somewhat artificial challenging of "qualifications" which in times past had been frequently used to cover up a desire by either a chairman or committee staff director to centralize power.

Of all the four chairmen, Fuqua's relations with the minority was perhaps the closest. To be sure, there was an initial brush with ranking minority member Wydler, when the latter charged he had not been sufficiently consulted on the committee budget to be presented to the House Administration Committee early in 1979. But gone were the days when the chairman and ranking minority member engaged in shouting matches over diametrically opposing views on staffing. No longer did the corridors of the committee reverberate with the colorful epithets which Teague and Wydler used to exchange on minority staffing. Although Wydler remained studiously independent in his beliefs and substantive views, a very close personal working relationship developed between Fuqua and Winn—the second ranking Republican on the committee.

RELATIONS WITH EXECUTIVE AGENCIES

Fuqua's relations with the agencies and their administrators also differed from the mode of operation of his predecessors. Brooks in his short tenure was noted for always trying to push, push, push for faster action, and keeping the agency heads at arm's length. Miller developed social friendships with, and frequently became almost an apologist for the agencies and their top personnel. Teague was a restless traveller who sponged up information on frequent field trips, asked hard-nosed questions where necessary, told agency heads bluntly when he thought they were wrong, and was an indispensable ally when it came to working toward mutual goals. Fuqua is determined to work more closely with high officials like Dr. Frank Press, head of the Office of Science and Technology Policy in the White House. At the same time, he will not relax his constitutional responsibility to exercise vigorous oversight when necessary. To illustrate this point, when NASA revealed to the committee surprising cost over-runs in Fuqua's No. 1 favorite program, the Space Shuttle, Fuqua addressed these crisp words to NASA Administrator Dr. Robert A. Frosch on June 28, 1979:

> The timing of this subsequent budget amendment raises questions with regard to the accuracy and candor of testimony and response to questions at the February hearings. It is most difficult to understand how a problem of this magnitude developed between March and May.

One could hardly imagine Chairman Miller addressing that kind of a critical statement to any NASA official.

A NEW EXECUTIVE DIRECTOR

Col. Harold A. Gould was a natural to move up to become executive director of the committee when Fuqua ascended to the chairmanship at the beginning of 1979. In the early 1960's he had served two uniformed tours of duty with the committee on assignment from the Army Corps of Engineers (see pages 118–121), specializing in construction and budgeting matters. He moved up slowly but steadily in the committee hierarchy.

When Teague became chairman, in 1973, he wanted an astronaut as his executive director. But former astronaut Swigert, recognizing Colonel Gould's administrative ability, persuaded Teague to name him as his deputy director in mid-1975. Then when Swigert went off to make his unsuccessful race for the U.S. Senate in Colorado, Mosher became the third executive director in September 1977, with Colonel Gould remaining as deputy director. Teague's decision in 1977 indicated a faith in Colonel Gould's competence in administration, but also a desire to extend the concept that the top post required someone with closer ties with the scientific community.

MUTUAL TRUST

Colonel Gould started his first tour of duty with the committee the same year that Fuqua began his service in Congress and on the committee—in 1963. Because Fuqua's central interest and specialty on the committee had been NASA, over a period of 16 years he had developed a good working relationship with Colonel Gould. Both men could anticipate how they each reacted. It was not unexpected when Fuqua tapped Colonel Gould to move up to become the committee's fourth executive director early in 1979.

At the age of 61, an avid golfer, Colonel Gould believes in keeping the troops happy by having all channels open and insuring that harmony reigns within the chain of command. His transition from deputy director to his new post was easy and painless; he did not even change offices. Colonel Gould announced at a staff meeting that one of his first decisions was to abolish his own former job of deputy director.

GENERAL COUNSEL

After having served on the staff of the select committee which preceded the Science Committee, and also for many years as staff director of the Science, Research and Technology Subcommittee, Philip B. Yeager (see page 133) was promoted in 1979 to the post of general counsel of the full committee. Unlike the connotation of the title, the new office did not make Yeager the chief "legal officer" of the committee. Rather, his responsibilities included principal authority for following and perfecting legislation as it moved through the committee, and its relationship with executive agencies—especially the

Office of Science and Technology Policy and OMB—the Senate and public input. He was given broad supervision over the committee rules, jurisdictional relationships within the committee and with other House committees, and general inquiries which involved committee policy. He also coordinated relationships with the legislative service organizations—GAO, OTA, CBO and CRS.

When Yeager became general counsel in 1979, his new job was defined in much different form than the old position of deputy director which Colonel Gould had held from 1975 to 1979. The general counsel performed advisory rather than administrative duties, and by personal preference Yeager stayed clear of such issues as personnel and staff relationships. He was well-equipped to undertake new initiatives in the area of "foresight" for the committee, a function which had been somewhat overshadowed by the emphasis on "oversight." For example, in the summer of 1979, Yeager undertook a study of those subjects of emerging importance on which the committee might concentrate its efforts in the nonenergy field.

SUMMARY OF CHAPTERS

From the days in 1958 when the Select Committee on Astronautics and Space Exploration created the Committee on Science and Astronautics and also the National Aeronautics and Space Administration, the character and influence of the Science Committee have changed markedly. Starting out as a committee which was set up to respond to the launching of the Soviet Sputnik, at the beginning it was looked upon by some as a repository of far out concepts. The committee quickly established a reputation for breadth of vision and responsibility by moving into areas such as the relation of space to military weapons development, and the need to invest in advanced research and the education and training of scientific and engineering talent to enhance the future strength of the Nation.

Chapter I

In the first year of its existence, the committee in 1959 began to forge links with the scientific community through the Panel on Science and Technology. These were later materially strengthened under the Subcommittee on Science, Research and Development chaired by Congressman Daddario. For the first time in history, Congress afforded to science and technology an open door for consultation and mutual interchange of ideas. Despite the headline-hunting pressures of the space race with the Soviet Union, this relationship stimulated the consideration of future issues and mature advanced planning beyond the next budget.

Chapter II

When House Armed Services Committee Chairman Carl Vinson succeeded in spiriting out the second-ranked member of his committee, Overton Brooks, by maneuvering to have him named Chairman of the Science and Astronautics Committee, there were those who concluded that this doomed the new committee because of poor leadership. To be sure, Chairman Brooks made mistakes—he operated all over the lot, his adrenalin was too strong for his small and overworked staff, and he angered the subcommittee chairmen by centralizing power without delegation. Yet he firmly and fearlessly pressed forward to instill a sense of urgency into the space program, while encouraging broader relationships throughout the scientific arena. In getting the committee off to a fast start, Brooks was aided considerably by the presence of Majority Leader—and later Speaker—John W. McCormack, who had chaired the select committee in 1958. Former Speaker and Minority Leader Joseph W. Martin Jr. also helped smooth Republican support for the committee's efforts. Later, Carl Albert's service on the committee helped weld stronger leadership support as Albert advanced through the ranks to become Speaker of the House.

Once the committee had firmly established its authority to conduct annual authorizations for NASA, the next step was to insure that the intent of the Congress was carried out in the policies and programs of this rapidly expanding agency. In a larger sense, the committee repeatedly prodded NASA into speedier action on propulsion, training, spacecraft development, and the timetable on a manned lunar landing. At the same time, the committee pushed for a speedup in programs for communications, weather and navigations satellites. In addition to its role as an accelerator, the committee actively worked to protect NASA from the intrusions of the military attempting to invade NASA's jurisdiction. This included committee initiative to expand the Cape Canaveral launch area for NASA's future operations.

Chapter III

When George P. Miller became chairman after Brooks' death in 1961, the subcommittees were delegated expanded authority, and morale generally rose among the staff. But the size of the staff never exceeded a dozen professionals during the Miller regime and this handicapped the scope of investigations and oversight. Despite the agitation of the minority for staff representation, not until 1968 was the minority allowed to have one staff member and it was 1971 before the minority had its own unit including more than one staffer.

Chapter IV

Avuncular and at times irascible, Chairman Miller developed a closer rapport with the scientific community, the National Science Foundation, and the National Bureau of Standards. The committee took the initiative to cut down the size and eventually terminate the super booster, Nova, which had originally been designed to make direct manned ascent to the Moon. The committee recognized that Nova was no longer needed when the lunar orbit rendezvous was picked as the technique of manned lunar landing.

Chapter V

While Miller was building international relationships in the scientific field, Teague and his Manned Space Flight Subcommittee were crisscrossing the country, dropping in on aerospace contractors, NASA installations, and laboratories to check on contracts, expenditures, and timetables. With the full support and encouragement of the committee, Colonel Gould was on the road frequently to spur NASA to adopt stricter design criteria and construction standards. Meanwhile Daddario, along with his key staff man Yeager, was broadening the legislative use of scientific talent and advice through several formalized panels, the encouragement of the Science Policy Research Division of the Library of Congress, and generally throwing out the congressional welcome mat to science and technology throughout the world. The committee rendered powerful support for the expansion of education in both the natural and social sciences, primarily through the National Science Foundation. Dr. Philip Handler, President of the National Academy of Sciences, put it this way, in a letter to Teague on July 13, 1978:

> Not only did the creation of your committee provide a formal institutional arrangement for legislative promotion and oversight of science and technology, it also gave to the Nation's scientific and technological community a valuable forum at our national seat of government for interaction with the political process.

Chapter VI

In addition to his concentration on oversight of the NASA program, Teague placed heavy stress on educating the Congress and the public on the practical values of space. He encouraged a steady stream of congressional visitors to Cape Canaveral for manned space launches, pioneered the establishment of a visitor's center at the Cape, stressed the development of a more practical public affairs program for NASA, and repeatedly needled NASA to give more attention to the spinoffs or industrial and human applications of the space program. Fuqua and Frey also reiterated this point, especially as NASA's budget declined.

During the 1960's, although Chairman Miller remained an almost uncritical supporter of NASA, numerous committee members spoke their own minds and engaged in open internecine warfare on various issues. At the start of the space program, there were deep splits within the committee over such issues as solid versus liquid propellants, how much emphasis should be placed on the development of nuclear rockets—the Nerva program—and whether or not to build an Electronics Research Center. The propellant battle was won by the liquid advocates, the Nerva fight roared on into the early 1970's and only ended when its No. 1 tub thumper—Senator Anderson—left the Senate in 1972, and the opposition forces finally won their point in the closing of the Electronics Research Center after President Nixon took office. Karth, Wydler and a determined group of Republicans fought the building of ERC in Cambridge, Mass., lost the early battles to superior forces, but won the war when ERC was shut down and turned over to the Department of Transportation in 1970.

The committee basked in the glory of the string of successful Mercury and Gemini flights of the early and midsixties. When tragedy struck on the launching pad as astronauts Grissom, White, and Chaffee were killed in the 1967 fire, Teague helped rescue the program through the thorough, fair, and constructively searching investigation he chaired on the causes of the fire and steps which had to be taken to protect the future safety of the astronauts. Astronaut Frank Borman, a member of NASA's review board on the Apollo fire, reflected in 1978 on the committee probe:

The investigation was tough, impartial, and a positive factor in the ultimate success of the Apollo program.

Chapter VII

In the early 1960's, Karth and his subcommittee set a high standard for hardnosed oversight investigations of the management, scheduling and performance of such programs as the Centaur launch vehicle, Ranger and Surveyor lunar probes, and Advent military communications satellite. He also furnished leadership for NASA's planetary programs, and applications such as Earth resources satellites. Hechler's subcommittee helped focus NASA's attention on aeronautics, the need to build a reservoir of basic and advanced scientific research, and the necessity for training a younger crop of future scientific talent. As the 1960's progressed, Pelly, Wydler, and Goldwater pitched in to stress aeronautical research and development, and bring more sense and coordination into the multiagency programs in this field where the United States had once maintained world superiority. Hechler, and later Karth, also concentrated their subcommittees on the speedier development of weather, communications, navigation, and geodetic satellites and radio astronomy.

Chapter VIII

Future planning was a recurrent theme which the committee pushed in every area of its activity. The committee pressure for post-Apollo space planning began several years before the first Moon flight. A deep bipartisan split developed in 1970, the first year NASA sought funds for the Space Shuttle. The efforts of Karth and Mosher to slash funding for the Shuttle lost only by a tie vote that year, although both Members later swung around to support this new space transportation system. Throughout the 1970's, the committee took a particular interest in the expansion of Earth resources and educational satellites, and Fuqua exerted leadership in the areas of space industrialization, the future development of solar power from satellites, and the establishment of an operational Earth Resources Information System.

When Teague assumed the chairmanship in 1973, the committee took a new lease on life. As chairman of the Democratic Caucus and longtime Veterans' Affairs Committee chairman, Teague had built a unique reputation in the House which was well-described by John Walsh in the Science magazine of January 12, 1979:

> The House has its own hierarchy of values, and Teague's perceived virtues were cardinal ones—personal integrity, concern for the House as an institution, and fairness in exercising power. There was also respect for his toughness: the nickname "Tiger" has stuck with him since high school and gives some inkling of one dimension of his personality. And there was the direct, unassuming manner and the invisible ribbons on his chest.

Chapter IX

Steady progress was made in the development of the Space Shuttle to carry numerous payloads into space with a recoverable booster. In 1973, three groups of astronauts had successfully completed flights of 28, 59, and 84 days during which they performed valuable experiments on Earth resources, astronomy, in medical, environmental and other areas while visiting the orbiting laboratory named Skylab. But by 1979, unforeseen cost and scheduling problems surfaced with the Shuttle, delaying its first launch. The effect of sun spots caused Skylab to fall into the Earth's gravity sooner than expected, on July 11, 1979. The pieces of Skylab fell harmlessly to Earth, primarily in isolated areas of Australia.

The committee continued to encourage NASA applications, spin-offs, and technology utilization, and always funded more in these areas than NASA requested. In addition to the direct benefits through the use of weather, communications, and Earth resources satellites, the spinoffs included such items as the following:

—Solar cells to convert sunlight into electrical energy.
—Rechargeable electric pacemaker for heart patients.
—Voice-activated wheelchair for paralyzed patients.
—Domed fabric roofs, originally developed from fiber glass fabric used for astronauts' space suits.
—Satellite video transmission for medical and educational use.
—"Image enhancement" through computers to enhance photographs and old documents.
—"Intruder detectors" for use in homes and industrial plants, developed from highly sensitive seismic monitors used on the Moon.
—Microminiaturized transistors and electronic equipment.

Chapter X

The year 1974 was a watershed year for the Science Committee. For it was during that year that the Select Committee on Committees, headed by Representative Richard Bolling (Democrat of Missouri) recommended vastly expanded authority and responsibility for the Science Committee—principally through legislative jurisdiction over all energy research and development, plus oversight authority over all Federal R. & D. and broadened jurisdiction over civil aviation and environmental R. & D. There were also other refinements, such as abolition of the Joint Committee on Atomic Energy and transfer of its jurisdiction over nonmilitary nuclear R. & D. to the Science Committee, which did not go through until 1977. The Democratic caucus scuttled the Bolling recommendations and referred the entire package to a caucus committee headed by Representative Julia B. Hansen (Democrat of Washington). The Hansen committee, with the significant exception of the nuclear R. & D. jurisdiction, was equally generous to the Science Committee. It took months of careful staff work, numerous drafts of testimony and help solicited from outside witnesses who testified. These efforts were supplemented by many strategy sessions within the staff and committee. The presence of two friendly staff members on the Bolling committee—Dr. Charles S. Sheldon II and Robert C. Ketcham—certainly helped. But above all, Teague's tremendous prestige in the House, his position as chairman of the Democratic caucus, and his personal phone calls and conferences with Bolling, Mrs. Hansen, and other Members were worth their weight in gold.

When the Bolling committee's recommendation to abolish the Merchant Marine and Fisheries Committee went down the tube, it looked as though the Science Committee would lose its chance to obtain jurisdiction over oceanic and atmospheric sciences. But a clarifying colloquy between Hechler and Mrs. Hansen established the intent of the Hansen committee that this jurisdiction be shared. This gave the Science Committee the green light to proceed in that area, in collaboration with the Merchant Marine Committee.

In December 1974, Teague took the senior committee Democrats out for a boat ride one evening on the Potomac River. After drinks and dinner, Teague asked each member to express his preference on the jurisdiction of subcommittees. McCormack argued it made sense to centralize authority over energy R. & D. in one subcommittee, and Hechler argued for two. After extended discussion, a show of hands indicated, by a narrow margin, that two energy subcommittees were preferred. McCormack subsequently chose to head up a subcommittee which encompassed solar, geothermal, conservation, and advanced energy technologies, while Hechler chose coal, oil, and gas (fossil fuels). In 1977, Flowers exercised his seniority to choose fossil and nuclear R. & D. (By 1977, with the abolition of the Joint Committee on Atomic Energy, nuclear jurisdiction passed to the Science Committee.) McCormack then chose advanced energy technologies—about the same jurisdiction he had had the two prior years. In 1979, the energy subcommittees were rescrambled, with McCormack getting nuclear and geothermal, while Ottinger was given fossil, solar, and conservation.

The minority, after many years of agitation, finally got their first staff member, Richard E. Beeman in 1968. Not until 1971 did they have a "minority unit" including more than one designated staff member. By 1973, when Teague became chairman, the Republicans had only two professionals and one secretary to call their own.

Teague strongly resisted the concept of a minority staff. He frequently called attention to the fact that his executive director, Jack Swigert, had confessed to being a Republican, and Teague had warned Swigert he would fire him if he ever repeated that fact— because Teague stressed the staff should be nonpartisan and serve all members, Republicans and Democrats alike. The Republicans would not accept that argument, and continued to claw away, bolstered by strong encouragement from Republicans everywhere. The new rules of the House adopted in January 1975, authorized the ranking minority member of up to six subcommittees "to appoint one staff person who shall serve at the pleasure of the ranking minority party member." The House rules also authorized a total of six professional and four clerical personnel, as statutory members of the standing committee staff, to be assigned to the minority when requested by a majority of the minority members. The minority on the committee lost no time in attempting to implement these new rules and managed eventually to do so after some heated arguments over "qualifications." Paul Vander Myde became the first officially designated minority staff director in 1977.

In 1976, Teague established the Ad Hoc Subcommittee on Special Studies, Investigations and Oversight and hired Dr. Robert B. Dillaway as its staff director. Only one study was published, a report on EPA's Community Health and Environmental Surveillance System (CHESS). A large amount of the work on this study was actually done by the Brown Subcommittee on the Environment and the Atmosphere, with assistance from the Congressional Research Service of the Library of Congress. The reaction to Dr. Dillaway's work was mixed, there was a great deal of argument concerning the overlap of his work with the existing subcommittees, and he left the committee staff in 1977. Several unpublished studies on NASA's aeronautical R. & T. and NASA's energy R. & D. proved useful.

Teague also took the initiative to have the committee conduct research and help coordinate programs to aid the handicapped. Early in 1976, Sherman Roodzant was placed in charge of this special oversight program. Roodzant was assisted by John G. Clements, who succeeded him in 1979. A panel of experts studied the problems of the handicapped in 1977, followed by hearings chaired by Brown and conducted under the aegis of the Subcommittee on Science, Research and Technology. A second panel reported in 1978, and some of its recommendations were incorporated in legislation which the Congress passed in 1978, establishing a National Council on the Handicapped and a vastly increased research program for the handicapped, placed within the HEW Department.

Despite the failure of the Dillaway oversight operation in the 94th Congress, in 1979 the committee Democratic caucus voted to set up a new Subcommittee on Investigations and Oversight. Representative Jim Lloyd (Democrat of California) was elected chairman. In conjunction with the Science, Research and Technology Subcommittee, an investigative trip to Mexico examined the potential for the transfer of technology and energy resources between the United States and Mexico. The Lloyd subcommittee also held hearings on the aeronautical design of the DC–10, and those technical aspects which might require future modifications, in light of the worst tragedy in U.S. aviation history on May 25, 1979.

On July 19–20, the committee held two successful joint hearings with the Select Committee on Aging on applications of space technology for the elderly and handicapped.

Supplement to Summary of Chapters

Following are summaries of chapters not included in this abridged edition. These summaries essentially cover material not directly related to space or astronautics. Readers interested in this material should refer to Chapters X - XIV, XVI - XX in the original publication:

Toward the Endless Frontier, History of the Committee on Science and Technology, 1959-79, U.S. House of Representatives, U.S. Government Printing Office, Washington, D.C. 20402, 1980 (35-120 O)

Please note that the chapter numbers X - XIV and XVI - XX do not correspond to chapter numbers in the present volume but to the original unabridged edition.

Chapter X

Working through the United Nations and other international organizations, the committee gave strong support to the international aspects of science and space. This extended from encouraging agreements with other nations on satellite launchings, to the establishment of the worldwide tracking network, and exchange of scientific information with many nations to the 1967 Treaty on the Peaceful Uses of Outer Space. The committee sponsored the codification and publication of international space treaties and space law. In 1971, the committee formalized some of these activities through the establishment of the Subcommittee on International Cooperation in Science and Space, which lasted in one form or another until 1979, when its activities were re-incorporated primarily into the Science, Research and Technology Subcommittee. Chairman Miller and other committee members made countless journeys to speak at international scientific meetings and lend support to the scientific endeavors of other nations and groups of nations.

Teague, who vehemently opposed President Kennedy's 1963 advocacy of a joint Soviet-U.S. manned flight to the Moon, also had initial doubts about the rendezvous and docking of American and Russian spacecraft known as Apollo-Soyuz. He insisted that a sufficient number of experiments be placed on the Apollo flight to justify it in case the Soviets should back out at the last minute. The Apollo-Soyuz link up in 1975 proved to be successful in every respect.

In the late 1970's, the committee extended its interest in international areas, including oversight of the Law of the Sea Conference, comparative criminal justice, joint work with the House International Relations Committee on applying science and technology to foreign policy, and oversight on technology transfer to OPEC countries. Through many trips to other countries and attendance at international meetings, the committee members devoted an increasing amount of effort to fostering international scientific cooperation.

Fuqua exerted strong leadership toward the establishment of the Institute for Scientific and Technological Cooperation in 1979. The committee joined in the preliminary planning for the U.N. Conference on Science and Technology, which Fuqua and six other committee members attended in August 1979.

Chapter XI

From the early years of its establishment in 1959, the committee has consistently nudged the Congress and the Nation toward an eventual, voluntary conversion to the metric system. In 1961, the committee reported legislation to study the feasibility and problems involved in possible conversion to metric. Similar legislation was finally enacted in 1968. The Secretary of Commerce reported in 1971, recommending that the change be made "deliberately and carefully" through a coordinated national program with a specific target date for the U.S. to "become predominantly, though not exclusively, metric." Legislation to implement the report did not obtain the required two-thirds majority in 1974 needed for suspension of the rules, and ran into labor and right wing opposition. But a compromise bill passed in 1975, setting up a Presidentially appointed Metric Board, emphasizing any conversion to the metric system would be strictly voluntary, but would be coordinated through the Board and information distributed for schools and industries. By the time the 1975 law was signed, Trinidad, Tobago, Tonga, and the United States were the only nations in the world which have resisted the worldwide trend to adopt the metric system. Every Science Committee chairman, including Fuqua, has publicly endorsed the concept that the United States should voluntarily move toward eventual adoption of the metric system.

Chapter XII

In its review and authorization of the National Science Foundation, the committee fought to increase funding for basic research, high school summer institutes, and general support for higher education in the sciences. Mosher in particular called attention to developments such as the decline in science education funding from 36 percent to 10 percent between 1970 and 1974. The committee was torn by an emotional fight over the MACOS ("Man, a Course of Study") program which had been funded by NSF for use in anthropology courses for students in the fifth through seventh grades. Conlan charged that films and readings on the habits of the Netsilik Eskimos included "predominant emphasis on sex, pragmatic respect for life, shocking film segments displaying gore and immoral acts." Committee members reviewed the materials and films. Mosher, Symington, and Ottinger argued that the Federal Government should not interpose its judgment against the thousands of school boards throughout the country, which had the full and free right as elected bodies to accept or reject the materials. Teague, Wydler, and Mrs. Lloyd contended that it was shocking that the Federal Government should spend tax money on materials which were degrading. The House voted down by 218 to 196 Conlan's efforts to require Congressional committee review and approval of all MACOS materials before their release for use locally. A number of review committees were appointed, including a citizens committee which Teague appointed, headed by Dr. James M. Moudy, chancellor of Texas Christian University. The Moudy Report recommended that the MACOS materials be used only with care, with thorough training of the teachers who might use the materials, and also added:

From reports reaching us, we believe that the surest success of MACOS has come in those schools where ample preparations were made, including conferences with parents to show them in advance the MACOS materials, and to explain the purposes and methods of the course.

Starting in 1976, McCormack and Harkin had an annual clash over the "Science for Citizens Programs," inserted into the NSF authorization bill by Senator Kennedy to improve public understanding of science and technology issues. McCormack and many of the committee members looked on the measures as funding environmental extremists and intervenor groups to file delaying lawsuits, while Harkin viewed it as a natural extension of enlightening more people through the exercise of the democratic process. Generally, the result was to fund the program at a compromise level halfway between the House and Senate figures.

Until the end of the decade, the committee continued to stress the need for greater emphasis on science education by NSF, and increased the funding in this category. The perennial flap occurred over the "silly-sounding projects" nearly every time the NSF authorization bill was considered on the House floor. Teague cited the studies of the sex life of the fly which led to the elimination of the dreaded screw-worm which afflicted cattle. Harkin related that NSF grants to research "The Excretion of Urine in the Dog" and "The Excretion of Insulin in the Dogfish" led to the discovery of "vital information on the function of the human kidney and the relationship of hormones to kidney functions."

Among the major pieces of legislation originating in the committee, and passed exclusively through the committee's initiative, was the establishment of the Office of Technology Assessment. The first bill was cosponsored in 1969 by Daddario and Mosher, growing out of a phrase, popularized by Daddario, that Congress needed a "technological early warning system." The legislation set up a Technology Assessment Board with a staff to make studies and appraisals projecting the impact of technology in various fields. Although the committee-reported legislation, strongly backed by Miller, Mosher, Davis, and Symington, initially included a mixed Board of both congressional, executive agency, and public appointees, a floor amendment by Representative Jack Brooks (Democrat of Texas) made the Board an all-Congress affair of five House and five Senate Members. The President signed the OTA legislation on October 13, 1972. The OTA was only the third independent service organization created for Congress in the Nation's history—the first being the Library of Congress in 1800, and the second the General Accounting Office in 1921.

OTA Board members from the Science Committee included Teague (chairman, 1975–76), Mosher (vice chairman, 1973–74), Winn (vice chairman, 1977–78), Davis, Esch, Brown, and Wydler. The Subcommittee on Science, Research and Technology exercised fairly extensive oversight over OTA, and in 1977 and 1978 held hearings on its operation, noting some of its ongoing problems yet encouraging the continuance of its role in making technological evaluations and assessments for Congress.

Among other legislation enacted on the initiative of the Science, Research and Technology Subcommittee were the following laws:

> —Fire Research and Safety Act of 1968, adding new responsibilities to the National Bureau of Standards, including special training and demonstration programs in fire prevention, expanded by 1971 act to establish Fire Research and Safety Center. This was further supplemented by the 1974 legislation which set up a National Fire Prevention and Control Administration within the Department of Commerce, and authorized a U.S. Fire Academy. The legislation was further strengthened in 1976, and again in 1978.
>
> —Updating and strengthening the Standard Reference Data Legislation, originally passed in 1968, in legislation passed in 1972. This statute helped develop world-wide scientific and engineering standards for such elementary items as how much heat is given off when a substance is burned, how fast methane will react with air, or how soluble mercury is in water.
>
> —Earthquake Hazards Reduction Act of 1977, coordinating Federal research, prediction, and warning systems. Brown was the chief sponsor of this legislation.
>
> —Native Latex Commercialization and Economic Development Act of 1978. Popularly known as the "guayule bill," this legislation was also the result of Brown handiwork, authorizing the Agriculture and Commerce Departments, along with the NSF and Bureau of Indian Affairs to carry out research and development of the guayule plant as a possible source of natural rubber for commercial use.

Working with the National Bureau of Standards, the committee assisted in the development of voluntary industrial standards, helped draw up and write into law a new organic act for the National Bureau of Standards, and laid the groundwork for several searching studies of national materials policy. Led by Thornton, the committee also carried forward extensive oversight and public information hearings, and published useful reports on DNA and genetic engineering—popularly known as "gene-splicing." In 1977 and 1978, Thornton and Fuqua led hearings on the science policy questions and benefits which might be achieved by DNA research. In addition, inquiries were made in the effective utilization of Federal laboratories, fuller employment of scientists and engineers, the encouragement of science policy developments on the State and local level, river basin planning, water resources, agricultural research, world food problems, and Federal patent policies as they related to scientific and technological matters. In 1979, the House passed a materials policy R. & D. bill.

Chapter XIII

When President Nixon in 1973 abolished the scientific machinery in the White House, the committee started a long campaign to re-establish the presence of scientific advice at the highest levels of government. It was one of the major achievements of the committee, demonstrating congressional initiative at its finest. Under the bipartisan leadership of Teague and Mosher a thorough record was established through exhaustive hearings held in the three-year period from 1973 through 1975. With the assistance of a broad representation of the scientific community, including all of the former Presidential science advisers, the committee carefully proceeded toward drafting acceptable legislation which would give science and technology a strong voice in the White House structure. Watergate and President Nixon's resignation at first seemed to divert attention from the careful work the committee had accomplished, but the net effect was to speed up the timetable considerably. Mosher found Vice President Ford receptive, and both Teague and Mosher realized that once Ford became President he was a key factor in reestablishing the science machinery in the White House. In fact, Mosher wrote President Ford a personal letter the day after he was sworn in, urging him to give additional thought to the issue. There were still long months of negotiation ahead, particularly involving Vice President Rockefeller. The committee kept pressing toward perfecting a Teague-Mosher bill which was acceptable to the White House. Rockefeller provided a dramatic spectacular when he visited and endorsed the efforts of the committee in a public hearing on June 10, 1975.

As the Teague-Mosher bill finally evolved, it set up an Office of Science and Technology Policy with a Director who would also serve as the President's Science Adviser. These features survived in the progress of the legislation. A major breakthrough occurred when President Ford wrote Teague and Mosher on October 8 endorsing the legislation. The committee reported the bill unanimously, although Brown appended "additional views" to the committee report criticizing the lack of long-range planning language in the legislation. After long and difficult negotiations between the Senate and Science Committee staffs, the conference committee met to work out the final details. President Ford signed the legislation in the East Garden of the White House on May 11, 1976, before about 200 guests. There were some differences which Teague expressed to President Carter's Director of the Office of Science and Technology Policy over interpretation of the 1976 act. But when Fuqua became chairman he took the position that the new President and OSTP Director were entitled to devise the kind of operation they found most comfortable—so long as they did not, like President Nixon had, dismantle the machinery or violate its central goals.

Chapter XIV

Starting in 1971, the focus of the committee began to broaden with the establishment of the task force and later the Subcommittee on Energy under McCormack's leadership. Three landmark pieces of legislation were developed by the Energy Subcommittee and enacted in 1974: the Solar Heating and Cooling Act of 1974; the Solar Energy Research, Development, and Demonstration Act of 1974; and the Geothermal Energy Research, Development, and Demonstration Act of 1974. McCormack's influence added another dimension to the committee's jurisdiction as he coined the concept of "demonstration" to stretch the committee's activity beyond R. & D.

The task force and Subcommittee on Energy firmly established the reputation of the Science Committee in the energy field, and its expertise was recognized both in and out of the Congress. The successful work accomplished under McCormack's leadership had a direct relationship to the expansion of the committee's jurisdiction in the energy area in 1974.

Chapter XVI

Continuing the work which he had started in 1963 as chairman of the Subcommittee on Advanced Research and Technology, Hechler put increasing emphasis on basic research and an expansion of R. & D. work in aeronautics. With the strong support of Pelly, Wydler, and Goldwater, and the interested participation of Davis and Symington, Hechler's subcommittee succeeded in concentrating more of NASA's attention on safety, general aviation, and airport and airways congestion, short takeoff and landing planes, as well as the training of young aeronautical engineers. For several years, Hechler pressed for the upgrading of aeronautics in the NASA hierarchy, and also agitated for an Associate Administrator for Aeronautics and a separate office to handle General Aviation. In 1972, the name of NASA's Advanced Research and Technology Office was changed to the "Office of Aeronautics and Space Technology," opening the way for the Hechler subcommittee to change its name to the "Subcommittee on Aeronautics and Space Technology." Roy P. Jackson was then named NASA's Associate Administrator for Aeronautics and Space Technology. In 1973, NASA, following the insistence of the Hechler subcommittee, set up a separate General Aviation Technology Office.

Overflow crowds and high public interest attended the Hechler subcommittee hearings in 1972 reviewing the joint NASA–DOD study of "Civil Aviation Research and Development." The hearings and oversight which the subcommittee held in 1972 and 1973 laid the basis for later proposals by both the executive branch and the Congress for reducing aircraft noise levels by significant amounts through retrofitting the existing civil aviation fleet. The subcommittee also achieved improvement in communication and coordination among agencies engaged in the noise problem by summoning witnesses from NASA, FAA, EPA, and DOD around the table at one time.

In 1975, when Milford took over the subcommittee chairmanship, the jurisdiction was expanded to include all civil aviation R. & D. (part of which had been lodged in the Interstate and Foreign Commerce Committee) and ground transportation R. & D. Although the Milford subcommittee annually authorized FAA R. & D., the Senate would not go along with this initiative. The subcommittee became embroiled in an internal controversy over the validity of an FAA decision to adopt a microwave landing system which differed from the British "doppler system" and involved ultimately a decision by the International Civil Aviation Organization. The Milford subcommittee held constructive hearings on the future of aviation, the future needs and opportunities of the air traffic control system, and supersonic technology. In the fall of 1978, the subcommittee recommended legislation to provide a basic charter for the National Weather Service, pulling together bits and pieces of prior legislation going back over 100 years and consolidating the duties which had been previously authorized. The 1978 legislation passed the House but not the Senate.

Starting in 1974, Brown was joined by Symington and other members who sponsored automotive research legislation to devise more efficient auto engines, including a possible alternative to the internal combustion engine. McCormack took up the fight in 1976, and his subcommittee shepherded through a bill which President Ford vetoed, but was repassed in the next Congress and signed by President Carter in 1978. The legislation provided for a 5-year R. & D. program to develop a new and more efficient automobile engine. When Harkin took over the Transportation, Aviation and Communications Subcommittee in 1979, he put high priority on development of more fuel-efficient propulsion for automobiles and expansion of the use of electric vehicles. The Harkin subcommittee also held hearings on aviation collision avoidance.

Chapter XVII

In 1975, the Hechler subcommittee put increased funding into in situ low Btu coal gasification, coal mining research, health studies of the high incidence of nonpulmonary diseases among coal miners, and the environmental effects of coal mining. The subcommittee succeeded in line-iteming the numerous features of the fossil fuels R. & D. authorization, including funding for five coal liquefaction plants, a clean boiler fuel demonstration plant, and additional R. & D. in four different methods of converting coal to liquids for energy. The subcommittee also supported expansion of research in magnetohydrodynamics, oil shale, coal combustion, methods of recovery of oil and gas, methanol, gasohol, and fluidized-bed technology for coal.

A Senate plan for loan guarantees for synthetic fuels was embraced by the Ford administration, but later defeated in the House in both 1975 and 1976. Hechler contended that insufficient attention had been focused on the community impacts of the plan, and Ottinger led the opposition in 1976 on environmental grounds and the subsidies to large energy companies. Hechler suggested that the proper approach would be either to expand the existing Federal expenditures to bring the synthetic fuels plants to commercialization, or else to exercise a firmer control in the public interest such as in the World War II synthetic rubber program. The issue split the committee, and Teague, McCormack, Fuqua, and a majority of the committee favored the synthetic fuels loan guarantee proposals. The advocates contended that the opportunity should be seized when it presented itself, since time was of the essence in the energy crisis.

The supporters also stressed that any environmental, community impact or other problems could be identified and then corrected as a consequence of the demonstration process. Supported by the administration and encouraged by a topheavy 80–10 majority in the Senate, those advocating loan guarantees and other supports for synthetic fuels also received widespread industrial support and in articles and editorials. Strategically, however, a critical editorial in the Wall Street Journal proved very influential in mobilizing conservative support against loan guarantees. On December 11, 1975, the issues were debated in the House, which soundly defeated the loan guarantee provision by 243–140, and went on to vote down the leasing of public lands for oil shale demonstration, 283–117.

The following year the advocates of loan guarantees, price supports and construction grants for synthetic fuels regrouped their forces and extensive hearings were held in both the House and Senate. On September 23, 1976, the House, after one hour of debate on the resolution from the Committee on Rules to bring up the synthetic fuels bill, defeated the rule by 193–192.

In 1977, Flowers became chairman of a newly constituted subcommittee with jurisdiction over fossil and nuclear R. & D. Among the issues with which the Flowers subcommittee dealt were the role of the national laboratories in energy R. & D., clean air standards, support for magnetohydrodynamics, and management of the MHD program, and construction oversight on coal liquefaction (at Cresap, W. Va., and Catlettsburg, Ky.)—at both of which remedial management action was recommended and carried out. The Flowers subcommittee also recommended the building of a second solvent refined coal (SRC) liquefaction plant, a recommendation which was also made in 1979 when Ottinger took over the subcommittee. The Ottinger subcommittee generally supported the DOE effort in 1979 to reorganize the coal mining R. & D. program, and reorient it toward meeting productivity and environmental standards. In 1979, the subcommittee added funds for anthracite mining, fuel cells, combustion systems, heat engines and heat recovery.

Chapter XVIII

A no-win controversy developed between the committee and President Carter over building the $2.6 billion Clinch River Breeder Reactor (CRBR). A majority of the committee strongly favored the CRBR, which had been supported by both Presidents Nixon and Ford but opposed by President Carter. President Carter's early position was that the production of plutonium would lead to nuclear weapons proliferation, a fact disputed by McCormack, who contended that it was cheaper and easier to make weapons outside of the fuel cycle. A small group of antinuclear Democrats, led by Ottinger, contended that the CRBR was outdated. Wydler and Mrs. Lloyd also helped lead the fight to build the CRBR.

Teague led an eight-member delegation to Europe in May and June 1977, where they inspected France's operating breeder reactor, and visited, among other places, the International Atomic Energy Agency in Vienna, Austria.

In 1977, Brown attempted to amend the ERDA authorization bill to reduce the $150 million of funding for the CRBR to the level recommended by President Carter: $33 million. Brown's amendment went down by 246 to 162 in the House. Committee Democrats voted 15 to 11 against the amendment and for full funding for the CRBR. Fish and Pursell were the only committee Republicans (out of 13) who supported the Brown amendment.

Although the President vetoed the ERDA authorization bill in 1977, it did not kill the CRBR, because Congress put funds into a supplemental appropriations bill and then arranged for the funding bill to become effective without a specific authorization. Included in the same supplemental bill were many other items the President wanted, such as the final death blow to the B–1 Bomber, so he was pretty much forced to sign the bill.

In 1978, Teague came up with a compromise negotiated with Secretary Schlesinger, which involved delay in the construction of the CRBR pending a study. McCormack, Mrs. Lloyd, and strong supporters of the CRBR rejected the compromise, which Flowers sponsored. Teague arranged for the President to meet with nuclear industry representatives at the White House, but it failed to shake their position and the Flowers compromise was defeated on June 14, 1978, by a vote of 187 to 142. Once again, in 1979, efforts to resolve the issue failed. The committee stuck to its position in support of the CRBR. The House defeated a Fuqua-Brown compromise effort on July 26, 1979, by a vote of 237–182.

Chapter XIX

In 1975, McCormack became chairman of the Energy Subcommittee which handled solar, geothermal, conservation, and advanced energy technologies. Hechler, Ottinger, Hayes, and Dodd supplemented McCormack's efforts to make increases over the President's budget, and the committee voted boosts of over 100 percent in solar, geothermal, and conservation R. & D. Although the committee increase for solar energy was from $70.3 million to $143.5 million, an amendment on the House floor by Representative Frederick W. Richmond (Democrat of New York) added $50 million on top of that. In a March 20, 1975, letter to Dr. Seamans, McCormack called for more aggressive management than "the low key, academic management style that was characteristic of the NSF."

Brown and McCormack, in 1975, introduced the electric vehicle research, development and demonstration bill to enable 8,000–10,000 electric vehicles to be demonstrated by Government, industry, and individuals throughout the Nation. Teague, Ottinger, Mosher, and Goldwater joined in supporting the bill. President Ford vetoed the bill, but Congress voted to override the veto.

Once again in 1976 and 1977, the committee voted hefty increases for solar, geothermal, and conservation, and on both occasions the House decided to make further increases in solar energy R. & D.

Another committee initiative was the enactment of the Energy Extension Service legislation, the brainchild of Thornton, who patterned the statute after the successful Agricultural Extension Service. The Energy Extension Service helped answer questions and give advice to individuals, businesses, and State and local government officials on energy conservation measures and alternate energy systems. President Carter signed the Thornton bill into law on June 3, 1977.

The committee, led by McCormack and Goldwater, sponsored and piloted through to enactment the Solar Photovoltaic Energy Research, Development, and Demonstration Act of 1978. The bill provided for a 10-year program, doubling the total production of photovoltaic systems each year. In 1978, the committee added $101.5 million to the President's budget request of $291.8 million for solar R. & D. operating expenses, and also voted add-ons for geothermal and conservation.

Beginning in 1973, McCormack's Subcommittee on Energy had teamed up with the Space Science and Applications Subcommittee for a joint hearing on the possibilities for solar satellite power. In subsequent years, Fuqua's Space Science and Applications Subcommittee had carried the ball, usually in conjunction with one of the energy subcommittees. A solar satellite power system bill passed the House in 1978, but failed in the Senate, and the committee voted out a new SSPS bill in 1979.

In 1979, Fuqua endorsed the President's goal to generate 20 percent of the Nation's energy by solar means by the year 2000. The committee continued to give strong support to solar and conservation initiatives in 1979. The House passed a wind energy R. & D. bill in December 1979.

Chapter XX

From 1975 through 1978, Brown headed the Subcommittee on Environment and the Atmosphere. One of the first steps the subcommittee took was to help beef up the environment and safety programs of ERDA, after which the subcommittee went to work on their first authorization bill for the Environmental Protection Agency (EPA). The subcommittee adopted an amendment by Hayes requiring a 5-year research plan, which proved to be an excellent management tool. The Brown subcommittee conducted a wide-ranging series of oversight hearings in such areas as depletion of the ozone layer through use of aerosols, sulfates in the atmosphere, waste disposal polluting the oceans, chronic exposure to low-level pollutants, and environmental research centers.

The subcommittee's first enactment was the Resource Recovery Act, which was incorporated into legislation being developed by the House Interstate and Commerce Committee. This legislation included the R. & D. portions developed in the Brown subcommittee relating to the use of waste to generate energy. Next, the subcommittee turned to the National Weather Modification Policy Act of 1976, which authorized the Secretary of Commerce to pull together data on scientific knowledge and technological developments concerning weather modification. President Ford signed the bill into law on October 13, 1976.

In 1977, James W. "Skip" Spensley transferred from the Merchant Marine and Fisheries Committee to become Brown's staff director. A vast amount of the subcommittee's work and achievements were in the environmental and safety features of coal burning and the safety and nuclear waste features of the development of nuclear energy. Among other legislation sponsored by the subcommittee which was enacted into law were the following statutes:

Marine Protection, Research and Sanctuaries Act Reauthorization

This bill, signed in 1977 and developed in conjunction with the Merchant Marine Committee, mandated the end of all ocean dumping of sewage sludge not later than December 31, 1981. A special environmental research program in the oceans was developed as part of the legislation by the Brown Subcommittee.

National Climate Program Act

This legislation, principally sponsored by Brown and signed by the President in 1978, designated the Commerce Department as the lead agency to coordinate Federal plans and research in climate analysis, information and forecasting.

Marine Pollution Environmental Research, Development and Monitoring Act

This legislation, approved by the President in 1978, set priorities for research by the National Oceanic and Atmospheric Administration in areas of ocean pollution.

Antarctic Conservation Act of 1978

This bill, signed by the President in 1978 provided for the conservation and protection of the animals and plants in Antarctica and of the ecosystems on which they depend.

The Brown subcommittee also held a wide number of foresight and oversight hearings and made reports on such issues as ground water quality research, health and safety implications of the President's national energy plan, nuclear waste management, environmental research reserve networks, oil spill recovery technology R. & D., health effects of ionizing radiation, environmental monitoring, and special urban air pollution problems. In conjunction with Representative Paul G. Rogers (Democrat of Florida), a conference was held on August 15, 1978, on the environment and health care costs.

Ambro took over as chairman of the subcommittee at the beginning of 1979, at which time it was renamed the "Subcommittee on Natural Resources and Environment." At that time, the jurisdiction of the subcommittee was expanded to include materials policy, and also all aspects of weather—except aviation-weather services.

*   *   *   *   *   *   *

On July 20, 1979, appropriate ceremonies were held to commemorate the 10th anniversary of the first small step taken by Neil A. Armstrong when he set foot on the Moon. On the same day, the committee indicated that its sights were focused on the long future rather than the past. In collaboration with its opposite number in the Senate, the committee had just completed a symposium which, instead of glorifying the past, was devoted to "Next Steps for Mankind—the Future in Space." Preparations were being made to enact legislation for space industrialization and the development of solar power from satellites. The way was being paved for the first Space Shuttle flight. The applications of space technology for the elderly and handicapped were being examined.

A new burst of activity in the two energy subcommittees was concentrated on both old and new sources of energy—solar, geothermal, conservation, biomass, synthetic fuels, and overcoming the problems of low level ionizing and nonionizing radiation and nuclear waste disposal while pointing toward the future uses of nuclear power, including the exciting possibilities of nuclear fusion. The committee was deeply involved in planning for the United Nations Conference on Science and Technology, as well as focusing on legislation to stimulate innovation in science and technology, the role of Federal laboratories in transferring technology to State and local governments, nutrition, decisionmaking in such areas as approving new drugs, and the complex area of risk/benefit analysis and R. & D. policy.

There was no slowdown in the committee's efforts to stress the development of a fuel-efficient and safer automobile, and to push forward the frontiers of research in aeronautics and aviation. The oceans, the climate, the atmosphere, and the total environment occupied the continued attention of the committee as it pointed its sights toward the future.

Even as this is written, the winds of change are still blowing. In the 20 years since 1959, the committee had vastly broadened its horizons. The early concentration on space now constituted only one portion of the committee's mission which in 1979 encompassed energy, transportation, natural resources and the environment, and the use of science and technology toward the solution of present and future problems faced by human beings on Earth. The committee's long string of legislative and other achievements affecting public policy, detailed in these chapters, should not obscure the primary focus on the future of mankind. This was the central concern of the committee as it provided the leadership in humanity's inexorable progress toward the endless frontier.

Photo Credits

The bulk of the illustrations are from NASA, the White House, Department of Energy, Navy Department, National Science Foundation, National Bureau of Standards, and the Democratic and Republican photographers at the U.S. Capitol.

Specific credits are for other photographs which appear on the following pages:

Page	*Credit*
27	Swann Studio.
61	Department of Archives and Manuscripts, Louisiana State University.
62	McDonnell Aircraft Corp.
315	David C. Greenwald.

Publishers' Credits

The White House Years: Waging Peace 1956–1961 by Dwight D. Eisenhower. Copyright 1965 by Dwight D. Eisenhower, Reprinted by permission of Doubleday & Co., Inc.

Sputnik, Scientists, and Eisenhower by James R. Killian, Jr. Copyright 1977 by The Massachusetts Institute of Technology. Reprinted by permission of The MIT Press.

The Stately Game by James Symington. Copyright 1971 by James Symington. Reprinted by permission of Macmillan Publishing Co., Inc.

We Propose: A Modern Congress, edited by Mary McInnis. Copyright 1966 by McGraw-Hill, Inc. Reprinted by permission of McGraw-Hill Book Co.

Firsthand Report by Sherman Adams. Copyright 1961 by Sherman Adams. Reprinted by permission of Harper & Row.

Appendix

Committee Chairmen

Name	Official dates of service
Overton Brooks	Jan. 7, 1959–Sept. 16, 1961.
George P. Miller	Sept. 21, 1961–Jan. 3, 1973.
Olin E. Teague	Jan. 24, 1973–Jan. 3, 1979.
Don Fuqua	Jan. 24, 1979–

Ranking Republican Members

Name	Official dates of service
Joseph W. Martin, Jr	Jan. 7, 1959–Jan. 3, 1967.
James G. Fulton	Jan. 26, 1967–Oct. 6, 1971.
Charles A. Mosher	Oct. 26, 1971–Jan. 3, 1977.
John W. Wydler	Jan. 19, 1977–

Longevity Record of Committee Members Serving Over 10 Years
(As of End of 1979)

Name	Years served
Olin E. Teague	20
Ken Hechler	18
Don Fuqua	17
John W. Wydler	17
Charles A. Mosher	16
Alphonzo Bell	16
Thomas N. Downing	15
George P. Miller	14
John W. Davis	14
George E. Brown, Jr	13
Joseph E. Karth	13
James G. Fulton	13
Larry Winn, Jr	13
Thomas M. Pelly	12
Emilio Q. Daddario	12
Barry M. Goldwater, Jr	11
Louis Frey, Jr	10
Richard L. Roudebush	10
J. Edward Roush	10

Source Notes

Key to quotations and citations:
CF—Correspondence or memoranda in committee files
CR—Congressional Record
Hg—Published committee hearing
Int—Interview (copy in committee historical files)
Rpt—Published committee report
U—Unpublished committee transcript in committee files

Quotation or citation

Page

1 Rayburn, McCormack, Maurer, CR, 3–5–58
2 Select committee charter, CR, 3–5–58
3 Ducander, Int.
6 Natcher quote, Natcher Diary
7 Fulton, Hg, 4–15–58
8 Brooks, CR, 3–5–58
9 Ford, Int; McCormack, CF
10 Brooks and McCormack, U
11 Dryden, Hg, 4–16–59, 4–22–59; McCormack, Brooks and Fulton, *Ibid.*
13 Boushey, Hg, 4–23–58
13 Natcher quote, Int
14 Senate committee jurisdiction, CR, 7–24–58; House committee jurisdiction, CR 7–21–58; McCormack and Albert quotes, Int
15 Rayburn and Albert quotes, Int
16 Quotation from committee report, Rpt, 5–24–58; McCormack quote, CR, 6–2–58
17 Johnson quote, CR, 6–16–58; Yeager quote, Int
18 O'Neill Report on H. Res. 580, Rpt, 5–29–58; Bolling quote, CR, 7–21–58
19 McCormack and Martin quotes, CR, 7–21–58
20 Feldman, Int
21 Arends quote, CR, 6–2–58; quote from H. Rept. 1770, Rpt, 5–24–58
22 Feldman and Senator Johnson quotes from Feldman Int
23 McCormack quotes, CR, 7–16–58
24 Ford, Keating and McDonough quotes, CR, 8–20–58; Ford quote, Ford Int
25 Sisk, Ford, Judd, Thomas quotes, CR, 8–20–58; Sheldon, Int
26 Ford, Keating, Cannon quotes, CR, 8–21–58
31 Ducander, Int
32 Brooks quote, CF; Teague quote, Int
33 Letters from Feldman and select committee staff to McCormack and all members of select committee, 7–21–58, CF
33–4 Feldman, Int
34 Ducander, Int
35–7 Quotation from Aviation Week and Space Technology, 2–1–60; Brooks response to Hotz, *Ibid.*, 2–22–60
38 Brooks comments, U, 1–20–59
39 Ducander, Int; Brooks, press release, 1–31–59; *Ibid.*, Hg, 2–2–59
40 Brooks and Fulton, Hg, 2–2–59
41 Quotations from "The Next Ten Years in Space, 1959–1969", Rpt
42–3 Brooks letter to Vinson, 5–9–59, CF; Vinson's response, 5–11–59, CF; Further Brooks-Vinson correspondence, CF
44 Fulton quote, Hg, 2–27–59
45 Brooks quote, Hg, 8–25–59; Brooks letter to Miller, 9–1–59, CF; Ducander memorandum, 4–18–60, CF

Quotation or citation

Page

111   Teague and von Braun quotes, Hg, 3–18–63; Daddario, Holmes and Shea quotes, Hg, 5–8–63;
        Fulton quote, Hg, 2–20–64

112   Teague and Gray quotes, Hg, 2–20–64; Teague, Int

113   Teague quote, Hg, 3–29–62; Teague, Int

114   Teague quote, 4–16–62, U; Petrone quote, Hg, 7–24–62

115   Teague, Int; H. Rept. No. 1959, 7–2–62, Centaur quote

116   Karth quotes, 6–27–62, U; Hg, 5–15–62

117   Karth quotes, Hg, 3–8–63

118   Teague quote, 5–10–62, U; quote from conference report, H. Rept. No. 2038, 7–26–62;
        Ducander, Int; Gould, Int

120–1 Gould, Int

124   Gross quote, CR, 5–23–62

127   Albert, Int

129   Miller quote, 2–5–63, U

130   Galloway, Int

131   Miller quote, CR, 9–11–63; Daddario, Int

132   Miller and Daddario quotes, Yeager, Int; Daddario, Int

134   Kistiakowsky quote in letter to Daddario, 9–11–63, reprinted in "Government and Science" Hg,
        p. 421; Stever letter, *Ibid.*, p. 425

135   DuBridge letter, *Ibid.*, p. 455; Revelle letter, *Ibid.*, p. 427; Urey letter, *Ibid.*, p. 426; Seitz quote,
        Hg, 10–15–63

136–7 Daddario quote, Hg, 11–5–63; Miller quote, *Ibid.*, von Braun and Revelle quotes, *Ibid.*

142   Miller quote, Hg, 1–26–65; McCormack quote, *Ibid.*

144   Letter, Wirths to Teague, 6–5–78, CF; Haworth to Teague, 6–30–78, CF

145   Mosher and Daddario quotes, CR, 4–12–67

147–9 Humphrey quote, Hg, 1–25–66

150   Letter, Brown to Miller, 9–23–69, CF

151   Daddario, letter of transmittal, 6–7–67, Rpt; Haworth to Teague, 6–30–78, CF

155   Albert, Int

156   Teague, Int

157   Teague quote, Hg, 2–3–59; Teague, Int

158   Ford quote, remarks on 2–27–75, CF

160–1 Petrone quote, Int; Jensen and Teague quotes, CR, 5–23–62

162   Abelson quote, Science magazine, 4–19–63

163   President Eisenhower to Halleck, letter reprinted in CR, 4–2–63; The New York Times, 6–13–63

163–4 "Additional Views", minority, in H. Rept. 591, pages 201–5, 7–25–63

164   Teague quote, CR, 8–1–63

165   Miller, Martin and Teague quotes, *Ibid.*

166   Pelly and Fuqua quotes, *Ibid.*; Petrone quote, Int

167   Teague to Kennedy, 9–23–63, CF; Teague to O'Brien, 9–27–63, CF

168   Bundy to Teague, 10–4–63, CF

169   Teague to Webb, 1–10–64, CF

170   Subcommittee on Manned Space Flight, Report, 3–11–64, U; Teague quote, Hg, 2–18–64

172   Teague telegram to Dr. Milton Eisenhower, 5–28–64; Eisenhower to Teague, 6–2–64, CF

173   Teague to Adm. Burke et al., 6–10–64, CF; Teague quote, CR, 5–2–68

175–6 Ducander, Int

176–7 Miller quote, CR, 2–16–65

177   Roush quote, 4–29–65, U

178   Glennan, Int

179   Webb to Johnson, 5–23–61, copy in NASA Historical Files

180   Miller quote, 3–17–64, U

181   Quote from conference report, H. Rept. 514, 6–15–65, page 7

182   Fuqua quote, CR, 5–6–65; Teague quote, Hg, 3–10–66

183   "Future National Space Objectives," Rpt, committee print, 7–25–66; Webb to Teague, 12–1–66, CF

184   Notes on 1–20–66 NASA meeting with Subcommittee on Manned Space Flight, copy in NASA
        Historical Files

185   Teague quote, CR, 5–3–66; Lt. Col. Edward H. White to Teague, 8–17–66, and Teague to White,
        8–26–66, in Teague personal correspondence file; Teague observation to Silverstein, Hg, 2–18–60

187–8 Miller, Teague, Albert, Pelly, Roush, Fuqua quotes, CR, 1–30–67

188   Teague letter to members of Subcommittee on NASA Oversight, CF, 3–22–67; Winn and Webb
        quotes, Hg, 4–10–67

Quotation or citation

Quotation or citation

Page

252 Quotation from "The National Space Program—Present and Future", committee print, Rpt, Dec. 10, 1970

253 Letters to Teague and Miller, cited, CF; Fletcher interview in Washington Sunday Star, 2-28-71

254 Teague and Fulton telegram to Fletcher, 3-1-71, CF; Casey quote, CR, 6-3-71

257 Mosher quote, 3-30-71, U; Wydler, Karth, Teague quotes, 3-30-71, U

258 Wydler quote, *Ibid.*; Fuqua-Frey statement, H. Rept. No. 92-143, 4-1-71

259 Fletcher, Int

260-1 Fuqua and Frey quotes, Hg, 2-8-72

261 Teague quote, Hg, 2-17-72

262 Teague, Abzug quotes, Hg, 3-14-72

263 Abzug, Winn, Wydler, Fuqua quotes, *Ibid.*

264 Teague quote, *Ibid.*

265 Mosher, Wydler, Boggs quotes, CR, 4-20-72

266 Miller to Fletcher, 1-28-72, CF

269 Abzug quote, CR, 5-23-73; Teague quote, *Ibid.*

270 Fuqua and Myers quotes, Hg, 2-19-74; Mosher quote, CR, 6-12-74; Winn quote, CR, 4-9-75

271 Abzug, Badillo, *Ibid.*; Fuqua, Winn quotes, CR, 3-22-76

272 Frey quote, *Ibid.*; Stafford quote, Hg, 1-26-78; Lloyd quote, *Ibid.*

273 Winn quote, Hg, 2-23-78

275 Fuqua and Winn quotes, CR, 3-28-79

276 Winn, Fuqua, Frosch, Yardley quotes, Hg, 6-28-79

277 Frosch quote, *Ibid.*; Fuqua and Winn quotes, statements on release of report of Subcommittee on Space Science and Applications, 8-30-79

279 Roudebush quote, CR, 4-23-70

280 Teague to Nixon, 11-5-69, CF; Bell, Goldwater and Fulton quotes contained in H. Rept. 92-143, 4-1-71

281 Quote from S. Rept. 92-146, 6-8-71; Myers quote, 2-17-72

283 Esch quote, Hg, 7-17-73; Fuqua quote, Hg, 8-1-73

284 Fuqua quote, *Ibid.*; Bell and Fuqua quotes, CR, 6-12-74; Fuqua letter to Lindsey, 7-23-79, CF

285 Fulton and Wydler quotes, 3-5-70, U

286 Karth and Fulton quotes, *Ibid.*

287 Miller quote, Hg, 3-2-71; Miller address in Rome, 5-3-71, CF; Teague quote, CR, 6-3-71

288 Transcript of executive session, 3-16-72, U

290 Miller to Fletcher, 12-26-72, CF

292 Teague statement, 4-24-73, CF; Teague, Int

293 Karth quote, CR, 4-23-70

294 Quotes from Karth address to American Institute of Aeronautics and Astronautics, 4-1-71, CF

295 Karth quote, CR, 6-3-71; Downing to Miller, 12-10-71, CF

297-8 Downing subcommittee report, 3-23-72, U; Downing quote on funding space applications, CR, 4-20-72

300-1 Symington quote, Hg, 1-27-72; Symington quote from first subcommittee hearing, Hg, 3-1-73; Bergland, *Ibid.*

302-3 Esch, Symington, Camp, Mathews quotes, 9-24 to 10-4-73, U

303 Symington quote, opening Viking oversight hearings, Hg, 11-21-74

304 Symington, Int; Winn quote, Hg, 11-21-74

307 Winn quote, 3-5-75, U

308 Quote from report of Subcommittee on Space Science and Applications, 3-11-75, U; quote from conference report, H. Rept. 94-259, 6-4-75

309 Quotes from Rpt "Future Space Programs," 9-75

311 Fuqua quote, Hg, 6-21-77; Fuqua to Press, 11-29-77; Press to Fuqua, 12-6-77, CF

312 Fuqua quotes, Hg, 5-2, 5-3-79; Brown quote, 5-3-79; Frosch quote, Hg, 1-26-78

313 Wydler, Press, Winn, Fuqua quotes, Hg, 1-26-78

313 Fuqua quote, CR, 4-25-78

313 Fuqua to Press, 10-20-78, CF

315 Fuqua, Wydler quotes, Hg, 2-14-79

316 Fuqua quote, Hg, 5-23-79

317 Rangel quote, CR, 4-20-72

318 Bell, Pelly, Rangel, Teague, Miller, Dellums quotes, *Ibid.*

319 Teague, Hechler quotes, CR, 5-23-73

320 Teague quote, 4-17-73, U; Hechler, Downing quotes, *Ibid.*

321 Milford, Winn, Hechler, Teague quotes, *Ibid.*

322 Hechler to Fletcher, 6-18-73, CF; Fletcher to Hechler, 6-29-73, CF; Hechler to Teague, 7-31-73, CF

Quotation or citation

Page

326 Report of Monroney-Madden Joint Committee on the Organization of Congress, 1966

328 Symington, Int.; McElroy, Int

329 Karth, Int

330-2 Davis, Mosher, and Martin quotes, Hearings before House Select Committee on Committees, 5–11–73; Bolling quote, *Ibid.*, 6–8–73

333 Brown quote, House Select Committee on Committees, 9–13–73

334-5 Mosher quote, *Ibid.*, 5–11–73

337 Swigert memo to all staff members, 2–8–74, CF

339 Hansen, Int

341 Swigert summary of Hansen and Bolling reforms, with charts, 9–26–74, CF

342 Teague quote, CR, 9–30–74

344-7 No documentary record has been kept on the boat trip, and this account is drawn from a wide number of interviews with members and staff.

348 Bingham statement, record of Democratic caucus decision, 12–8–76, copy in CF; Memorandum of understanding on nuclear R. & D. jurisdiction, CR, 1–14–77; McCormack-Wright exchange, *Ibid.*

349-50 Teague, Int; Teague-Price exchange, CR, 9–13–77; Teague to Price, 5–9–78

351 Brown quotes, H. Rept. No. 95–1166, Part 3, page 79

353-4 Swigert memo, 7–18–73, CF; Mosher to Teague and Swigert, 7–18–73, CF

355-6 Teague-Mosher exchange, 1–23–75, U

358 Mosher quote, Int; Wydler testimony before House Administration Committee, 3–1–78, CF

359 Teague to Thompson, 4–19–78, CF

360-1 Miller, Fulton, Karth, Hechler, Teague, Wydler, Frey quotes on subcommittee chairman's power to appoint a staff member, 2–23–71, U

362 Ducander to Miller, 2–24–71, CF

364 Swigert memo to Teague on "Special Investigations and Oversight Task Team," 10–9–75, CF

365 Teague instructions on how Dr. Dillaway should operate, from notes by Colonel Gould, CF; Teague memo to all committee members on methods of operation for the investigations subcommittee, 2–3–76, CF

366 Gould to Swigert memo, 5–25–76, CF; Mosher quote, Hg, 8–10–76; Goldwater quote, *Ibid.*

367 Teague, Int; memorandum to House Government Operations Committee on decision to disband the Subcommittee on Special Studies, Investigations, and Oversight, 2–18–77, CF

368 Choice of subcommittee chairmanships and subcommittee assignments are detailed in record of committee Democratic caucus, 2–1–79, U

373 Roodzant testimony before committee, 8–10–76, U

374 Teague statement at news conference, 4–7–77, CF

376 Teague statement on violation of Federal laws covering the handicapped, 2–24–78, CF

376-7 Teague to Carter, 3–22–78, CF

377-8 Watkins, et al., letter to Fuqua on space-age technology to aid the elderly and handicapped, 2–13–79, CF

378 Fuqua to Frosch, 5–15–79, CF

381 Fuqua, Int; Fuqua, on Department of Energy authorization, CR, 7–26–79

383 Fuqua, Int

384 Fuqua quote in remarks to Frosch, Hg, 6–28–79

Selected Bibliography

1. DOCUMENTARY SOURCES

The voluminous files of the committee and its subcommittees constitute the major source to detail and interpret the events covered in this volume. In addition, each individual member maintained personal correspondence files, many of which were removed when members left the committee or left office. For example, some papers of the late Representative Overton Brooks, the first chairman of the committee, are deposited at Louisiana State University, Baton Rouge, La.

The papers of Representative George P. Miller, the second chairman of the committee, have been offered to St. Mary's College, Moraga, Calif. Texas A. & M. University is receiving the papers of the third chairman of the committee, Representative Olin E. Teague of Texas.

Since 1975, when the jurisdiction of the committee was greatly expanded, the most useful file is entitled the "Chairman's Reading File," which is maintained for the full committee and each of the subcommittees. The committee minutes are useful to record basic decisions, but include little of the debate which led up to them or their significance.

The unpublished stenographic transcripts of committee meetings and caucuses are very useful, especially for markup sessions. During the 1970's, these markup sessions, although public, were not published as a general rule. A great deal of revealing material is included in the unpublished transcripts of executive sessions which the committee frequently held during the 1960's.

As noted in the "Acknowledgments" (pages XXI–XXVI), a large number of personal interviews have been recorded, and these have both enlivened the history with anecdotal material and assisted in background interpretations of events.

2. COMMITTEE PUBLICATIONS

Important sources for the official actions of the committee are the printed hearings and reports. The reports include committee prints and House documents, as well as legislative reports. The latter cover analyses of each bill as it passes the committee prior to consideration by the House, plus the official decisions of conference committees, and the text of public laws which originate in the committee.

3. OTHER GOVERNMENT PUBLICATIONS

The Congressional Record is the central source for the stenographic record of the debate by the House of Representatives and the U.S. Senate, in addition to a potpourri of commentary, articles and editorials and other interpretive data supplied by Members each day the Congress is in session. Sometimes the key to unlock the secret of why certain actions were taken is contained in miscellaneous material inserted into either the body of the Record or Extensions of Remarks (formerly termed "Appendix").

The annual reports of the agencies with which this history deals, the official budget documents produced annually, the reports of the General Accounting Office, Office of Technology Assessment and Congressional Research Service are vital sources for an understanding of what happened and why. In recent years, the publications of the Congressional Budget Office have become increasingly important. Needless to say, the hearings and reports of the House and Senate committees, dealing with subject matter closely related to the work of the Science Committee, cannot be overlooked. This is particularly true of the actions of the House and Senate appropriations committees, and their conference reports, dealing with authorization legislation initially voted by the Science Committee and also dealing with related legislation.

Presidential statements, messages, texts of news conferences, addresses, and other official actions are contained in the "Weekly Compilation of Presidential Documents," and annually published by the National Archives and Records Service in "Public Papers of the Presidents of the United States." The "Federal Register" is the official source for executive orders, official regulations and other announcements.

4. DOCTORAL DISSERTATIONS

Two doctoral dissertations were particularly helpful in compiling this history: James R. Kerr, "Congressmen as Overseers: Surveillance of the Space Program," Stanford University, 1963; and Thomas P. Jahnige, "Congress and Space," Claremont Graduate School, 1965.

5. ARTICLES AND JOURNALS

For a running commentary on action in Congress and its committees, Congressional Quarterly and the National Journal were very useful, as well as magazines such as Science, Aeronautics and Space Technology, International Science and Technology, Daedalus, Chemical and Engineering News, Saturday Review, Nature, Bulletin of the Atomic Scientists and Physics Today. Among newspapers most frequently consulted were Washington Post, Wall Street Journal, Washington Evening and Sunday Star, The New York Times, Chicago Tribune, Louisville Courier-Journal, St. Louis Post-Dispatch and Los Angeles Times.

6. BOOKS

Adams, Sherman. *Firsthand Report: The Story of the Eisenhower Administration.* New York, N.Y.: Harper and Brothers, 1961.

Benson, Charles D., and Faherty, William Barnaby, *Moonport.* Washington, D.C.: National Aeronautics and Space Administration, 1978.

Bergaust, Erik, *Murder on Pad 34.* New York: G. P. Putnam's Sons, 1968.

Berkner, Lloyd V. *The Scientific Age: The Impact of Science on Society.* New Haven, Conn.: Yale University Press, 1964.

Boone, Adm. W. Fred. *NASA Office of Defense Affairs.* Washington, D.C.: National Aeronautics and Space Administration, 1970.

Brooks, Courtney G., Grimwood, James M. and Swenson, Loyd S., Jr. *Chariots for Apollo—A History of Manned Lunar Spacecraft.* Washington, D.C.: National Aeronautics and Space Administration, 1979.

Brooks, Harvey. *The Government of Science.* Cambridge, Mass.: The M.I.T. Press, 1968.

Byers, Bruce K. *Destination Moon: A History of the Lunar Orbiter Program.* Washington, D.C.: National Aeronautics and Space Administration. 1976.

Cochrane, Raymond C. *Measures for Progress—A History of the National Bureau of Standards*. Washington, D.C.: National Bureau of Standards, 1966.

Corliss, William R. *Histories of the Space Tracking and Data Acquisition Network, the Manned Space Flight Network, and the NASA Communications Network*. Washington, D.C.: National Aeronautics and Space Administration, 1974.

Eisenhower, Dwight D. *Waging Peace: 1956–1961*. Garden City, N.Y.: Doubleday & Co., Inc. 1965.

Emme, Eugene M. *A History of Space Flight*. New York, N.Y.: Holt, Rinehart and Winston, 1965.

Ezell, Edward Clinton and Neuman, Linda. *The Partnership—A History of the Apollo-Soyuz Test Project*. Washington, D.C.: National Aeronautics and Space Administration, 1978.

Gibney, Frank B., and Feldman, George J. *The Reluctant Space-Farers: The Political and Economic Consequences of America's Space Effort*. New York, N.Y.: New American Library, 1965.

Gilpen, Robert and Wright, Christopher, ed. *Scientists and National Policy Making*. New York, N.Y.: Columbia University Press, 1964.

Greenberg, Daniel S. *The Politics of Pure Science*. New York, N.Y.: New American Library, 1967.

Griffith, Alison. *The National Aeronautics and Space Act: A Study of the Development of Public Policy*. Washington, D.C.: Public Affairs Press, 1962.

Hacker, Barton C. and Grimwood, James M., *On the Shoulders of Titans: A History of Project Gemini*. Washington, D.C.: National Aeronautics and Space Administration, 1977.

Hall, R. Cargill. *Lunar Impact: A History of Project Ranger*. Washington, D.C.: National Aeronautics and Space Administration, 1977.

Hartman, Edwin P. *Adventures in Research—A History of Ames Research Center*. Washington, D.C.: National Aeronautics and Space Administration, 1970.

Holmes, Jay. *America on the Moon: The Enterprise of the Sixties*. Philadelphia, Pa.: J. B. Lippincott Co., 1962.

Hughes, Patrick. *A Century of Weather Service*. New York, N.Y.: Gordon and Breach, Science Publishers, Inc., 1970.

Johnson, Lyndon B. *The Vantage Point· Perspectives of the Presidency, 1963–1969*. New York, N.Y.: Popular Library, 1971.

Killian, James R., Jr. *Sputnik, Scientists, and Eisenhower*. Cambridge, Mass.: The M.I.T. Press, 1977.

Kistiakowsky, George B. *A Scientist at the White House*. Cambridge, Mass.: Harvard University Press, 1976.

Lambright, W. Henry. *Governing Science and Technology*. New York, N.Y.: Oxford University Press, 1976.

Logsdon, John M. *The Decision to Go to the Moon*. Chicago, Ill.: The University of Chicago Press, 1970.

Lomask, Milton. *A Minor Miracle—An Informal History of the National Science Foundation*. Washington, D.C.: The National Science Foundation, 1975.

McInnis, Mary, ed. *We Propose: A Modern Congress*. New York: McGraw-Hill Book Co.

Medaris, John B. *Countdown for Decision*. New York, N.Y.: G. P. Putnam's Sons, 1960.

Reedy, George. *The Twilight of the Presidency*. New York, N.Y.: World Publishing Co., 1970.

Rosenthal, Alfred. *Venture into Space: Early Years of the Goddard Space Flight Center*. Washington, D.C.: National Aeronautics and Space Administration, 1968.

Rosholt, Robert H. *An Administrative History of NASA, 1958–1963*. Washington, D.C.: National Aeronautics and Space Administration, 1966.

Shannon, James A., ed., *Science and the Evolution of Public Policy*. New York, N.Y.: The Rockefeller University Press, 1973.

Swenson, Loyd S., Jr., Grimwood, James M. and Alexander, Charles C. *This New Ocean—A History of Project Mercury*. Washington, D.C.: National Aeronautics and Space Administration. 1966.

Symington, James W. *The Stately Game*. New York, N.Y.: Macmillan, 1971.

Wenk, Edward, Jr. *The Politics of the Ocean*. Seattle, Wash.: University of Washington Press, 1972.

Wiesner, Jerome B. *Where Science and Politics Meet*. New York, N.Y.: McGraw-Hill Book Company, 1965.

Index

Dr. Eugene M. Emme, Chairman
CLIO Research Associates
11308 Cloverhill Drive
Wheaton, MD 20902 (593-2938)

Pres. Philip H. Bolger
Cranston Research Associates
6060 Duke Street
Alexandria, VA 22304

Mr. Stephen E. Doyle
Director, Space and Weapons Systems
Aerojet Liquid Rocket Company
P.O. Box 13618
Sacramento, CA 95813

Mr. Frederick C. Durant, III
Astronautics History Consultant
109 Grafton Avenue
Chevy Chase, MD 20815

Dr. Albert E. Eastman
Hq. A.O.P.A.
4616 Winston Place
Alexandria, VA 23310

Prof. Richard P. Hallion
Department of History
University of Maryland
College Park, MD 20742

Mr. Frederick I. Ordway, III
Director, Special Projects
U.S. Dept. of Energy
Home: 2401 N. Taylor Street
Arlington, VA 22207

Col. George Ryan Weinbrenner, USAF
Aerospace Technology Historian
P.O. Box 35342
San Antonio, TX 78235

LIAISON ASSOCIATES

Prof. Stephen G. Brush
Institute for Physical Science
  and Technology
University of Maryland
College Park, MD 20742

Dr. Tom D. Crouch
Historian of Flight
National Air and Space Museum
Washington, DC 20560

Mr. R. Cargill Hall, Director
Office of Research
USAF Simpson History Center
Maxwell AFB, AL 36112

Mr. George James
Communications (ISPT)
National Science Foundation
Washington, DC 20550

Prof. John M. Logsdon
Program in Science, Technology,
  and Public Policy, Graduate Col.
George Washington University
Washington, DC 20052

Mr. Mitchell R. Sharpe
Historian of Rocketry and Space
Alabama Space and Rocket Center
Huntsville, AL 35807

Mr. Frank H. Winter
Historian, Dept. of History of
  Space Science and Exploration
National Air and Space Museum
Washington, DC 20560

AAS HISTORY SERIES

Series Editor, Eugene M. Emme

Other Titles in the Series

Vol. 1 *TWO HUNDRED YEARS OF FLIGHT IN AMERICA: A BICEN-TENNIAL SURVEY*, edited by E. M. Emme, 1977, 326p, third printing 1981.

Vol. 2 *TWENTY-FIVE YEARS OF THE AMERICAN ASTRONAUTICAL SOCIETY: HISTORICAL REFLECTIONS AND PROJECTIONS, 1954-1979*, edited by E. M. Emme, 1980, 248p.

Vol. 3 *BETWEEN SPUTNIK AND THE SHUTTLE: NEW PERSPECTIVES ON AMERICAN ASTRONAUTICS, 1957-1979*, edited by F. C. Durant, III, 1981, 350p.